# Between Sanity and Madness

# Between Sanity and Madness

## *Mental Illness from Ancient Greece to the Neuroscientific Era*

ALLAN V. HORWITZ, PH.D.
*Board of Governors Distinguished Professor of Sociology, Emeritus*
*Rutgers University*

OXFORD
UNIVERSITY PRESS

OXFORD
UNIVERSITY PRESS

Oxford University Press is a department of the University of Oxford. It furthers the University's objective of excellence in research, scholarship, and education by publishing worldwide. Oxford is a registered trade mark of Oxford University Press in the UK and certain other countries.

Published in the United States of America by Oxford University Press
198 Madison Avenue, New York, NY 10016, United States of America.



Library of Congress Cataloging-in-Publication Data
Names: Horwitz, Allan V., author.
Title: Between sanity and madness : mental illness from ancient greece to the neuroscientific era / Allan V. Horwitz.
Description: New York : Oxford University Press, [2020] |
Includes bibliographical references and index. |
Identifiers: LCCN 2019032581 (print) | LCCN 2019032582 (ebook) |
ISBN 9780190907860 (hardback) | ISBN 9780190907884 (epub) |
ISBN 9780190907877 (updf) | ISBN 9780190907891 (online)
Subjects: LCSH: Mental illness. | Psychology, Pathological. | Neurosciences.
Classification: LCC RC469 .H68 2020 (print) | LCC RC469 (ebook) | DDC 616.89—dc23
LC record available at https://lccn.loc.gov/2019032581
LC ebook record available at https://lccn.loc.gov/2019032582

*To my daughters: Rebecca, Jessica, and Stephanie*

# Contents

# Preface

My interest in mental illness began as an undergraduate during the late 1960s when questions about this topic were prominent aspects of the youth culture. At the time, the work of such figures as Ken Kesey, R. D. Laing, Thomas Szasz, and Erving Goffman were widely read and discussed. When I entered graduate school in sociology at Yale in 1970, I encountered far more conventional, but equally fascinating, views of psychiatric disorders. I was exposed to the rigorous and pathbreaking sociological research of Sandy Hollingshead and Jerry Myers about how factors such as social class and family structure profoundly shaped definitions of and responses to those who were perceived as mentally ill. In addition, Donald Black's innovative studies of law and social control, which relied on a huge range of cross-cultural and historical material, provided a model of inquiry that I have tried to emulate ever since. My own work focused on the then-new topic of how social network structures influenced the processes through which people came to enter psychiatric treatment. My first book, *The Social Control of Mental Illness* (1980) attempted to generalize the findings from my thesis about how individuals in New Haven defined their conditions and sought psychiatric help to a far broader range of settings.

After graduate school, I joined the faculty of the Sociology Department at Rutgers University, which I never left. A few years after my arrival, I experienced an unusually rare stroke of luck. David Mechanic, probably the foremost medical sociologist in the world, moved to Rutgers and we became close collaborators. I was well-versed in David's scholarship but had no idea that he was also an extraordinary institution builder. In 1985 he established the interdisciplinary Institute for Health, Health Care Policy, and Aging Research (IFH), which strongly influenced my subsequent work.

One of the core members of IFH was the historian Gerald Grob, whose prolific studies on American psychiatric history and policy had a major impact on my own writings. I was fortunate to collaborate with Gerry on a number of projects before his untimely death in 2015. This book is deeply indebted to Gerry's scholarly vision and perspective. The postdoctoral program in mental health at IFH was another major inspiration for my research. Gerry's presence attracted many leading young historians of mental illness including Elizabeth Lunbeck, Nancy Tomes, and the late Jack Pressman, all of whom became major figures in this field. The postdoctoral program also led to my collaboration with the extraordinary theorist of mental illness Jerry Wakefield, with whom I co-authored two books, *The Loss of Sadness* (2007) and *All We Have to Fear* (2012) as well as a number of articles and chapters. Jerry's brilliant thought has infused my writings for the past quarter-century. More recently, I have been fortunate to work with another former postdoc, historical sociologist Owen Whooley, who is producing dazzlingly insightful work on the history of psychiatric knowledge in the United States. IFH also brought me into contact with a wide range of interdisciplinary researchers including Deborah Carr, Stephen Crystal, Peter Guarnaccia, Ellen Idler, Howard Leventhal, Jane Miller, Kathleen Pottick, Sarah Rosenfield, Keith Wailoo, and Jamie Walkup. My colleague in the Rutgers Sociology Department Eviatar Zerubavel provided a model of creative sociological thinking. All of these scholars not only provided an incredibly congenial work environment but also models for imaginative and innovative scholarship.

In addition to the many wonderful faculty and postdocs that I encountered at Rutgers, I have been fortunate to work in a field—the social and historical study of mental illness—that is filled with exceptional scholars. Those who have had especially important influences on this book include David Healy, George Makari, Andrew Scull, Edward Shorter, and the late Roy Porter.

This book is the result of these many diverse sociological, historical, anthropological, and philosophical influences. Consequently, it is unusual in a number of ways. Although the volume is organized chronologically, it does not adhere to a traditional concentration on a single culture or historical era. Instead, it considers how various societies have defined madness and sanity, viewed the relationship between psychic and physical disorders, and understood the origins and appropriate responses to mental illnesses. Bookended by an analytic introductory chapter and a synthetic concluding one, the first three substantive chapters involve understandings of mental illness in Ancient Greece and western European societies while the following five chapters focus on American psychiatry. Many of these chapters draw upon my previous writings, which are noted in the Acknowledgments section in each chapter.

Terminology represents a great challenge in writing about mental illness. Terms such as "mad" and "insane" that were historically descriptive labels have become highly pejorative. I continue to use them, despite the considerable baggage they carry, because they convey meanings that have traditionally accompanied understandings of mental illness. In addition, I use the more general terms "mental illness" and "mental disorder" interchangeably although others employ them more distinctively.

Finally, I have been fortunate to continue my affiliation with Oxford University Press. I am especially grateful to my editors, Andrea Knobloch and Jacqueline Buckley, for their support throughout this project. I am also indebted to the anonymous reviewers of this manuscript for their many insightful suggestions, which have resulted in an immensely improved book.

# 1
# Puzzles of Mental Illness

> [In ancient Israel] it was proposed that a person who wandered about alone at night, who spent the night in a cemetery, or who tore his garments and destroyed what was given to him might be considered deranged—if such behavior appeared irrational. However, it was pointed out that otherwise normal persons could also behave in this way, e.g. one who spent the night in a cemetery might have done so to practice magic, or that another who tore his clothes might have done so in a fit of anger, or because he was a cynic philosopher exhibiting his contempt for material things.
>
> —George Rosen, *Madness in Society* (1968)

Every society, regardless of time or place, regards some of its members as mad. Typically, this label has been reserved for a small number of seemingly senseless behaviors. For example, the Native Alaskan term for crazy, "nuthkavihak," refers to "such phenomena as talking to oneself, screaming at someone who does not exist, believing that a child or husband was murdered by witchcraft when nobody else believes it, believing oneself to be an animal . . . killing dogs, and threatening people."[1] Similarly, the Kalumburu people in Australia use the word "wambaba" to mean mad or crazy; it refers to "an excited or assaultive individual who talks strangely, swears a lot, forgets his manners, opens his bowels where people live, knocks over his water holder, or runs off in fright to the bush. He behaves in a highly disturbed and incomprehensible way."[2] Early 20th-century German psychiatrist Karl Jaspers comparably isolated "incomprehensibility" as the most distinctive aspect of madness: "All the sudden impulses, the unfathomable affect and lack of affect, the sudden pauses in conversation, the disconnected ideas. . . . Some call these actions eccentric, silly. But with all these terms we are at the end of the day saying the same thing: The common factor is 'incomprehensible.'"[3]

The inexplicable aspects of madness contrast with the socially shared conceptions of reality that mark sanity.

Present views of what constitutes a mental illness encompass such unfathomable conditions but far more as well. Despite general agreement regarding the pole of madness, huge disparities exist on where dividing lines should be placed between it and sanity and even if there is any clear demarcation at all. The range of current mental disorders includes, among much else, depression, anxiety, post-traumatic stress disorder (PTSD), various addictions, attention-deficit/hyperactivity disorder (ADHD), and numerous personality disorders. The most highly regarded surveys indicate that a quarter of the U.S. population experiences some mental illness in any particular year and well over half at some point in their lives.[4] Other, more comprehensive, studies show that two-thirds of people will become mentally ill over their lifetime.[5] Recent cohorts of young people report especially vast amounts of disorder: in the early 21st century, half of American adolescents seemingly have some mental illness by age eighteen.[6] Global data from the World Health Organization indicate that mental illnesses now account for more years of disability than any other disease category.[7]

Enormous historical and cultural disparities also exist regarding what sort of authority should respond to mental illnesses. Dominant current practices—following a long tradition—regard these conditions as brain-based defects that medical and other mental health professionals should treat. In other times and places, however, similar behaviors have been viewed as sins that require confession, possessed states in need of exorcism, crimes that entail punishment, or creative inspirations that should be cultivated.[8]

While various groups provide an extensive array of answers to questions about the nature of mental illness and its boundaries with sanity, all have confronted the same issues. What distinguishes mental illnesses from other sorts of devalued conditions and from normality? Should medical, religious, psychological, legal, or no authority at all respond to the mentally ill? Which factors lead people to become mad? What treatments might help them recover? The various responses that societies have provided to these puzzles are both widely divergent and surprisingly similar to current understandings.

## Distinguishing Sanity from Madness

All groups face the challenge of defining what aspects of madness distinguish it from normal actions on the one hand and other sorts of deviant behaviors on the other. A debate among Talmudic scholars two millennia ago illustrates the difficulties that arise in isolating the distinctive qualities of insanity:

> It was proposed that a person who wandered about alone at night, who spent the night in a cemetery, or who tore his garments and destroyed what was given to him might be considered deranged—if such behavior appeared irrational. However, it was pointed out that otherwise normal persons could also behave in this way, e.g. one who spent the night in a cemetery might have done so to practice magic, or that another who tore his clothes might have done so in a fit of anger, or because he was a cynic philosopher exhibiting his contempt for material things.[9]

For these ancient observers, the characteristic feature of mentally ill people did not lie in their behavior. Instead, it involved the absence of contextually appropriate reasons for their actions. The mad often wandered at night, slept in cemeteries, or tore their clothes off. Yet, sorcerers, angry people, or philosophers also did the same things. The essential difference was that the latter had socially comprehensible motivations—practicing magic, fits of anger, philosophical principles—for their behaviors. In contrast, observers could not connect the activities of the mentally ill to any socially explicable purpose. Inferences about sanity and madness, therefore, involved whether or not actions could be linked to understandable motives.

Two thousand years later, acclaimed sociologist Erving Goffman made a similar point: "The delusions of a private can be the rights of a general; the obscene invitations of a man to a strange girl can be the spicy endearments of a husband to his wife; the wariness of a paranoid is the warranted practice of thousands of undercover agents."[10] Judgments about whether some behavior is deluded or justified, obscene or romantic, and paranoid or reasonable require knowledge of relevant contextual information. When observers can find a socially plausible reason for some action, they do not define it as a sign of mental disturbance.

The Sebei people of East Africa provide an example of how groups differentiate mental illness from other devalued behaviors such as criminality, drug use, intellectual deficiency, bad manners, or ignorance. Anthropologist Robert Edgerton recounts a dialogue among a group of Sebei who are discussing the behavior of a young man that they observed talking nonsense, making abrupt and jerky movements, flapping his arms like a bird, and giggling like a child:

SALIMU: He is a strange boy.
SAYEKWA: He is a foolish boy. Why does he behave that way?
SALIMU: He may be mad or he may be foolish (mentally retarded).
SAYEKWA: It could also be bewitchment or a fever or something like a fit. . . .
INTERPRETER: There is something wrong. People here smoke bhang. Perhaps he may be a bhang man.
SAYEKWA: No. I don't think that. He is more like he is crazy.
SALIMU: I think he may be a fool.
INTERPRETER: It is impossible to know without knowing about his family. We could ask one of these people about him.
ANTHROPOLOGIST: What is the difference between a fool and a madman?
SALIMU: A fool was born without sense. A madman becomes senseless because of a disease or witchcraft. We would have to know his history to tell about this young man.[11]

This debate shows the typical constructions involved in making attributions of insanity. Observers use factors such as family history, biographical circumstances, and situational contingencies to decide among the various possible interpretations of inappropriate behaviors. Before participants use mental illness to explain some strange behavior, they rule out other designations such as foolishness, bewitchment, psychotropic drug use, or epilepsy. The answers to the questions that the Sebei pose also imply the kind of response that should be made to the strange behavior. If he is a "bhang man," then the village equivalent to the police might be called. If he is a fool, he was born that way and no remedies are available. If, however, the boy is deemed to be mad, they might summon a witchdoctor to expel the possessing force. Similar distinctions persist in current psychiatric diagnostic manuals.

## The DSM Definition of Mental Disorder

Discussions of the nature of mental illness mark all societies, including our own. Consider the "official" designation of mental disorder that is found in the American Psychiatric Association's *Diagnostic and Statistical Manual of Mental Disorders* (fifth edition; DSM-5):

> A mental disorder is a syndrome characterized by clinically significant disturbance in an individual's cognition, emotion regulation, or behavior that reflects a dysfunction in the psychological, biological, or developmental processes underlying mental functioning. Mental disorders are usually associated with significant distress or disability in social, occupational, or other important activities. An expectable or culturally approved response to a common stressor or loss, such as the death of a loved one, is not a mental disorder. Socially deviant behavior (e.g. political, religious, or sexual) and conflicts that are primarily between the individual and society are not mental disorders unless the deviance or conflict results from a dysfunction in the individual, as described above.[12]

Much like the Sebei, the DSM strives to isolate the particular qualities that distinguish mental disorders from other kinds of devalued conditions and from normality. Its characterization has four central aspects, two that try to express what mental illnesses are and two that attempt to delineate them from non-disordered behaviors.

### Dysfunctions

At the heart of the way that the DSM identifies a mental disorder—comparable to the various definitions of premodern groups—is the presence of a "dysfunction in the psychological, biological, or developmental processes underlying mental functioning." Although the definition does not specify what a "dysfunction" is, the term presumably indicates that some process is not simply undesirable or rare but that something has *gone wrong* with mechanisms involving cognition, emotional arousal, perception, memory, and the like.[13] The resulting symptoms are neither explicable responses to circumstances nor culturally normative. The definition does not stipulate whether dysfunctions involve defective brain or psychic processes but allows for either sort of account. Nor does it mention what qualities mark the *appropriate* functioning of psychological, biological, or developmental processes.

The DSM's placement of dysfunction at the heart of its definition of mental disorder echoes a long historical tradition. For example, a renowned Ancient Greek physician, Aretaeus of Cappadocia (~2nd century AD), described melancholics in these terms: "Sufferers are dull or stern: dejected or unreasonably torpid, without any manifest cause."[14] Here, "without any manifest cause" implies that melancholic symptoms result from some inner dysfunction as opposed to reasonable grounds for melancholy such as the loss of an intimate relationship, economic catastrophe, or diagnosis of a serious physical disease. Two thousand years later, one of the founders of 19th-century biological psychiatry, German psychiatrist Wilhelm Griesinger (1817–1868), made a comparable distinction between identical symptoms that result from a dysfunction or from some appropriate external cause:

> The melancholia which precedes insanity sometimes appears externally as the direct continuation of some painful emotion dependent upon some objective cause . . . , e.g. grief, jealousy; and it is distinguished from the mental pain experienced by healthy persons by its excessive degree, by its more than ordinary protraction, by its becoming more and more independent of external influences, and by the other accessory affections which accompany it.[15]

Symptoms that are rooted in external situations are "healthy"; those with disproportionate severity or duration to their generating contexts indicate insanity. The DSM follows this enduring line of thought, emphasizing how only symptoms that reflect an internal dysfunction rather than some appropriate situational cause mark some disorder. If nothing is wrong with someone's inner functioning, they do not have a mental illness.

As the debate among Talmudic scholars shows, it is often difficult to distinguish conditions that arise from a dysfunction or from some other reason. Consider the case of Michael Phelps, who has won more Olympic medals than any athlete in history. When he was nine years old, Phelps received a diagnosis of ADHD, which is made when someone is unable to focus, sit still, or concentrate, presumably because of some inner dysfunction. "I was told by one of his teachers that he couldn't focus on anything," his mother recounts.[16]

This suggests that Phelps had some—possibly biological—inability to pay attention. Yet, he had no comparable problems of focus when it came to his swimming. "For the past 10 years, at least, he's never missed a practice," Phelps's mother recalled. "Even on Christmas, the pool is the first place we go, and he's happy to be there." Did Phelps have a dysfunction or, instead, a compelling competing interest that preoccupied him?[17] Alternatively, he could have been one of the youngest children in his class so that his relative immaturity put him at greater risk for receiving an ADHD diagnoses.[18] All groups have faced the intrinsic uncertainties in distinguishing behaviors that stem from dysfunctions or from other reasons.

Pederasty—sexual relations involving adult men and prepubescent boys—provides another example of the problems involved in making decisions about the presence of some dysfunction. Our culture views pederasty as perhaps the most reviled form of sexual behavior. The DSM-5 defines adults who are aroused by having sex with children (generally thirteen years or younger) as having a pedophilic disorder, which involves inappropriate targets of sexual preferences.[19] In our society, some inner dysfunction is likely to motivate pedophiliacs' anomalous sexual preference because their desires persist despite strong social sanctions.[20]

Yet, many cultures have institutionalized sexual relationships between older men and prepubescent and adolescent boys. Most notably, the ancient Greeks esteemed intergenerational erotic ties among males, which were deeply embedded in their civic life: "A whole philosophy was built up round [pederasty], based on the idea that the lover was the educator and military trainer and partner of the beloved, and would do everything to earn his admiration," historian Michael Grant notes.[21] Involvement in culturally approved man–boy sex among the ancient Greeks (and other societies that do not condemn such relationships) would rarely stem from defective sexual arousal mechanisms but instead are learned and culturally approved practices. The Siwan group in North Africa provides an example: "All men and boys engage in anal intercourse. Males are singled out as peculiar if they did not do so."[22] In such groups, a dysfunction is unlikely to underlie pederasty. Therefore, similar modes of sexual attraction are likely to stem from dysfunctions in groups where they arise in the face of harsh disapproval but not in others where they are customary and, sometimes, even mandated.

The example of pederasty illustrates some of the many issues raised by the requirement that a mental disorder must stem from some dysfunction: is it possible to know what the appropriate functions of psychological mechanisms are without reference to cultural values; can a mechanism that is dysfunctional in some groups be functional in others; do many negatively valued behaviors such as extreme jealousy, hatred of outgroups, fear of strangers, or intercourse with pubescent adolescents stem from natural, rather than dysfunctional, processes?[23] Such concerns raise the question of the objective or value-laden nature of dysfunctions. "The most fundamental issue, and also the most contentious one," psychiatrist Robert Kendell contends, "is whether disease and illness are normative concepts based on value judgments, or whether they are value-free scientific terms; in other words, whether they are biomedical terms or sociopolitical ones."[24] The DSM introduces a normative component into its definition through invoking the distress or disability associated with some dysfunction.

## Distress or Disability

The DSM implicitly deals with the issue of whether definitions of mental disorder are objective or value-laden through stating that a dysfunction in itself typically does not provide sufficient grounds to infer the presence of a mental disorder. Only dysfunctions that are "usually associated with distress or disability" are disorders.[25] Therefore, adequate characterizations of mental disorder contain elements of *both* a dysfunction that involves incapacity to perform a natural function *and* a cultural judgment that defines the dysfunction as undesirable. Jerome Wakefield's term "harmful dysfunction" captures this dual quality of mental disorder.[26]

The necessity of harm means that dysfunctions are not mental disorders when they are neither distressing nor disabling. Many psychological dysfunctions—for example, persistent depressions, crippling compulsions, or inexplicable anxiety—intrinsically entail distress. Yet, many others do not involve any misery at all. For example, anosognosia—the unwillingness of people to believe that anything is wrong with their minds—can characterize people with schizophrenia.[27] Or, the manic stage of bipolar disorder is marked by euphoria, seemingly exceptional ideas, highly pleasurable behaviors, and often unusual degrees of creativity.[28] The DSM takes this into account by saying that dysfunctions that do not involve distress must entail "disability in social, occupational, or other important activities" if they are to be considered as disorders. In contrast to distress, which refers to the feelings of the disturbed individual, disability often involves problems with other people such as parents, spouses, teachers, bosses, or social control agents.

The DSM's distress or disability constraint means that, although all mental disorders result from some inner dysfunction, all dysfunctions are not disorders. Only dysfunctions that also have impairing consequences for selves or others are mental disorders, presumably because it would be pointless to be concerned with conditions that are unproblematic for both individuals and those around them. One important consequence of this aspect of the DSM definition is that, while dysfunctions can be grounded in universal organic or psychic processes, mental disorders can be cross-culturally relative. Because the degree of distress or disability that any dysfunction entails varies considerably across cultures, the definition implies that dysfunctions can be disorders in some cultures but not in others.

Anthropologist Ruth Benedict's "Anthropology and the Abnormal" (1934) presented the classic statement that the concept of mental disorder is culturally relative. Benedict (1887–1948) questioned the validity of applying Western definitions of normality and abnormality to different cultures. Instead, she asserted that various groups around the world consider as normal and appropriate the sorts of behaviors—paranoia, seizures, trances, and the like—that Western psychiatry defined as abnormal. She used as an example the Dobuans of Melanesia, who display a constant fear of poisoning that they saw as normal rather than as paranoiac. Or, she noted that the Shasta Indians in California and the native people of Siberia did not view seizures as dreaded illnesses but instead as signs of special connections to supernatural powers that singled out people for authority and leadership. She also discussed catalepsy—a state that involved trance-like states, hearing voices, losses of voluntary motion, and rigid limbs—to illustrate how some other cultures treated as valued conditions what Western psychiatry considered as the mental disorder of catatonic schizophrenia. "There are," Benedict asserted, "well-described cultures in which these abnormals function at ease and with honor, and apparently without danger or difficulty to the society."[29]

Conversely, Benedict described behaviors that were normalized and even rewarded in our culture that would be considered mad in other cultures. Dobuans, for example, would regard a person who was always cheerful, happy, and outgoing as crazy. Divisions between justified suspicions or paranoia, ritualistic invocations of spirits or hallucinations, and unnaturally prolonged or appropriate periods of grief were often blurry, tied to context, and subject to value judgments. Normality thus resided in culturally approved conventions, not in universal biological or psychological standards of appropriate functioning. "All our local conventions of moral behavior and of immoral are without absolute validity," Benedict concluded.[30]

Advocates of the DSM definition could use the distress or disability requirement to challenge Benedict's assertion that there is no validity to Western conceptions of schizophrenia because some groups place people who hallucinate, have delusions, fall into trance states, and speak incoherently into valued and honored social roles. From their perspective, such individuals have dysfunctions that their cultures do not evaluate as harmful.[31] The distress or disability requirement implies that the universal aspects of dysfunctions are compatible with the culturally relative nature of mental disorders.

Like decisions about the presence of a dysfunction, judgments about distress or disability can be difficult to make. Take the example of the poet Emily Dickinson who, by the time she was forty, would not leave her home and hid in her room, unwilling to see even her longtime friends.[32] Nothing about Dickinson's circumstances could account for her refusal to walk outside or to meet with people she had known for extended periods of time. Her extreme social isolation seems likely to have stemmed from some inner dysfunction. Yet, it was arguably neither distressing nor disabling. Indeed, many people regard her as the greatest American poet, largely because of the poems she wrote during her period of seclusion. Whether her extreme fear of going out in public or seeing friends should count as disabling or, on the contrary, a precondition for her genius has no easy answer.

The distress or disability stipulation is a complicating factor in setting boundaries between sanity and madness. It also raises challenging issues related to the value-laden component of definitions of mental disorder. Both feelings of distress and social impairments often arise because of negative societal attitudes rather than any intrinsic qualities of some dysfunction. For example, the DSM-III diagnosis of "ego-dystonic homosexuality," which applied to homosexuals who were distressed by their sexual orientation, seems to result more from individual responses to oppressive social norms than from a dysfunction.[33] It shows how the harm that some condition engenders can stem from cultural values rather than from the dysfunction. Separating the normative from the biomedical elements of mental disorder is often an imposing task.

## Mental Disorders Are Not Explicable Responses to Circumstances

Like ancient Talmudic scholars, the DSM tries to specify what sorts of phenomena do *not* constitute internal dysfunctions and so are not mental disorders. The DSM's definition distinguishes dysfunctions from "an expectable or culturally approved response to a common stressor or loss, such as the death of a loved one." This means that symptoms resembling dysfunctions and entailing intense suffering or disability are nevertheless not disordered because they are natural responses to circumstances.[34] The definition uses "the death of a loved one" as a model for highly distressing states that are not disordered because they are contextually explicable. People naturally become intensely sad after someone close to them dies; their internal functioning is appropriate, not pathological. Although such symptoms can be very distressing and, often, disabling, they are reasonably commensurate to the severity of the situation that led to them and, unlike dysfunctions, typically dissipate with the passage of time. In contrast, as Griesinger noted, mental disorders arise in the absence of a suitable context or initially appear after some precipitating event but become disengaged from their initial trigger, take on a life of their own, and are immune to changes in external conditions.

The intense grief that psychiatrist and anthropologist Arthur Kleinman suffered after the death of his wife illustrates this aspect of non-disordered yet highly distressing conditions.

> In March, 2011, my wife died and I experienced the physiology of grief. I felt greatly sad and yearned for her. I didn't sleep well. When I returned to a now empty house, I became agitated. I also felt fatigued and had difficulty concentrating on my academic work. My weight declined owing to a newly indifferent appetite.[35]

Kleinman's symptoms easily meet the DSM criteria for major depressive disorder. Yet, as long as emotions have been recorded, experiences of grief—feelings of deep sadness that follow the death of an intimate—have been central to portrayals of basic human nature. Kleinman's grief, as the DSM definition recognizes, is a normal response to an intense loss. Accordingly, his distress gradually eased over time in the absence of any treatment: "This dark experience lightened over the months, so that the feelings became much less acute by around 6 months."[36] The intense depressive symptoms that Kleinman and other grievers experience are not products of a dysfunction but are explicable in light of their circumstances of loss.

Like Kleinman's grief, anxious emotions that could indicate a dysfunction when no danger is present are appropriate in perilous situations. Psychiatrist Kenneth Kendler uses the example of a mountain climber who suffers a full-blown panic attack when he loses his grip and falls before his rope catches him. "A panic attack," Kendler explains, "is not—in and of itself—psychopathological. It only becomes pathology when it occurs in certain contexts—at times and in places when it should not. Thus the diagnostic status of panic disorder is inherently contextual. It is not a disorder in and of itself but only in certain contexts."[37]

As is the case with grief, not panic symptoms alone but their relationship to the circumstances in which they emerge accounts for the distinction between "expectable" panic attacks that are not disorders and "unexpectable" ones that are disorders. Identical symptoms that are explicable in one context, such as falling down a mountain, are signs of a mental disorder when there is no contextually appropriate explanation for them. "Most negative mental states such as sadness, despair, anxiety, fear, agitation, and anger," Wakefield and psychiatrist Michael First observe, "are not abnormalities but normal responses to life's vicissitudes."[38]

PTSD illustrates some of the nuances involved in distinguishing "abnormalities" from "normal responses to life's vicissitudes." It definitionally arises from exposure to some traumatic event and so might seem to be an "expectable" response to a stressor and, therefore, not a dysfunction. However, the characteristics that Griesinger noted in regard to depression—"excessive degree," "more than ordinary protraction," and "becoming more and more independent of external influences"—suggest that continuing symptoms of PTSD have become detached from their initial generating context. Their intrusive, recurrent, and involuntary nature indicates that an extreme environmental stressor has led the natural functioning of memory mechanisms to breakdown.[39] Because such symptoms are also almost invariably associated with distress and/or disability, PTSD would be a genuine mental disorder and not simply a contextually explicable response to circumstances. Needless to say, it is often difficult to distinguish behaviors that are "dysfunctional" from those that are "expectable" reactions to given contexts.

## Mental Disorders Are Not Forms of Social Deviance

The DSM definition strives to separate mental disorders not just from contextually explicable behaviors but also from "socially deviant behaviors" and "conflicts between individuals and society." Often, drinking heavily, persistently using culturally devalued drugs, engaging in unusual sexual activities, or offending the wrong kind of person are types of deviant social behaviors: violations of social norms that define standards of proper behavior. Likewise, people who disregard social conventions, hold divergent political beliefs, or have disreputable lifestyles might develop conflicts with society. The DSM definition distinguishes such nonconformists from those who have dysfunctions that lead them to suffer negative effects such as persistent abuse of addictive substances, compulsive pursuit of activities that harm them, or an inability to refrain from engaging in socially disvalued activities. The latter would have mental disorders only when some internal dysfunction leads to their distressing or disabling conditions.

This aspect of the definition again contains echoes of the Sebei, who divide madness from deviant behavior: "Respondent after respondent qualifies his description of a psychotic behavior by saying 'without reason.' That is, murder as such is not psychotic—only murder *without some good reason is psychotic*. The same thing is true of every other behavior cited."[40] The DSM strives to separate rule-violating behaviors that entail individual responsibility from those that stem from a dysfunction and therefore should be treated rather than punished.

Such decisions are often controversial. Consider the case of Adam Lanza, who killed his mother, twenty elementary school children, six staff members, and then himself in Newtown, Connecticut, in 2012. Like Emily Dickinson, his life was marked by extreme social isolation: "Lanza seemed to have no friends or people he could turn to for support or assistance and did not appear to have any enjoyment of life."[41] Unlike Dickinson, Lanza was obsessed with previous mass shootings and spree killings. Many people would agree with Lanza's father, who stated "You can't get any more evil."[42] For him, Lanza's actions were not the senseless acts of a madman but those of a depraved killer. Current clinicians would be hard pressed to make more informed decisions than the Sebei about whether Lanza was a mass murderer or a victim of some inner dysfunction.

Hoarding disorder, which was added to the DSM in 2013, shows the often value-laden nature of decisions that try to distinguish mental disorders from conflicts between individuals and society. The essence of this condition is "persistent difficulties discarding or parting with possessions regardless of their actual value."[43] Hoarders typically accumulate so many items that they and others have difficulty navigating their living spaces. Individuals themselves are typically not troubled by their condition unless someone tries to *stop* their hoarding. Interventions typically arise after complaints from family members, neighbors, or public health departments. The justification for calling their behaviors "dysfunctions" or "mental disorders" as opposed to "conflicts between individuals and society" is unclear.

Precise lines rarely exist between dysfunctional and deviant conditions. "No other specialty of medicine deals with diseases whose initial signs can be so easily confused with moral lapse," an officer at the Rockefeller Foundation wrote in 1944.[44] Indeed, many particular DSM diagnoses seem to encompass activities that were traditionally regarded as sins. Examples include distinctions between binge eating disorder and gluttony, hypersexual disorder and lust, intermittent explosive disorder and wrath, or narcissistic personality disorder and pride.[45] To say the least, the grounds for separating dysfunctions and, therefore, mental disorders from such traditional vices are rarely precise. Other diagnoses, such as oppositional defiant disorder or conduct disorder are extraordinarily difficult to separate from "conflicts between individuals and society" as well as from delinquency and criminality.[46] Moreover, such definitions diverge widely across cultures. American psychiatrists might diagnose an unmarried woman (but not a single man) who engages in unrestrained sex as hypomanic; the same behavior might elicit harsh punishment in many Muslim societies and indifference in Scandinavian ones.[47] Nevertheless, all groups strive to distinguish people who are mad from those who are normal, deviant, sinful, criminal, or prophetic, however difficult this splitting might be in practice.

The poles of madness and sanity are usually obvious; a remarkable consensus exists across various cultures regarding what are clear cases of mental illness.[48] Talmudic scholars, Sebei observers, Australian Native People, 19th-century psychiatrists, and the promulgators of the DSM would concur that certain kinds of unreasonable, inexplicable, and incomprehensible actions constitute one extreme of distinctly disordered behaviors. They would also agree that actions that are understandable responses to particular contexts represent the opposite pole of normality.[49] Yet, all of these definitions—including the DSM's—recognize that the boundaries between the extremes of mental health and mental illness are typically vague and fuzzy. The concept of mental disorder, therefore, is often indefinite because the phenomena that it applies to are themselves indefinite.[50]

Answers to the central questions involved in defining mental disorder—what is a dysfunction; to what extent are the negative consequences of mental disorders direct results of dysfunctions or of social attitudes; what standards distinguish contextually explicable from dysfunctional symptoms; and how is it possible to separate mental disorders from other forms of social deviance?—seem no more definitive now than they were thousands of years ago. Perhaps the reason for this lack of progress does not lie as much in the flaws of the DSM definition as in the intrinsically elusive and possibly insolvable nature of the question of what a mental disorder is. While the extremes of madness and sanity are usually clear, different societies have placed boundaries for the vast terrain of conditions that fall in between these poles in very diverse places.

## To What Extent Do Mental Illnesses Resemble Physical Ones?

One perennial issue groups have faced has been how to specify the distinctive qualities of mental disorders. A second recurrent concern regards what kind of authority defines their nature and controls their treatment. A particularly contentious issue has been the extent to which mental illnesses involve somatic or psychic processes. The answer to this question typically dictates whether mental disorders fall within medical or, alternatively, religious, legal, philosophical, psychological, or some other jurisdiction.

The view that some organic defect lies behind madness has persisted for millennia.[51] This belief implies that mental illnesses are comparable to bodily diseases: both are biologically grounded, subject to the laws of nature, and have qualities that are independent of individual or cultural characteristics. Hippocratic physicians in 4th-century BCE Greece provided perhaps the first statement of this perspective: "Men ought to know that from the brain, and from the brain only, arise our pleasures, joys, laughter, and jests, as well as our sorrows, pains, griefs and tears."[52] The idea that mental disorders are brain diseases also dominates contemporary portrayals of mental illness. A 1999 report from the U.S. Surgeon General echoes the Hippocratics: "It is, in fact, a core tenet of modern science that behavior and our subjective mental lives reflect the overall workings of the brain. Thus, symptoms related to behavior or mental lives clearly reflect variations or abnormalities in brain function."[53] The DSM, too, implies that the ultimate benchmark for health and disorder is the same in psychiatry as it is in physical medicine, namely, whether the individual's mental processes are performing the functions they are naturally designed to perform.

In the biomedical view mental disorders, like organic diseases, have properties that cannot be reduced to individual idiosyncrasies or cultural values. "This modern history of diagnosis," according to historian Charles Rosenberg, "is inextricably related to disease specificity, to the notion that diseases can and should be thought of as entities existing outside the unique manifestations of illness in particular men and women: during the past century especially, diagnosis, prognosis, and treatment have been linked ever more tightly to specific, agreed-upon disease categories."[54] The symptoms of, say, depression, anxiety, schizophrenia, or bipolar disorder, no less than those of cancer, diabetes, asthma, or malaria, can be abstracted from the individuals who have them and studied as objects with distinct causes, courses, and outcomes. Nosologies such as the DSM thus do not classify individuals: "A common misconception is that a classification of mental disorders classifies people, when actually what are being classified are disorders that individuals have," the manual asserts.[55]

Because in the biomedical view mental as well as organic diseases reflect physical laws, they operate independently of time, space, and social evaluations. For example, in 1882 famed neurologist Jean-Martin Charcot (1825–1893) proclaimed of hysteria that "everything unfolds according to the rules, which are always the same. . . . They are valid for all countries, for all epochs, for all races, and are, in short, universal."[56] Similarly, Emil Kraepelin (1856–1926), the foremost psychiatric diagnostician in the late 19th and early 20th centuries, visited Singapore to see if Asian patients differed from European ones. He concluded that they had the same disease: "we must therefore seek the real cause of dementia praecox [schizophrenia] in conditions which are spread all over the world, which thus do not lie either in race or in climate, in food, or in any other general circumstances of life."[57] Although local factors typically lead to divergent prevalence in different settings, the core features of each mental disorder are invariant results of natural processes.

Medicine is only one, although often the most prominent, form of jurisdiction over madness. Benedict, for example, showed how many premodern groups place individuals who enter possessed states that were similar to madness into valued social roles that entail exceptional spiritual connections to supernatural powers. For a millennium in the West, religious authorities dictated responses to the insane, which typically strove to eliminate the evil influences that polluted them. Legal definitions, too, which place rule-violating behaviors within frameworks of responsibility and guilt, often compete with medical authority.

Other views contend that madness should not be subject to any form of social control. Perhaps the best-known statement of this position is psychiatrist Thomas Szasz's assertion that the concept of mental illness itself is a "myth."[58] Szasz (1920–2012) claimed that terms such as "illness" and "disease" only apply to physical lesions. In contrast, mental illnesses deviate from psychosocial norms that have no objective reality in the natural world and so reflect culturally grounded value judgments. Hence, the notion of a "mental illness" is an inherent contradiction. Unless they have violated some legal standard, there is no justification for exerting any form of involuntary social control over people who are thought to be "mentally ill." A variety of competing philosophical, religious, legal, and other perspectives continues to challenge the primacy of medical authority over mental illness.

## What Factors Lead People to Become Mentally Ill?

A third continuing issue regards the perceived reasons for why some people become mad. Observers have sometimes found the causes of mental illness in inner biological or psychological forces and, at other times, in environmental, social, and cultural factors. Often, the two types of explanations have been thoroughly intertwined.

One type of internal characterization emphasizes how some organic defect in the brain, nerves, or other bodily systems including hormones, the digestive system, or the heart accounts for madness. Depending on the particular time period, physiological foundations have included the humors, constricted blood vessels, weak nerves, or deficient or excessive levels of neurochemicals among many others. For example, 18th-century physician Nicholas Robinson (*c.* 1697–1775) declared that all mental disorders "from the slightest Symptoms of the Spleen and Vapours, to the most confirm'd Affections of Melancholy Madness and Lunacy . . . are no imaginary Whims or Fancies, but real Affections of Matter and Motion, whenever the Constitution of the Brain warps from its natural Standard."[59] Current philosopher Paul Churchland echoes this view: "The victims of mental illness are the victims primarily of sheer chemical circumstances, whose origins are more metabolic and biological, than they are social or psychological."[60] Such biological explanations typically entail medications or other manipulations of the brain as treatments of choice. One of the few constants across history has been the use of drugs, particularly opium and alcohol, to alleviate distress among people with mental disorders.

While many inner interpretations ground mental disturbances in defective brain processes, many others use psychological understandings. The Freudian emphasis on repressed memories of childhood sexuality as causes of a variety of neuroses among adults is one prominent example of a psychic explanation.[61] Another is when cognitive distortions in information processing about the self, the world, and the future lead to persistent negatively biased thoughts and appraisals and consequent depression and anxiety. Current cognitive methods strive to change distorted patterns of thinking that are presumably responsible for the presence of mental illnesses.[62] They are descendants of ancient philosophies such as Stoicism and Epicureanism, which have taught people to use psychological techniques that distance themselves from worldly concerns as ways to establish inner peace of mind.

Throughout history, explanations of mental disorders that rely on external stimuli stand alongside those emphasizing internal factors. Such influences have included supernatural interventions, witchcraft, climate, traumatic events, or problematic interpersonal relationships to explain the emergence of mental illnesses. Many treatments of mental illness as well have manipulated environments, often aiming to promote restrained modes of living through changing diets, sleep patterns, and exercise routines. For example, the mental institutions that developed in the early 19th century in Europe and the United States were based on treatments that assumed changing surroundings would alleviate or cure madness.

Internal and external explanations often have not been regarded as mutually exclusive. Hippocratic conceptions of mental illness, for example, focused on imbalances among bodily humors. Yet, they also emphasized how environmental disturbances led to excesses or deficiencies of these physical fluids. Similarly, since the 17th century observers have associated the pressures of civilization, stressful lifestyles, domestic woes, and unhealthy living conditions with mental illness while simultaneously indicating that biologically and psychologically vulnerable people are the most likely ones to succumb to these burdens. Likewise, although 19th-century psychiatrists commonly viewed heredity as the chief source of insanity, they also invoked triggering causes such as financial ruin, family troubles, or masturbation to explain who among predisposed people actually developed some mental illness.[63]

Many current neuroscientists as well are beginning to focus on gene–environment relationships rather than on isolated brain or genetic processes.[64] Such interactive conceptions do not view internal and external forces as competing explanations but seek to understand the ways in which various kinds of social environments provoke or suppress biological predispositions. Such explanations emphasize how the causes of mental disorder resist reduction to any single factor, whether within or outside of individuals. Accounts and treatments featuring internal or external dynamics or their combinations have waxed and waned throughout history.

A number of the questions that engage the modern mental health professions—what constitutes a mental disorder; are mental disorders continuous with or discrete from normal behaviors; do holistic cultural and personality factors or specific diseases underlie mental symptoms; and how do biological and psychological liabilities interact with external stressors?—would be familiar to practitioners across the ages. Although the technologies available to mental health specialists at present vastly exceed those available in prior periods, it is an open question whether extant definitions of, explanations for, and responses to mental disorders surpass those of preceding eras.

## Plan of the Book

The chapters that follow examine the answers that have arisen in various historical periods about the nature of mental illness, the degree to which it resembles or is distinct from bodily diseases, what type of experts have the legitimate authority to treat it, and which factors cause and might cure it. The following chapter considers how these questions were formulated and answered in the West before the psychiatric profession was established in the 19th century. It begins by describing three distinct views that developed among the ancient Greeks: the Homeric, Platonic, and Hippocratic conceptions that developed external, mental, and organic views of madness, respectively. During the long period between the fall of Rome and the 17th century, faith-based outlooks supplanted organic views and medicine became subordinate to the church. The subsequent European Enlightenment resulted in two distinct biological and psychological conceptions of mental disorders. At the same time, definitions of mental illness expanded to encompass nervous conditions that had little resemblance to traditional stereotypes of madness.

Chapter 3 considers the model of mental disorder that arose in the 19th century alongside the creation of a new profession dedicated to the treatment of mental illness. This period marked the advent of psychiatrists as the socially legitimate arbiters of what constituted sanity and madness. Their conceptions embraced biomedical tenets that grounded mental disorders in defective brains, viewed mental illnesses as comparable to physical illnesses, and focused on inner, organic causes. Nevertheless, most believed that madness usually arose in conjunction with some socially based triggering event. A new type of response to the insane—inpatient mental asylums—also became widespread during this era.

The fourth chapter discusses Sigmund Freud's revolutionary model of mental illness that he developed at the beginning of the 20th century. Upending previous conceptions, Freud's psychodynamic approach emphasized the similarities of normal and neurotic behaviors, the differences between mental and physical symptoms, and the conflict between inner instincts and social forces of repression as the major generator of both healthy and pathological expressions. The dynamic model, which became especially influential in the United States, expanded the realm of psychiatry to outpatient practices, the control of deviance, and understandings of everyday life. It was far less influential in affecting the treatment of psychotic forms of madness, which it generally ignored.

While the previous three chapters primarily concern European developments, the remaining chapters focus on the United States. Chapter 5 examines how American dynamic psychiatry evolved into a more environmental approach after World War II. In collaboration with the military and the National Institute of Mental Health, a new socially oriented psychiatry applied therapeutic frameworks to a wide array of distressing conditions and strove to prevent, as well as treat, mental disorders. Its psychosocial model embraced expansive definitions of what constituted mental illness, separated mental from physical conditions, and focused on external rather than interior causes of psychic disturbances. During this era, psychiatric influence gained unprecedented scope and reach.

The next chapter examines the crisis that arose in American psychiatry during the 1960s and deepened in the 1970s. An anti-psychiatry movement developed that denied the very existence of mental disorders. An influential counterculture emerged that regarded psychiatry as upholding outmoded and oppressive social norms. At the same time, non-medical mental health professionals challenged psychiatry's dominion over psychotherapy. Third party insurers that paid for mental health treatment came to question the efficacy of dynamic therapies. Psychiatry's sponsors in the federal government also became skeptical of expansive programs of prevention and social change. The field's very existence came into peril in this period.

Chapter 7 describes how the appearance of the DSM-III in 1980 resolved psychiatry's crisis of legitimacy. This new diagnostic system used a biomedical model that equated mental with physical disorders to reclassify all of the numerous conditions that had concerned dynamic psychiatrists. But, in contrast to the psychodynamic model, diagnostic psychiatry sharply split mental disorders from normal behaviors. At the same time, it was agnostic about the causes of mental illnesses. The manual's great accomplishment was to provide the specific diseases that psychiatry required for it to seem to be a genuine medical specialty. The DSM-III allowed psychiatry to reinvent itself as a scientifically grounded discipline, a portrayal that persists to the present.

The following chapter surveys the rise of neuroscientific studies of mental illness that have dominated research since the 1980s. Much like its 19th-century predecessors, this model regards understanding the brain as the key to unlock the mysteries of mental illnesses. It combines the DSM's classifications with a militantly biological approach to explaining various conditions. Neuroscientific findings, however, thoroughly conflict with the DSM model: they indicate that mental illnesses reflect overlapping general vulnerabilities more than isolatable, specific disorders, gradations rather than discrete entities, and a complicated array of inner and external causes that interact with brain-based mechanisms.

The penultimate chapter examines the paradox of the enormous institutional, social, and cultural successes of the DSM revolution with its inability to validly capture the nature of mental disorder. It focuses on three consequences of the DSM nosology. First, epidemiological studies produce vast, but useful, overestimates of the amount of mental disorder in the general population. Second, the DSM system has produced a divide between clinicians who find the extant system of great practical value and researchers who see it as hindering progress in understanding the nature of mental disorders and in developing targeted treatments for them. This led psychiatric researchers to attempt to replace the DSM's categorical taxonomy with a dimensional system in the DSM-5, published in 2013. Finally, the chapter discusses the DSM-5's elimination of the bereavement exclusion in the criteria for major depression and its implications for psychiatric diagnosis.

The concluding chapter surveys the arc of answers observers have provided to questions about the division between sanity and madness, the resemblances and distinctions of mental and physical illnesses, and the causes of and treatments for mental illnesses. It considers the extent to which current conceptions of mental disorders represent advances or, in some cases, regressions compared to historical understandings. It ends by speculating about future possible developments in responses to mental health and illness and in the psychiatric profession.

All societies have grappled with questions about the nature of mental illness, its commonalities with and differences from physical illness, and the forces that cause and might overcome it. Examining the changing answers they have provided allows us to see the extent to which current explanations surpass or are equivalent to—or in some cases, are even deficient compared to—prior understandings.

## Notes

1. Murphy, 1972.
2. Cawte, 1974, 56–57.
3. Jaspers, 1946, quoted in Shorter, 2015, 122.
4. Kessler et al., 2003a; https://www.nimh.nih.gov/health/statistics/mental-illness.shtml.
5. Moffitt et al., 2010.
6. Rohde et al., 2013.
7. World Health Organization, 2001, 37.
8. E.g. Rosen, 1968; Porter, 1997.
9. Rosen, 1968, 67.
10. Goffman, 1971, 356.
11. Edgerton, 1969, 59.
12. American Psychiatric Association (APA), 2013, 20.
13. Wakefield, 1992 1999. See also Klein, 1999.
14. Jackson, 1986, 45.
15. Griesinger, 2000, 226.
16. https://www.additudemag.com/michael-phelps-adhd-advice-from-the-olympians-mom/.
17. Wakefield, 1999.
18. Layton et al., 2018. Phelps was born on June 30; he would have been about ten months younger than the oldest students in his class.

19. APA, 2013, 697–700.
20. Spitzer & Wakefield, 2002.
21. Grant, 1978, 32.
22. Green, 2002, 467.
23. See especially Cosmides & Tooby, 1999.
24. Kendell, 1986, 25.
25. The distress or disability requirement emerged out of the controversy, recounted in Chapter 6 of this volume, over whether homosexuality should be seen as a mental disorder. The outcome was to eliminate homosexuality from the diagnostic manual unless it was "regularly accompanied by subjective distress and/or 'some generalized impairment in social effectiveness of functioning'" (Bayer, 1981, 127). This solution was illogical because if homosexuality was not a dysfunction, it could not be a mental disorder regardless of the degree of disability or disability it entailed. Later psychiatric manuals completely excised this condition from consideration as a mental disorder.
26. E.g. Wakefield, 1992; 2006.
27. It is, of course, possible that people who deny they have a serious mental illness are correct.
28. Jamison, 1993.
29. Benedict, 1934, 60.
30. Benedict, 1934, 79.
31. In fact, Benedict's own work indicates that at least some of the shamans she described had seriously impairing dysfunctions. She notes that Siberian shamans could be "violently insane for several years, others irresponsible to the point where they have to be watched constantly lest they wander off in the snow and freeze to death, others ill and emaciated to the point of death, sometimes with bloody sweat" (Benedict, 1934, 62).
32. Garbowsky, 1989.
33. Ego-dystonic homosexuality appeared in the DSM-III in 1980 but was removed from the next edition of the DSM in 1986.
34. Although the DSM uses the statistical term "expectable" to distinguish appropriate from disordered conditions, "explicable" seems to more accurately convey the intended meaning. See Busfield, 2011, 155.
35. Kleinman, 2012, 608.
36. Kleinman, 2012, 608.
37. Kendler, 2008.
38. Wakefield & First, 2013.
39. APA, 2013, 271–72.
40. Edgerton, 1966, 419, italics in original.
41. https://www.upi.com/FBI-docs-reveal-Sandy-Hook-shooters-interest-in-mass-murder/4481508889851/.
42. https://www.vanityfair.com/news/2014/03/peter-lanza-adams-father-newtown-shooting.
43. APA, 2013, 248.
44. Alan Gregg, quoted in Whooley, 2019, 120.
45. Sadler, 2013a, 455.

46. E.g. Wakefield, Kirk, & Pottick, 2006.
47. Ghaemi, 2013, 811.
48. Horwitz, 1982; Wakefield, 1999, 379.
49. The DSM-5 criteria for adjustment disorder provide a notable exception. This condition arises after a stressor and terminates within six months of the stressor, virtually the definition of a contextually appropriate response (APA, 2013, 286–87).
50. Wakefield, 1999.
51. Freidson, 1970, 208.
52. Quoted in Porter, 2002, 37.
53. U.S. Department of Health and Human Services, 1999.
54. Rosenberg, 2007, 13.
55. APA, 1994, xxi. Similarly, mental health advocacy groups encourage the use of language such as "a person with schizophrenia" instead of "a schizophrenic" to separate disorders from the individuals who have them. E.g. https://www.ncbi.nlm.nih.gov/books/NBK333029/.
56. Quoted in Shorter, 1992, 181.
57. Kraepelin, 1919.
58. Szasz, 1961.
59. Quoted in Scull, 2015, 171.
60. Churchland, 1984, 145.
61. E.g. Freud, 1900/1965, 1905/1962.
62. E.g. Beck, 1967, 1991.
63. See especially Porter, 2018.
64. E.g. Conley & Fletcher, 2018.

# 2
# Before Psychiatry

> Medicine is a science which hath been more professed than labored, and yet more labored than advanced; the labour having been, in my judgment, rather in circle than in progression. For I find much iteration, but small addition.
>
> —Francis Bacon, *The Advancement of Learning* (1605)

A dedicated profession of psychiatry did not emerge until the late 18th century in Europe and the mid-19th century in the United States. Nevertheless, before that time all groups confronted questions about how to define mental illness, which authority had jurisdiction over the mad, what caused madness, and the best ways to treat it. Their answers show both divergences from and similarities to more recent conceptions.

## Sanity and Madness in Ancient Greece

The Greeks produced most of the extant written sources regarding views of mental disorder in the ancient world. Their earliest literary works from about 800 to 700 BCE, the *Iliad* and the *Odyssey*, portray many characters who are driven mad through supernatural interventions. Greek physicians during the Classical period in the 5th and 4th centuries BCE produced the first sustained medical discussions about sanity and madness.[1] Hippocratic medicine rooted mental illnesses in the physical composition of brains and bodily fluids, developing sophisticated theories about mental disturbance that persist in various guises to the present. During the same era, Platonist philosophers, in certain ways foreshadowing Freud's theories, viewed madness as reflecting some inner psychic conflict or weakness. The widely varying external, organic, and psychological theories of mental illness that the Greeks developed, as channeled through Galenic medical culture in

Rome, provided the major templates for thought about mental illness for the following two millennia. Indeed, a "substantial unity" still exists between ancient and modern views of mental disorder.[2]

## Distinguishing Madness from Sanity

For the Greeks, mental illness referred to behaviors that clearly fell outside the boundaries of sanity and so were "mad," "crazy," or "lunatic." Socrates (c. 470–399 BCE) summarized this view when he stated that people

> Do not call those mad who err in matters that lie outside the knowledge of ordinary people: madness is the name they give to errors in matters of common knowledge. For instance if man imagines himself to be so tall as to stoop when he goes through the gateways in the Wall, or so strong as to try to lift houses or to perform any other feat that everybody knows to be impossible, they say he's mad. They don't think a slight error implies madness, but just as they call a strong desire love, so they name a great delusion madness.[3]

The mad acted in ways that diverged in extreme ways from commonly accepted social norms. The famed historian Herodotus (c. 484–425 BCE) used the Persian king Cambyses as an example of someone who is a "madman," "lunatic," "completely out of his mind," and "far from sound in his mind."[4] Cambyses, among other acts, murdered his brother and had incestuous relations with two of his sisters before killing both. For some philosophers, the irrational thoughts of madmen stood so far apart from commonly shared beliefs that they more closely resembled dreams than consciousness in waking reality. The pre-Socratic Heraclitus "stated that whosoever does not recognize the world in common is not of sound mind but insane, and he acts and speaks like a sleeper."[5]

Madness was a social judgment applied to behavior that violated expectations of what behaviors were sensible in given situations. "I see," Hippocrates observed, "insane and delirious persons doing inappropriate things for no obvious reason."[6] Aristotle (384–322 BCE) showed how such terms as "inappropriate" or "no obvious reason" were inherently contextual. He insisted that determinations of madness could only be made through examining the motives and reasons for why people behave the way they do.[7] Aristotle provided as examples of madness one man who killed and ate his mother and a second who murdered another man and then ate his liver.

Yet, Aristotle emphasized how it was not such behaviors themselves but only the context surrounding any action that provides the warrant for calling it a sign of "madness." Much like the DSM's general definition of mental disorder, Aristotle indicated that behaviors that sometimes indicate madness can be explicable and culturally approved when they occur as part of a social pattern. He provided the example of "some of the tribes about the Black Sea that have gone savage are said to delight in raw meat or in human flesh, or in lending their children to one another to feast upon."[8] In such settings, cannibalism was a learned behavior, not a form of mental disorder. A Black Sea tribesman ate human flesh because it was customary in his group; an Athenian who did the same had no culturally appropriate motivation for cannibalism and so was likely to have some mental disease. For Aristotle the dividing line between disordered and normal behaviors was inherently tied to the circumstances and cultures in which they arose.

The actions of the philosopher Diogenes (c. 404–323 BCE) illustrate the contextual nature of madness. Diogenes was an adherent of Cynic philosophy, which disdained conventional Greek norms, values, and institutions and advocated living according to the most basic laws of nature. He was well known for committing outrageous behaviors such as urinating on people who insulted him, defecating in theatres, and masturbating in public.[9] Yet, the Greeks regarded these activities, which they would undoubtedly have seen as signs of madness when they lacked any understandable motive, as ingenious demonstrations of the Cynic philosophy that held existing customs to be irrational. Therefore, they did not consider Diogenes to be mad.

Like philosophers, Greek physicians used context to distinguish who should be called "mad" from those who were sane or had other medical conditions. For example, Celsus (c. 25 BCE—50 AD), who wrote the influential *De Medicina* in the 1st century, emphasized how patients in the throes of some medical conditions could become delirious and talk nonsense. They were, however, distinct from the mad because their comparable symptoms disappeared as soon as their fevers dissipated.[10] Consequently, their acts were contextually explicable and not signs of madness.

### Diagnoses

Greek medicine in general was not diagnostically oriented. "It is more important to know what sort of person has a disease than to know what sort of disease a person has," Hippocrates reputedly said.[11] Accordingly, Greek nosology contained only a few basic categories of mental disorder, most importantly, mania, which was marked by delirium, frenzy and, sometimes, violence, and melancholy, which was associated with chronic isolation, brooding, fear, anxiety, and sadness without cause.[12]

Mania was an agitated form of insanity, which could appear in either a violent or a comic subtype. It embodied a stereotypical conception of mental disorder that persisted for thousands of years. According to Hippocrates, agitated madmen were "screamers, restless, troublemakers, and repeatedly doing something inappropriate."[13] Psychiatrist Bennett Simon summarizes:

> What was, as far as we can reconstruct it, the Greek stereotype of the madman? First, there are the physical signs: raving, roaming around or running wild, eyes rolling, sweating, drooling, foaming at the mouth. There is a greater emphasis on visual disturbances than auditory ones: terrifying visual images cause or accompany madness. Also, madmen do things that are contrary to all good Greek custom, such as deeds that are harmful to their friends and helpful to their enemies.[14]

While some madmen were angry and violent, others acted in incomprehensible yet inoffensive and often comic ways.[15] The celebrated physician Aretaeus, for example, contrasted one type "with whose madness joy is associated, laugh, play, dance night and day, and sometimes go openly to the market crowned, as if victors in some contest of skill; this form is inoffensive to those around," while in the other type "madness is attended with anger and these sometimes rend their clothes and kill keepers, and lay violent hands upon themselves."[16] Both the comic and the violent kinds were united by the fact that their behavior could not be accounted for within commonly understood cultural categories. Aretaeus anticipated diagnoses of manic-depressive or bipolar disorder when he described how "Some patients after being melancholic have fits of mania . . . so that mania is like a variety of melancholy."[17]

Melancholy, the second broad category of mental disturbance, entailed a quieter form of madness.[18] This condition, which is now called "depression," is probably the psychological disorder that is most easily recognizable throughout history; similar symptomatic descriptions occur over a 2,500-year span, representing what historian and psychiatrist Stanley Jackson calls a "remarkable consistency."[19] In the 5th century BCE, a Hippocratic text provided the first known definition of melancholia as a distinct disorder: "If fear or sadness last for a long time it is melancholia."[20] The Hippocratics routinely grouped sadness and fear together as symptoms of melancholia because melancholics were generally worried or morose about actual experiences as well as apprehensive about suffering from future negative events. In addition to fear and sadness, possible symptoms included "aversion to food, despondency, sleeplessness, irritability, [and] restlessness," indications of melancholia that are extraordinarily similar to current definitions of depression.[21]

Ancient medicine differentiated melancholia from normal loss responses because they were "without cause"; that is, they could not be explained by reference to the context in which they arose and so were incomprehensible to others. In addition, Hippocratic definitions indicated that it is not such symptoms alone but symptoms of unexpected duration that indicate disorder ("last for a long time"). Thus, melancholic disorders differed from normal reactions because they either arose in the absence of situations that would normally produce sadness or were of disproportionate severity or duration relative to their provoking causes. This insistence that melancholic sadness or fear must be prolonged is a first medical attempt to capture the notion that disproportion to circumstances is an essential aspect of mental disorder. Such conditions indicated that something was wrong in the individual, not in his or her environment.

Greek medical discussions of melancholy consistently used context to distinguish melancholic disorders from nondisordered types of deep sadness or fear that could have many of the same symptoms but that were normal, proportionate reactions to serious losses. Such losses included the death of intimates, reversals in fortune, failures to attain valued life goals, romantic disappointments, and the like. Just as external losses could lead to normal sadness, regaining what was lost could lead sadness to disappear. For example, Aretaeus described one case that featured a melancholic girl who "recovered when she had back her loved one . . . she was cured by the physician Love."[22]

Well before the Hippocratics, in the *Iliad* Homer (c. 8th–7th century BCE) provided the example of Achilles' unreasonable grief after the death of his friend Patroclus, which he contrasted to normal grief:

> A sane one may endure
> An even dearer loss: a blood brother
> A son; and yet, by heaven having grieved
> And passed through mourning, he will let it go.[23]

Men naturally grieve after the death of intimates, but their suffering typically diminishes with the passage of time. In contrast, Achilles' grief was not natural because it persisted for an unreasonable period and he could not "let it go." Similarly, the Greeks considered trance and possession states, seizures, and frenzied behaviors as appropriate when they occurred in the proper settings of religious rituals or among people in valued prophetic roles. The same behaviors, however, were likely signs of insanity when they arose in everyday interactions.[24]

A century after Hippocrates, Aristotle (or one of his students) in the *Problemata* elaborated the distinction between a variety of normal mood states of sadness on the one hand and disease states on the other. He clearly expressed the idea that disordered sadness is disproportionate to events because it was marked by "pathological fears and excessive imagination." The terms "pathological" and "excessive" distinguish melancholic disorders from normal forms of sadness. Aristotle also noted that, if the black bile "be cold beyond due measure, it produces groundless despondency."[25] Here "beyond due measure" refers to what is disproportionate to the circumstances, making the resultant sadness "groundless."

The key distinction in ancient definitions of melancholia was thus between states of sadness without cause and those with similar symptoms that arose from actual losses; only the former were mental disorders. But "without cause" did not mean uncaused, for throughout history melancholy has been attributed to postulated physical or psychological causes such as excessive black bile, disturbances in the circulation of blood, or depletion of energy. Rather, "without cause" meant that the symptoms of melancholia were not proportional to environmental events that would appropriately lead to sadness, such as bereavement, rejection in love, economic failure, and the like. Conversely, ancient physicians did not consider melancholic symptoms that occurred "with cause" as signs of a mental disorder because they were explicable reactions within their contexts.[26]

In addition, Aristotle and others acknowledged that variations in temperament (what we now call "personality") predispose some people to more readily or intensely experience sadness or fear but that these variations could be within a normal range of reasonably proportionate responses that were not disorders. The Hippocratics categorized people as having choleric, sanguine, melancholic, or phlegmatic temperaments depending on the relative mixture of the four bodily fluids of yellow bile, blood, black bile, or phlegm they possessed. Aretaeus noted that melancholy was more frequently found among those whose personalities are "already inclined to sadness" because of the particular mixture of humors within them.[27] In this vein, Stoic philosophers noted how "anxiety makes itself manifest in people who already were anxious, sadness in men inclined to that sentiment, fear in timid men."[28] The difficulties of distinguishing normally introverted or fearful personality characteristics from depressive or anxiety disorders persist to this day.

Galen, a Greek doctor residing in Rome during the 2nd century AD, was the most influential figure in psychiatric nosology for the next millennium and beyond. He was the towering physician in Roman medicine, refining Hippocratic notions and synthesizing them with Aristotelian philosophy. Following the Hippocratics, Galen distinguished normal grief that stems from such causes as the death of intimates or the devastation of war from grief without any adequate external cause.[29] He emphasized how melancholic disorders were marked by prolonged fear and discouragement without cause—"fear and sadness without a real reason."[30]

While most Greek discussions involved clear cases of madness, physicians occasionally recognized less severe forms of disorder. Anxiety provides an example. Anxiety disorders were not signs of madness but were marked by the Hippocratic criteria of being "prolonged" fears that endured "for some time."[31] For example, a medical text describes the case of Democles who had such a severe fear of heights that he could not walk on a bridge that was over even a very low ditch.[32] Later, Aretaeus observed that some people experienced symptoms of fright such as pounding heart and disturbances in the chest, although nothing alarming had occurred to them.[33] Those who were afflicted experienced intense fear despite knowing that they were not in danger. Therefore, their anxiousness was "unreasonable" to sufferers themselves as well as to others. Unlike natural fears that arose in dangerous contexts, these cases indicated that fear emerged from some internal derangement rather than from an appropriate external cause.

In essence, then, ancient medicine and philosophy took a *contextual* approach to the diagnosis of mental disorders; whether a condition was considered as disordered depended not just on symptoms, which might be similar to conventional behaviors, and not just on the condition's severity, for explicable responses to extreme situations can be severe, but on the degree to which the symptoms were understandable responses to circumstances. They also recognized that non-disordered aspects of temperamental qualities could be associated with sadness, fear, or mania. The doctrine that emerged in the period between Hippocrates and Galen, which distinguished disordered states that stemmed from emotions that were "without cause" or "lasted for a long time" from those that were proportional reactions to external circumstances or aspects of temperaments, persisted for thousands of years.

## Mental Illness and Physical Illness

The Greeks developed sophisticated theories of both the similarities and the differences between bodily and mental diseases. On the one hand, Hippocratic medicine provided a foundational account for how mental illnesses resembled physical ones. On the other hand, Platonic philosophy anticipated psychodynamic accounts about the distinctive nature of mental as opposed to physical illnesses.

### Hippocratic Medicine

The emergence of Hippocratic medicine during the 5th century BCE provided a physical foundation for the study of mental disease.[34] Its approach was single-mindedly empirical and thoroughly opposed to any invocation of supernatural, magical, or religious forces. The Hippocratic insistence on careful observation and description of symptoms profoundly influenced subsequent medical practice. Indeed, its rendering of bodily and psychological disorders in naturalistic, organic terms was a major and lasting turning point in the history of human thought. The Hippocratic School stressed the physiological basis of all human behavior, emphasizing how mental disorders were psychic reflections of brain disturbances:

> Men ought to know that from the brain, and from the brain only, arise our pleasures, joys, laughter, and jests, as well as our sorrows, pains, griefs and tears. . . . It is the same thing which makes us mad or delirious, inspires us with dread and fear, whether by night or by day, brings sleeplessness, inopportune mistakes, aimless anxieties, absentmindedness, and acts that are contrary to habit.[35]

Physiology, not mythology or psychology, was at the root of madness (and sanity) no less than of organic diseases: "the brain is the most powerful organ in man" underlying all mental and emotional states.[36]

The foundational principle of Hippocratic medicine was that health is a state of equilibrium between the body and its environment while disease—mental as well as physical—stems from some kind of disturbance to this balance.[37] The humors of blood, phlegm, yellow bile, and black bile were what were kept in or out of balance. Each humor possessed two of the four basic qualities of hot, cold, moist, and dry. When the humors were in line with each other, a healthy state resulted. Anticipating modern theories of chemical imbalance as the basis of mental disorder, diseases stemmed from an excess or deficit of one of these humors. A surplus of yellow bile or blood could result in mania; too much black bile led to melancholy.[38]

Hippocratic treatments for mental and physical disturbances also resembled each other, both featuring combinations of environmental and organic responses. Restrained lifestyles were particularly important facilitators of health. Greek physicians counseled that rest, dietary restrictions, pleasant walks, warm baths, and massages could alleviate mental agitation and help relieve bodily ailments.[39] They also used somatic measures, especially opium, to cool overheated brains and produce calm, tranquil states. Psychotherapeutic measures could also help relieve all forms of disorder.[40]

The Hippocratics left a lasting legacy that equated mental and physical illnesses. Both types of disturbances disrupted a holistic relationship between individuals and their surroundings. Healthy bodies were accompanied by healthy minds as well as vice versa. Conversely, disturbed minds resulted in somatic ills just as disturbed bodies distorted mental functioning.[41] The causes of both psychic and organic disorders—for example, foul air and water, toxins, poor interpersonal relationships, bad diets—were indistinguishable. Likewise, treatments for mental as well as physical ailments involved restoring states of equilibrium.[42] No sharp lines distinguished any aspect of psychic from organic conditions.

## Platonic Philosophy

Alongside Hippocratic medicine, philosophers in the 5th and 4th centuries BCE developed a new conception of the distinctively psychic nature of mental disorder that remains influential at present. By this time, a worldview dominated by scientific and rationalistic approaches had replaced the Homeric world of mythological heroes. The uncontrolled, impulsive, and unreasonable nature of madness contrasted with the ideals of moderation that characterized Greek culture in this period.

Plato (c. 428–c. 348 BCE) developed the most influential psychological account of madness. Unlike Hippocratic portrayals that unified minds and bodies, for Plato mental illnesses afflicted the soul, not the physical organism. He divided the soul into three components: reason, instinct, and emotion.[43] The ability to reason was the primary factor that elevated humans above other species that were dominated by their instincts. It also allowed people to comprehend the unchanging principles of reality that underlay transient and deceptive outer appearances.

Classical philosophy's focus on the conflict between reason and passion contrasted with the organic basis of the Hippocratic model. Presaging Freud, Plato developed a model of the mind as a battlefield where competing rational and irrational parts fought each other.[44] The mind contained irrational instincts and emotions that reason must struggle to contain. In the *Phaedrus* Plato famously developed a metaphor of the mind as a charioteer who drives a wagon pulled by two horses, one noble and the other wild.[45] Madness resulted when the irrational part of the mind overcame the rational component and led men to act against their own self-interest. People who became mentally ill had lost control of their passions to the extent that they behaved in ways that violated taken for granted customs.[46] In contrast, the sane were able to use their reason to control the expression of their passions. Rational thought was the major means through which humans could contain and master their underlying destructive urges. The Roman philosopher Seneca's maxim, "The wise man checks his passions and the fool follows them," perhaps best summarizes this rationalist approach.[47]

The inexplicable and unreasonable nature of madness sometimes had positive effects. For example, Plato indicated that no one could attain prophetic truths when they were in their right minds.[48] "Our greatest blessings," he had Socrates write in the *Phaedrus*, "come to us by way of madness."[49] Although Socrates recognized that some forms of madness could lead to poetic inspiration and prophesy, in general, rational forms of actions were superior to irrational forms.[50] The benefits of madness were only present when they "are given by divine gifts" that were distinct from madness of human origin, which was a disease.[51] Both the inspired and the pathological types of insanity differed from organic conditions.

## Internal and External Sources of Madness

Before the Classical period, the Greeks viewed madness as arising from external forces that were outside of individual souls or brains. All early civilizations attributed incomprehensible departures from normal behavior to the involvement of supernatural agencies. Divine intervention caused mental disorder, especially among those who had incurred the anger of the gods. This connection to the supernatural set the afflicted apart from ordinary people.

The earliest depictions of madness in the *Iliad* and *Odyssey* usually stem from supernatural intrusions. The afflicted were typically normal before the god intervened. For example, after her nurse Eurykleia tells Odysseus' wife that her husband has returned home, the incredulous Penelope declares: "Dear old nurse the gods have made you mad. They have that power, putting lunacy into the clearest head around or setting a half-wit on the path to sense. They've unhinged you, and you were once so sane."[52]

Madness was also a persistent theme in Greek tragedy in the Classical Age; many of its best-known characters including Oedipus, Phaedra, and Orestes become mad.[53] Like Homer, tragedians viewed divine intervention as the source of their madness. Sophocles' *Ajax* provides a prominent example.[54] It features a warrior who was reputed to be the greatest Greek fighter save only Achilles. After Ajax is passed over for receiving the armor of the slain Achilles in favor of Odysseus, he raises his sword against his superior officers but is blinded by the goddess Athena who drives him mad. He goes into a berserk rage and slaughters many cows, sheep, and goats, believing that they are men. Eventually, Ajax falls on his own sword and dies. Aeschylus' *Orestes* provides another example.[55] The Furies drive Orestes mad as vengeance for the murder of his mother. While in the grip of madness, he has terrifying visual hallucinations that mark him as insane. In *Bacchae* Euripides describes how the god, Dionysius, "stung these women into madness. . . . All the females, all the women of Thebes—I sent them crazy from their homes." Euripides describes how Agave "was foaming at the mouth, face twisted, eyes rolling, not thinking as she ought to think. She was possessed by Bacchus."[56] Agave tears off the head of her son, Pentheus, believing that he is a mountain lion.

The supernatural view also persisted in Herodotus' writings. The historian continued to recognize a form of madness that resulted from divine intervention, although he also noted a form that arose through natural causes such as epilepsy or heavy drinking.[57] He provided detailed descriptions of two mad kings, of whom one was internally deranged and the other the victim of divine retribution. Herodotus speculated that the Emperor Cambyses' murderous and incestuous behaviors stemmed from "the fact that a serious physical malady should have affected his brain." He also described the case of the Spartan king Cleomenes, who was always "a little strange in the head" but then "went quite mad," mutilating himself with a knife and "sliced his flesh into strips, working upwards to his thighs, and from them to his hips and sides, until he reached his belly, and while he was cutting that into strips he died."[58] In contrast to Cambyses, Herodotus attributed Cleomenes' derangement to divine punishment for his sacrilegious behaviors.

Because laypeople often believed that divine curses were the source of madness, the Greeks typically ridiculed and shunned those they regarded as mad.[59] The public simultaneously stigmatized the insane and viewed the mad as possessing unusual powers that allowed them access to an ethereal realm.[60] In the 5th century BCE:

> It is clear from various sources that the madman, even when regarded as in some way touched by the divine was a person to be shunned. Contact with holiness, like contact with its opposite, uncleanliness, was perilous and to be avoided. In fifth century Greece, madness was widely considered the consequence of a divine curse, and an insane person was therefore polluted and a thing of evil omen.[61]

Treatments for the afflicted similarly relied on invoking divine healing interventions, often through visits to religious shrines. The supernatural conception of madness persisted for many centuries, particularly among laypersons who believed that the insane suffered from demonic possession.

Classical medicine and philosophy dramatically changed the previous emphasis on external, supernatural reasons for madness. The Hippocratics fiercely rejected the notion that divine intrusions caused madness. They did not see any sharp dichotomy between internal and environmental forces. Melancholia, which literally means "black bile disorder," exemplifies the ancient belief that health and disease depended on the balance or imbalance between four bodily fluids, or "humors." It was connected to an excess of black bile—a humor thought to be produced in the spleen.[62] Some cases of this humoral imbalance arose from internal causes, such as heredity, while others stemmed from external factors including poor food, climate, air, or water. Terrifying external events could also lead to severe states of mental illness because they led the humors to become imbalanced.[63] The Greeks thus sometimes emphasized internal causes of mental disorder, sometimes environmental sources, and, often, the interactions between inner and outer forces.

## Conclusion

For the period that ran from the first literary documents through the end of Greco-Roman civilization, the ancients conceived of madness as a radical departure from commonly understood social norms. They used the criterion of whether any action had a rational, contextually relevant relationship with its perceived cause to distinguish the mad from the sane. Yet, around this consensual notion of what mental illness is, they developed a variety of theories for why some people became mad. The initial notions, which persisted in lay cultures for centuries, concentrated on divine or demonic possession as the cause of insanity. During the Classical Age, philosophers emphasized how internal conflicts led irrational forces to overcome reason with resulting mental disturbance. Finally, Hippocratic physicians focused on imbalances in physiological factors as reasons for mental illness. In his influential synthesis, Galen joined a focus on the nervous system with the Hippocratic emphasis on the humors.[64] The basic ideas of the Hippocratic corpus, as reflected through Galen's works, comprised the core of Western medical thought about mental disorder for the next 1,500 years. One or the other of the Greeks' pluralistic theories has continually resurfaced in Western thought through the present.

## The Long Pre-Modern Period

For the millennium and a half from the fall of Greco-Roman civilization through the establishment of the psychiatric profession in the 19th century, definitions of madness remained fairly stable. The term continued to refer to a narrow range of incomprehensible behaviors that fell outside of commonly understood cultural categories. Stereotypical madmen resembled those of Ancient Greece—people disconnected from reality who were unpredictable, violent, and outcast.[65] Others experienced the deep dejection that characterized melancholic states. Explanations and treatments for their conditions, however, profoundly changed.[66]

The Roman Emperor Constantine's conversion to Christianity in the early 4th century AD began a new era marked by a religious worldview that largely supplanted the empirical conceptions of Hippocrates and Galen as well as Plato's rationalist philosophy. The Christian emphasis on the immateriality and immortality of the soul and its popular counterpart of magical practices thoroughly transformed organic and material conceptions of disease to views grounded in faith, sin, divine will, and the potency of nonmaterial forces.

Early Christianity, rejecting previous naturalistic and rationalistic interpretations, returned to the archaic mystical explanations of madness that prevailed before the Classical period in Greece. Through the medieval era, lay accounts of mental illness focused on supernatural intervention, witchcraft, and sinful thoughts and behaviors as sources of insanity. Philosophers such as St. Augustine and St. Thomas Aquinas, too, explained madness as resulting from the demonic possession of spiritually lapsed souls. Prevention and cure were generally not medical concerns but consisted of true belief in Christian tenets supplemented by clerical exorcisms and visits to religious shrines.

St. Augustine (354–430 AD), the most influential theological writer in this era, emphasized that relief from mental suffering stemmed from faith in the teachings of Jesus Christ. After a period of profound agonizing, he recognized the importance of Christian principles: "I neither wished nor needed to read further. At once, with the last words of this sentence, it was as if a light of relief from all anxiety flooded into my heart. All the shadows of doubt were dispelled." Augustine's notion that "our heart is restless, until it repose in Thee" epitomized the idea that faith in God was the best therapy for mental disturbances.[67]

In this period, religion could cause, as well as remedy, distress. Belief in God and an eternal afterlife relieved psychic distress but at the same time led to tremendous uncertainty. Preoccupations with whether one was a member of the elect who would be chosen to enter the kingdom of heaven and guilt over the consequences of sinning were potent generators of emotional disturbance. Fear of perpetual damnation in the afterlife was a particular source of terror that persisted through the Renaissance and Reformation. Witches and devils also had powerful influences on illnesses in an era where mystical views predominated.[68]

The Christian epoch not only changed views of mental illness from an empirical to a spiritual focus but also more generally downplayed the status of medicine. As religious and magical concepts of healing reappeared, medicine itself was subjected to ecclesiastic control.[69] The Church emphasized the salvation of souls more than the healing of bodies so that priests gained primacy over doctors. The latter continued to emphasize Hippocratic notions of maintaining a balance of the humors and of restoring equilibrium when an excess or deficiency of some humor led to some disease. However, physicians had neither the cultural authority nor the practical skills to override the dominant spiritual interpretations of mental illness. Based on descriptions of faith-based cures in the Gospels, patients required religious piety while healers needed divine powers. Much of the populace believed the numerous accounts of miracle cures at saints' shrines.

Through the 17th century, demonic possession persisted as a major aspect of madness.[70] The possessed lost their autonomy and came under the control of external, often supernatural, powers. Exorcism, which was usually conducted by religious authorities, was the treatment of choice for expelling the occupying forces. Such states were strongly gender-linked: women constituted at least three-quarters of the afflicted. Beginning in the late medieval period, women were also linked to the practice of witchcraft, and many thousands of perceived witches were tortured and executed.

## The Persistence of the Hippocratic Corpus

While physicians became subordinate to theologians and medical thought itself languished for over 1,000 years, their basic beliefs did not change. Indeed, from Galen's time through the Renaissance, the psychiatric corpus was virtually undisturbed. The Hippocratic-Galenic tradition that featured a small number of mental diseases persisted in medical texts. For example, during the 11th century, the Persian physician Avicenna (980–1037), author of the influential *Canon of Medicine*, defined the signs of melancholy as "bad judgment, fear without cause, quick anger, delight in solitude, shaking, vertigo, inner clamor, tingling, especially in the abdomen."[71] Renaissance physicians also retained earlier Hippocratic "without cause" conceptions of melancholy. Sadness and fear were still the chief characteristics of melancholic conditions, which were mental disorders when they were "without cause." The notable Swiss physician, Felix Platter (1536–1614), for example, called melancholy a "kind of mental alienation, in which imagination and judgement are so perverted that without any cause the victims become very sad and fearful."[72] In 1745, an English medical dictionary continued to define melancholy as "alienation of mind, long continued dejection, dread and sadness without any manifest cause."[73]

The humoral theory of disease endured in medical understandings and treatments of illness until the end of the 17th century and, sometimes, beyond.[74] Humoral thought was foundational not only in the culture of physicians but also in the medical lore of common people and the lay healers who often treated them. Each humor was associated with a certain temperament, and each temperament was in turn associated with specific kinds of maladies. Diseases resulted from imbalances between the various humors: treatments aimed to correct such imbalances and restore the body to appropriate equilibrium. Fresh air; exercise; good sleeping, eating, and elimination habits; and control of the passions remained prominent treatments for melancholic and other mental conditions. Such treatments were typically intertwined with religious, magical, and folkloric methods.[75] The work of three men, Robert Burton, Richard Napier, and William Shakespeare, illustrate the prevailing conceptions of mental illness during the late 16th and early 17th centuries.

## Robert Burton

At the beginning of the 17th century, melancholia, a capacious category that emphasized a broad mixture of depressive and anxious symptoms, was the most prominent diagnosis of mental disturbance. English vicar Robert Burton's (1577–1640) encyclopedic *The Anatomy of Melancholy*, initially published in 1621, culminated a 2,000-year-old tradition that began with the Hippocratic corpus.[76] It was not only the major compendium of literature about this condition but also the most comprehensive description of melancholy ever amassed. Running to nearly 1,500 pages, Burton's magisterial work surveyed the entire sweep of writing on melancholy, beginning with the Bible, Ancient Greeks and Romans, and ending with contemporary 17th-century studies. Burton described the emotional, cognitive, and physical components that remain at the heart of definitions of depression. He followed the Hippocratics in emphasizing contextual distinctions between normal and disordered states, the similarities of mental and physical disorders, and the wide range of both internal and external causes of melancholy.

Burton's work illustrates the persistence of the Hippocratic tradition that distinguished normal from abnormal conditions through the contextual "without cause" criterion. He sharply distinguished those types of melancholy that were normal responses to circumstances from those that indicated mental disorders. Burton emphasized that a propensity to melancholy was often a normal and ubiquitous aspect of the human condition that was present in all people:

> Melancholy . . . is either in disposition or habit. In disposition, it is that transitory melancholy which goes and comes upon every small occasion of sorrow, need, sickness, trouble, fear, grief, passion, or perturbation of the mind, any manner of care, discontent, or thought, which causeth anguish, dullness, heaviness, and vexation of spirit. . . . And from these melancholy dispositions, no man living is free. . . . Melancholy, in this sense is the character of mortality.[77]

Transitory melancholic dispositions were explicable and, indeed, indicative of "the character of mortality" when they were responses to occasions of "sorrow, need, sickness" and the like.

However, Burton insisted that melancholic symptoms were not in themselves evidence of disorder; it was only when such normal reactions to specific events became established as an ongoing condition independent of events that Burton saw disorder:

> (I)t falleth out oftentimes that these Dispositions become Habits, and . . . make a disease. Even as one Distillation, not yet growne to custome, makes a cough; but continuall and inveterate causeth a consumption of the lungs: so doe these our Melancholy provocations. . . . This Melancholy of which we are to treat, . . . a Chronicke or continuate disease, setled humor . . . not errant but fixed, . . . growne to an habit, it will hardly be removed.[78]

In contrast to normal (or "ordinary") cases of melancholy that arose naturally in people who have suffered losses and disappointments, Burton held that melancholic afflictions were "contrary to nature."[79] Echoing the "without cause" tradition, he defined the disorder of melancholy as "a kind of dotage without a fever, having for his ordinary companions fear and sadness, without any apparent occasion."[80] This tradition persisted through the mid-20th century under the label "endogenous depression."

Although Burton occasionally separated the depressive and anxious components of melancholy, he typically combined fear and sorrow in his descriptions. "Cousin-german to sorrow is fear," he noted, "or rather a sister, *fidus Achates* [trusty squire], and continual companion, an assistant and a principal agent in procuring of this mischief; a cause and symptom as the other."[81] Clusters of symptoms marked by fear, anxiety, and apprehension were often indistinguishable from moods of sadness, despondency, and despair.

As much as sorrow, anxiety was also a fundamental aspect of human nature. "Great travail is created for all men, and an heavy yoke on the sons of Adam, from the day that they go out of their mother's womb, unto that day they return to the mother of all things. Namely, their thoughts and fear of their hearts, and their imagination of things they wait for, and the day of death." Moreover, anxiety was a universal affliction that touched everyone: "From him that sitteth in the glorious throne, to him that sitteth beneath in the earth and ashes; from him that is clothed in blue silk and weareth a crown, to him that is clothed in simple linen."[82] Similar to his separation of melancholic symptoms that were natural or pathological, Burton's Hippocratic distinction between fears that were with or without a cause provided a template to separate anxiety disorders from contextually grounded normal fears.

Like the Hippocratics, Burton assumed that both external and internal forces could bring about melancholic disorders. He postulated a huge range of possible causes of melancholy that spanned from supernatural interventions, magic, and witchcraft, to external life events, to internal forces such as excessive jealousy and anger or the ravages of physical diseases. In line with Hippocratic principles, Burton also noted how disturbances in factors such as diet, sleep patterns, and air quality could bring about melancholic diseases.

Burton astutely observed the extremes to which normal reactions to loss could go. He noted that the most extremely painful losses included separation from friends and bereavement following loss of a loved one ("in this Labyrinth of accidental causes . . . loss and death of friends may challenge first place"[83]) and compellingly described the extremes that nondisordered grief can reach:

> If parting of friends, absence alone, can work such violent effects, what shall death do, when they must eternally be separated, never in this world to meet again? This is so grievous a torment for the time, that it takes away

> their appetite, desire of life, extinguisheth all delights, it causeth deep sighs and groans, tears, exclamations, . . . howling, roaring, many bitter pangs, and by frequent mediation extends so far sometimes, they think they see their dead friends continually in their eyes, . . . Still, still, still, that good father, that good son, that good wife, that dear friend runs in their minds; a single thought fills all their mind all year long. . . . They that are most staid and patient are so furiously carried headlong by the passion of sorrow in this case, that brave discreet men otherwise oftentimes forget themselves, and weep like children many months together.[84]

In addition to grief, melancholy had an especially intimate link to love: "Every poet is full of such catalogues of love-symptoms; but fear and sorrow may justly challenge the chief place . . . love melancholy . . . 'Tis full of fear, anxiety, doubt, care, peevishness, suspicion." Burton went on to note: "Now if this passion of love can produce such effects if it be pleasantly intended, what bitter torments shall it breed when it is with fear and continual sorrow, suspicion, care, agony, as commonly it is, still accompanied! What an intolerable pain must it be!"[85]

Burton proposed a cornucopia of recommendations for the treatment of melancholy. Many echoed the tenets of Hippocratic medicine: restoring balance to diet, exercise, surroundings, sleep, and emotion. Talking with friends, physicians, and clergy; soothing music; and diverting activities could also serve as healing forces. Blood-letting and purges were often efficacious. Burton was also partial to alcohol. "A cup of wine or strong drink," he observed, "takes away fear and sorrow."[86]

Burton's work illustrates the long persistence of the Hippocratic tradition that used context to separate normal from pathological symptoms, did not clearly separate mental and physical conditions, and grounded melancholy in a broad combination of organic, psychological, and external forces. More scientific approaches that strove to specify a variety of psychic disturbances and root them in the operation of physiological processes would soon supplant these ancient tenets.

## Richard Napier

Most of what we know about the nature of mental disorder through the 17th century stems from medical, philosophical, or theological writings. At the time, no practitioners specialized in the care of the mentally ill.[87] Instead, a tremendous heterogeneity of healers emerged, most using treatments involving some combination of religious folklore, beliefs in demonic possession, and Hippocratic notions about the humors.

Few documents survive about mental disturbances and their treatment among ordinary healers and patients. A remarkable exception to this lack of sources are the 767 records of people who sought help from one 17th-century rural English physician and clergyman Richard Napier (1559–1634), the subject of historian Michael MacDonald's book *Mystical Bedlam*.[88] They vividly show the prominent role of normal and, more rarely, disordered mental conditions in the complaints that sufferers brought to this general practitioner. Few of Napier's cases—about 5%in all—were mad. Most had conditions that would currently be diagnosed as anxiety or depression; the vast majority of these sufferers "themselves frequently judged that their emotions were abnormal."[89]

Napier classified psychic conditions into two general sorts. The first stemmed from universal experiences of sorrow and grief, rejection in love, loss of fortune, severe illness, religious despair, and conflicts with spouses, lovers, or parents. Napier explicitly separated these sorts of ubiquitous adverse states from a second category of melancholic diseases because "not every gloomy person suffered from the disease of melancholy."[90] A few of Napier's patients' complaints went beyond ordinary suffering and seemed to reflect true disorders.

Like the Hippocratics and Burton, Napier used the criterion of disproportion to actual circumstances to distinguish the mad from the normally distressed:

> Everybody was afraid occasionally of the perils that many of Napier's melancholy patients complained about: devils, death, illness, accident, disgrace, robbery, and witches, for example. The apprehensions of patients . . . that they would die or be driven mad by disease were tokens of melancholy because they did not seem ill enough to justify such fears. . . . Ann Wilson thought that the medicine a physician had given her harmed her unborn child. The idea that women should avoid dosing themselves with physic during pregnancy was commonplace; Wilson's fearfulness was melancholy because it persisted even after the child's healthy birth had vindicated the careless physician. . . . Similarly, although violent crime was endemic, persons . . . who claimed that some unknown malefactor was going to kill them,

> turned a legitimate apprehension into a melancholy fear by detaching it from any plausible situation or dangerous enemy.[91]

These cases indicated melancholic disorders because they could not be encompassed within the commonsense assumptions village culture held about sanity. Such deranged conditions that were detached from their contexts seem to have been relatively rare among this group.

Napier considered two kinds of melancholic states as disorders. First, he used the term "baseless sorrow" for some of his disordered patients.[92] "Baseless" referred to cases that were unprovoked or delusional, thus wholly unexplained by external circumstances. Napier's records show that although melancholia often arose without situational provocations, it sometimes stemmed from a disproportionate response to actual losses. This second type of disordered condition came from sources such as "legitimate occasions in the death of loved ones and were revealed to be the sign of melancholy delusion by their unusual intensity and duration."[93] As MacDonald notes, "Contemporaries believed that the feelings experienced by melancholy and troubled people were exaggerations of normal states of mind. The sheer intensity of their moods was abnormal."[94] Many melancholic diagnoses, for example, resulted from bereavement, usually after the loss of a spouse or a child, where the sadness was of such intensity and duration that it led to states of madness.[95] Judgments of disease consequently required the physician to obtain knowledge of the relationship of the symptoms to the context of the situations in which they arose and persisted.

Common people during this period had many sources of loss and disappointment. Debt and the consequent fear of poverty was the greatest single generator of complaints that brought patients to Napier. Men, in particular, worried about the lack or loss of money. Fear of disease was also a prominent concern: "Seventeenth-century Englishmen were death's familiars, for epidemics, consumption, parasites and dysentery, accidents, infections, and botched childbirths killed children and adults, family and friends, earlier and more suddenly than the diseases we dread today." Nearly half the cases of deep depression that Napier handled involved intense grief that followed a child's death. Fears of bewitchment or witchcraft accusations were also acute sources of anxiety within the closed, intimate world in which these villagers resided. The perceived witch was likely be an everyday presence in one's life and impossible to avoid. Such fears bedeviled about two-thirds of Napier's patients. Anxiety over sins that could lead to eternal damnation and religious despair also plagued these sufferers.[96]

Problems involving courtship and married life, especially lover's quarrels, unrequited love, parental objections to a love object, and double dealing were another central concern. Such difficulties afflicted nearly 40% of Napier's clients. Among patients complaining of marital problems, 84% were women (then, as now, about two-thirds of all patients were women). Age was also related to treatment: Young adults between 20 and 29 comprised a disproportionate number of Napier's patients. Uncertainties over their futures were prominent reasons for seeking advice. "Many of these young people complained to their physician about the anxieties of courtship and marriage and the uncertainties of getting a living and bearing children, problems that accompanied the transition from youthful dependence on parents and masters to full independence as married adults."[97]

The lack of distinction between mental and physical illnesses that marked Hippocratic medicine also persisted into the 17th century. No distinct causes or prognoses separated mental and physical illnesses. Likewise, no discrete treatments were used for mental complaints. Napier's patients who suffered from life problems received similar responses as those with other problems:

> The regimen of treatment for mad and troubled patients differed little from the measures used to cure other patients. Regardless of their symptoms, almost every one of Napier's mentally disturbed patients was purged with emetics and laxatives and bled with leeches or by cupping.[98]

Like physicians more generally, Napier employed a "therapeutic eclecticism" that combined medical with astrological and psychological interventions.[99] Typical of medical practitioners at the time, Napier also used prayer and protective amulets to help heal his patients.

The ubiquitous distress and anxiety of Napier's patients were usually linked to the normal uncertainties of existence, not to madness or melancholia. Worries about courtship, marriage, death, disease, impoverishment, witchcraft, and damnation were realistic sources of concern among villagers facing both present threats and uncertain futures. While it is impossible to know how similar Napier's patients were to others during this period, it seems very likely that they reflected the central role of normal distress among patients of general practitioners more broadly.

## William Shakespeare

The substantial attention that William Shakespeare (1564–1616) paid to madness provides an additional source of information about premodern conceptions of this phenomenon. Instances of mental illness abound in Shakespeare's plays; many of his major characters were mad. Madness is a particularly central topic in three of his major writings: *Hamlet, King Lear,* and *Macbeth*. While these works illustrate the mixture of supernatural, religious, folkloric, and medical conceptions of mental disorder that prevailed at the time, their most striking aspect—analogous to the prosaic complaints of Richard Napier's patients—is the primacy they place on the *interpersonal* causes of madness.

*Hamlet* provides the best-known instances of madness (or its simulation) in Shakespeare's works. The play revolves around Hamlet's attempts to deal with the intense grief that follows upon his father's death and the rapid remarriage of his mother to his uncle. This situation involves a powerful disturbance in expectations regarding how the bereaved should act: "Things rank and gross in nature/Possess it merely. That it should come to this?/But two months dead—nay, not so much, not two."[100] Hamlet's resulting anguish has served to, according to his friend Horatio, "deprive your sovereignty of reason/And draw you into madness."[101] His symptoms are classic signs of depression: "How weary, stale, flat, and unprofitable/Seem to me all the uses of this world!"[102]

After Hamlet's father dies his mother marries her dead husband's brother, Claudius, who becomes the new king. Claudius tries to convince Hamlet that grief is normal: "'Tis sweet and commendable in your nature, Hamlet/To give these mourning duties to your father./But you must know your father lost a father,/That father lost his, and the survivor bound/In filial obligation for some term/To do obsequious sorrow." Yet, the king also makes a classic Hippocratic (and DSM-like) distinction between normal and unnatural grief, noting how natural grieving can "go wrong," as in Hamlet's unnatural mourning: "But to persever/In obstinate condolement is a course/Of impious stubbornness. 'Tis unmanly grief/It shows a will most incorrect to heaven."[103] Hamlet's "persever[ing]" grief is a sign that his enduring preoccupation with his father's death seems to be unmoored from its context and so is unnatural rather than expectable. Unlike normal grief, prolonged bereavement can indicate the presence of a mental disorder.

Hamlet was not prone to madness before his father's death. His bereavement and, especially, his mother's subsequent incestuous relationship with her dead husband's brother led to his bizarre behavior. Hamlet proclaims: "The time is out of joint. O cursed spite/That ever I was born to set it right!"[104] He disputes his mother's diagnosis that "This is the very coinage of your brain" and casts blame on her: "Mother, for love of grace/Lay not that flattering unction to your soul/That not your trespass but my madness speaks."[105] Hamlet believes that his unnatural interpersonal situation involving his father's death and his mother's ensuing "trespass," not his "madness," accounts for his behavior.

*Hamlet* also raises another issue regarding how to distinguish madness from sanity. Despite the indications that Hamlet's behavior results from some psychic disturbance—deep melancholy, suicidal thoughts, inappropriate affects and emotions—he also suggests that he feigns his madness for strategic reasons: "I am but mad north-north-west: when the wind is southerly, I know a hawk from a handsaw," he tells himself. Hamlet also observes "That I essentially am not in madness/But mad in craft." This allows him to deflect blame for his killing of Laertes: "Then Hamlet does it not, Hamlet denies it/Who does it, then? His madness." Polonius suspects that Hamlet's madness is contrived: "Though this be madness, yet there is method in't." Hamlet's symptoms seem to simultaneously imply true mental disturbance and premeditated calculation.[106]

The second mad character in the play, Hamlet's lover Ophelia, also illustrates the power of intense loss events to generate mental disorder. After Hamlet kills her father Ophelia suffers from: "The poison of deep grief; it springs/All from her father's death."[107] Hamlet's subsequent cruel treatment and rejection lead her to become psychotic and eventually commit suicide. The combined losses of a parent and a lover drive Ophelia mad.

*King Lear* provides a second example of the importance of interpersonal conflicts in provoking madness. The play invokes a variety of factors to account for Lear's insanity. Lear clearly suffers from age-related dementia. Lear's daughter, Goneril, recognizes this when she attributes her father's decline to "the unruly waywardness the infirm and choleric years bring with them" when "old fools are babes again."[108] Lear himself accepts his age-related problems: "I am a very foolish fond [senile] old man; Fourscore and upward, not an hour more or less; And to deal plainly; I fear I am not in my perfect mind."[109]

Nonetheless, the basic cause of Lear's madness, as he recognizes, lies in his daughters' betrayal of him after he decides to leave the throne and bequeath them his kingdom: "The tempest in my mind; Doth from my senses take all feeling else; Save what beats there—filial ingratitude."[110] His one faithful daughter Cordelia, too, notes her "child-changed father."[111] Lear's madness stems from his daughters' treachery and cruelty more than his demented condition.

*Macbeth* provides a third instance of mental disorder that arises from interpersonal causes. Macbeth and, especially, Lady Macbeth become seriously disturbed after they murder King Duncan so that Macbeth can inherent his kingdom. A noble notes that Macbeth's actions have led to "the insane root; That takes the reason prisoner?"[112] Overcome by guilt for her murderous deeds, which "will make us mad,"[113] Lady Macbeth feels agitated, is anxious, has terrible dreams, and is unable to eat, sleep, or rest.

His wife's behavior disturbs Macbeth, who sends for a doctor to cure her. The doctor arrives, quickly recognizes the murder as the source of Lady Macbeth's problem, and tries to reject Macbeth's effort to medicalize his wife's disturbance: "This disease is beyond my practice . . . unnatural deeds/Do breed unnatural troubles: infected minds/To their deaf pillows will discharge their secrets:/More needs she the divine than the physician."[114] Macbeth rejects this diagnosis and demands that the doctor cure his wife:

MACBETH: How does your patient, doctor?
DOCTOR: Not so sick, my lord,/As she is troubled with thick coming fancies,/ That keep her from her rest.
MACBETH: Cure her of that./Canst thou not minister to a mind diseased,/ Pluck from the memory a rooted sorrow,/Raze out the written troubles of the brain/And with some sweet oblivious antidote
Cleanse the stuffed bosom of that perilous stuff/Which weighs upon her heart?
DOCTOR: Therein the patient/Must minister to himself.[115]

The physician is helpless in the face of a condition that is rooted in interpersonal guilt rather than in disease.

Shakespeare's views contain echoes of the supernatural, physiological, and psychological aspects that were central to Ancient Greek conceptions of madness. But the essential source of madness in his work lies in deep ruptures of the social fabric: a son confronting one parent's death and another's incest,

**Figure 2.1** Before the establishment of mental hospitals, literary and artistic works often portrayed the mad as cast-offs from conventional societies who were passengers on a "ship of fools."

a father betrayed by his daughters, and murderers tormented by their assassination of a king. Shakespeare's penetrating insights had lasting impacts on conceptions and treatments of madness for centuries to follow.[116,117]

## Empiricism Reemerges in the 17th and 18th Centuries

The spiritual worldview that tightly gripped the West for a millennium slowly gave way to a secular view of social organization, cultural norms, and individual thoughts and feelings.[118] Religious views of madness came under assault from two directions. First, the rise of new empirical approaches to mental disorder was one aspect of the radical change in the Western intellectual tradition that occurred when the inductive and observational methods of Francis Bacon and Isaac Newton overturned the theological conceptions that had permeated previous Western thought. Divinely grounded principles yielded to the power of observation and inductively derived scientific findings. This led authority over the workings of minds to gradually shift from Christian beliefs to results that emerged from empirical studies. Correspondingly, prevailing understandings about the nature of madness and its treatment moved from the religious to the medical domain.

Second, a number of renowned philosophers emphasized how knowledge of the world stemmed from the power of reason and experience rather than from God. Rene Descartes' radical split of minds from bodies implied that mental disorders were produced through the mechanistic actions of physical organisms and were not spiritual properties of souls.[119] Thomas Hobbes also developed a highly influential materialistic theory of man and nature. In essence, these views produced the radical new view that "matter could think."[120] As the French philosopher Voltaire later summarized, the influence of secular philosophies reordered conceptions of all human behavior: "It would be very singular that all nature, all the planets, should obey eternal laws, and that there should be a little animal, five feet high, who, in contempt of these laws, could act as he pleased, solely according to his caprice."[121]

## Mental Illnesses Become Medical

In the 17th century, ideas regarding disease specificity emerged in general medicine, most notably in the work of English physician Thomas Sydenham (1624–1689). Rejecting holistic Hippocratic conceptions of disease, Sydenham proposed that each malady took specific forms with uniform presentations in different individuals. Foreshadowing Kraepelin and other 19th-century psychiatrists, he urged careful observation of symptoms as the basis of disease classifications. This inductive approach produced medical breakthroughs such as English physician William Harvey's discovery of how blood circulated through the body.

In line with broader trends in medicine, a new line of thought focused on the nervous system as the source of health and illness, emphasizing the importance of nerves, fibers, and organs. Accordingly, physiological causes accounted for mental as well as organic conditions. Thomas Willis (1621–1675), a 17th-century English doctor who coined the term "neurology," was the first physician who strove to localize various mental functions in particular brain regions. He developed a theory of neurological activity based on the notion of "animal spirits"—life-carrying fluids that passed through the nerves and transmitted information between the sense organs, brain, and muscles. Willis viewed nervous disorders as bodily conditions:

> The anatomy of the nerves revealed the true and genuine reasons for very many of the actions and passions that take place in our body, which otherwise seem most difficult to explain: and from this fountain, no less than the hidden causes of diseases and symptoms, which are commonly ascribed to the incantation of witches, may be discovered and satisfactorily explained.[122]

Willis enumerated the physiological components of fear including hair standing on end, a loosening of the nerves, involuntary excretion, and fainting.

Throughout the 18th century medical writing featured materialistic views of psychic disorders. The Welsh physician Nicholas Robinson asserted that "every change of the Mind indicates a Change in the Bodily Organs."[123] Robinson denied that "thoughts themselves can ever start from a regular Way of Thinking without . . . a Change in the Motions of Animal Fibres."[124] Such conceptions were instrumental in initiating the overthrow of the humoral theory and establishing a new organic paradigm of mental disorder grounded in the brain. By the end of the century, a fully scientific approach to mental illness had emerged in academic medicine as organic views displaced earlier faith-based, supernatural, and humoral notions.

The consensual biological model that was developing contrasted with the confusion that marked classifications of mental disorders. Most physicians accepted physiological explanations of nervous conditions but viewed such states in a broad and undifferentiated manner. One noted that: "indeed the Limits and Partitions that bound and discriminate the highest Hypocondriack and Hysterick Disorders, and Melancholy, Lunancy, and Phrenzy are so nice, that it is not easy to distinguish them, and set the Boundaries where one Ends, and the other Begins."[125] Prominent Scottish physician Robert Whytt (1714–1766) also emphasized the protean nature of nervous illness, noting that "those morbid symptoms which have been commonly called *nervous*, are so many, and so various, and so irregular, that it would be extremely hard, either rightly to describe or fully to enumerate them."[126]

In sharp contrast, others strove to develop highly specialized diagnostic systems. The most prominent was William Cullen (1710–1790), an 18th-century Scottish professor of medicine and the founder of psychiatric theory and practice in Great Britain. Cullen created the term "neurosis" to refer to the various ailments of the central nervous system. His thought had lasting impact through the many physicians he trained, including the founder of American psychiatry, Benjamin Rush.[127] Cullen might be the first "splitter" of psychiatric disorders: neurosis (or nervous disorder) was one of four classes of diseases that he subdivided into four subclasses. He placed hysteria in the category of spasmodic affections (irregular motions of the muscles) and hypochondriasis in the adynamiae category (deprivations of involuntary motions) and distinguished both disorders from melancholia, which he put in the Vesaniae category (disorders of intellectual functions).[128] Eventually, Cullen's nosology came to contain 150 different types of insanity.[129]

Medical conceptions dominated portrayals of mental illness through the end of this period, extending a naturalistic view into the innermost beings of humans.[130] This revolutionary development, however, did not lead to major changes in the generally harsh and coercive treatments accorded the mad. "Theories changed, but the medicaments remained the same," one French physician noted at the time.[131]

## "In-between" Conditions Appear

Through the 17th century, madness continued to be stereotypically tied to incomprehensible behaviors that were unreasonable, defied social expectations, and lacked good sense. At this time madhouses that specifically confined the mentally ill began to emerge, although the initial institutions contained only a small number of highly disturbed individuals. The major development influencing the boundaries between sanity and insanity was the appearance of nervous diseases that were sharply separated from madness. The newly dominant medical conceptions created a market for nervous patients whose minds were intact but who claimed to suffer from a genuine physical disorder.[132]

George Cheyne (1672–1743) was a society nerve doctor whose book, *The English Malady: or, A Treatise of Nervous Diseases of All Kinds* (1733), was the most influential popularization of the new conception.[133] The English

Malady was a capacious state that subsumed a cornucopia of physical complaints as well as various anxious and melancholic symptoms. It was the first nervous condition that *appealed* to a broad group of lay people. It became a desirable diagnosis that had been "gentrified and received into good society."[134] Because it carried no stigma and allowed individuals to partake in the advantages of the sick role, this label became a sought-after state. Cheyne moved nervous afflictions to the center of concern among a wider lay audience; the English Malady both normalized mental illness and legitimated it through its connection to physical illness.

The English Malady employed a mechanistic model of the human organism focused on the nervous system, tissues, and physical organs of the body. Cheyne rooted it in the body rather than the mind, emotions, or soul and so placed nervous conditions within the purview of the physician rather than the clergyman or philosopher. He claimed of nervous disorders that: "the Disease is as much a bodily Distemper . . . as the Smalpox or a Fever."[135] The nerves mediated between the mind and the brain, possessing attributes of both.[136] They conveyed sensations throughout the body and sent messages among the brain, internal organs, and limbs.

The new nervous condition by no means involved minor complaints. Cheyne described how in his twenties he himself was suddenly stricken by a condition of "Fright, Anxiety, Dread, and Terror" that led him to be bedridden for months.[137] Indeed, Cheyne stated, "of all the Miseries that afflict Human Life, and relate principally to the Body, in this Valley of Tears, I think *Nervous* Disorders, in their extream and last Degrees, are the most deplorable, and beyond comparison the worst."[138] Nevertheless, he explicitly distinguished nervous disorders from insanity and serious melancholia, which continued to be seen as types of madness. Nerves were not viewed as a less severe form of insanity but, instead, as a true organic condition akin to other medical illnesses.

Cheyne believed that huge numbers of well-bred Englishmen suffered from this malady, noting: "These nervous Disorders being computed to make almost one third of the Complaints of the People of Condition in England."[139] Like Cheyne, many observers believed that the 18th century was marked by an epidemic of nervous disease. Historian Edward Shorter notes a competition in the Dutch city of Utrecht for the best essay on the topic of "the causes of the increasing nervous disease in our land."[140] Leading Scottish physician Thomas Trotter (1760–1832) later observed: "(A)t the beginning of the nineteenth century, we do not hesitate to affirm, that *nervous disorders* have now taken the place of fevers, and may be justly reckoned two thirds of the whole, with which civilized society is afflicted."[141] Trotter also claimed that: "In the present day, this class of [nervous] diseases forms by far the largest proportion of the whole, which come under the treatment of the physicians."[142] Nervous conditions seemed to have become ubiquitous in the upper classes of Western countries.

Cheyne's book was widely read and helped promote the notion of nervousness as a fashionable feature of well-bred gentlemen and ladies. Through associating the English Malady with success, prosperity, and excessive consumption while at the same time providing it with a physiological grounding, nervous conditions became socially acceptable.[143] Cheyne's view also moved conceptions of mental illness away from the lonely, marginal, outsider features of melancholia toward sociable, integrated, and respectable insider status.

At the time, specialists in dealing with psychotic states (typically called "mad-doctors") did not treat nervous disorders. Indeed, labels of "nervous disorders" were so popular because they were *not* connected with a profession that dealt with madness. British King George III, for example, insisted that he was not "mad but only nervous."[144] General physicians, neurologists, and spa doctors as well as a plethora of nonmedical herbalists, faith-healers, and religious healers provided care to people with nervous conditions.[145] Despite the displacement of Hippocratic explanations of mental disturbance, treatments continued to emphasize moderation in all aspects of lifestyle. As in medicine more generally, opium—which was readily available and profusely used throughout this period—was also a common response to nervous maladies.[146] They paved the way for subsequent notions of nervous diseases including George Beard's neurasthenia and Sigmund Freud's neuroses that became not just tolerable but even fashionable illnesses among the bourgeoisie.

## Internal and External Causes of Mental Illness

In contrast to the organic focus of many physicians at the time, a radically new externalist model of mind and the emotions emerged in the last half of the 17th century in the extraordinarily influential work of philosopher John Locke (1632–1704). Locke proposed that minds were comprised of mental sensations derived from the external environment. They were blank slates at birth but then became shaped by experiences. Anticipating the 20th-century development of cognitive approaches, Locke placed misconceived *thoughts* at the heart of madness. People were mad when the pictures of the world they formed were incongruent with external reality. They came to act in accordance with ideas they believed were true but that were, in fact, false: "A madman fancies himself a prince; but upon his mistake, he acts suitable to that character; and though he is out in supposing he has principalities, while he drinks his gruel, and lies in straw, yet you shall see him keep the port of a distressed monarch in all his words and actions."[147] False ideas, not nerves, humors, or demons, provided the key for understanding madness. This notion provided grounds for optimism about the treatment of the mentally ill because they could be trained to think correctly. Accordingly, the application of reason rather than changing souls or bodies was the best way to counter illogical thoughts. "Reason," Locke declared, "must be our last judge and guide in everything."[148]

Despite the sharply conflicting somatic and mental models of mental illness, advocates of both views agreed that mental illnesses generally arose from a mixture of internal and external factors. Dominant models among both medical practitioners and laypeople mixed physiological defects and environmental stressors as ways of accounting for mental disturbances. For example, Cullen's writings grounded mental disturbances in neurophysiology, especially in states of increased and decreased activity of nervous power in the brain, which he called "excitement" and "collapse," respectively. Yet, his theory was not mechanistic, and he emphasized how both of these conditions could emerge in response to psychic and environmental stimuli.[149] His influential works thus served as a bridge between psychological and somatic views of mental illness.[150]

In his *A Treatise on Madness* (1758), English physician William Battie (1703–1776) proposed another influential distinction. Battie considered some forms of madness as "original." These cases stemmed from the inherited constitution of sufferers with oversensitive nerves and were unlikely to be curable. In contrast, most manifestations of insanity were "consequential," resulting from some "intolerable impulse of external objects" that created "too great or too long continued force."[151] While there was little hope of remedying original types, those who suffered from consequential ones could benefit from the Hippocratic prescription of simple food, clean air, nonstressful employment, and moderate lifestyles. Locke, too, remained squarely within this tradition, prescribing "Plenty of open Air, Exercise and

Sleep; Plain Diet, no Wine or Strong Drink, and very little or no Physick; not too Warm and strait Clothing."[152]

Cheyne provides another example of how a combination of organic, social, and cultural factors produced mental illnesses. On the one hand, he firmly rooted nervous conditions in the physical properties of brains and nerves. On the other hand, he held that the English Malady represented the clearest case of an externally caused ailment that stemmed from pathologies of wealth, weather, land, diet, and sedentary lifestyles. It was the result of:

> the Variableness of our Weather, (from our situation amidst the Ocean), the Rankness and Fertility of our Soil, the Richness and Heaviness of our Food, the Wealth and Abundance of Inhabitants (from their universal Trade), the Inactivity and Sedentary Occupations of the Better Sort (among whom this Evil mostly rages) and the Humor of living in great, populous and consequently unhealthy Towns, have brought forth a Class and Set of Distempers with atrocious and frightful Symptoms, scarce known to our Ancestors and never rising to such fatal Heights, nor afflicting such Numbers in any other known Nation.[153]

Social factors, especially prosperity, rapid social change, and high living, led to the epidemic of nervous afflictions among the English.

Cheyne attributed the English Malady to the progress of civilization and especially to the stressful lifestyles of the best-off members of this country. Conversely, those lower in the social hierarchy were largely immune to nervous disorders: "Fools, weak or stupid Persons, heavy and dull Souls, are seldom much troubled with Vapours or Lowness of Spirits."[154] The English Malady was the first mental illness that arose from collective conditions that were shared by large portions of a population. It initiated a trend that continues to the present to see modern life as a threat to mental health.

## Conclusion

Over the course of the 17th and 18th centuries, mental illness became increasingly likely to be viewed within medical, as opposed to spiritual or moral, frameworks. Within medicine, the influence of humoral pathology, which had dominated medical thinking since Hippocratic times, gradually waned. In its stead, mechanistic notions grounded in nerves, fibers, and organs displaced ancient conceptions associating mental disorders with excessive or deficient levels of fluids. By the end of this period, while humoral conceptions remained influential in popular culture and among general physicians, academic medicine emphasized physiological accounts of how mental disturbances resulted from malfunctioning nervous systems. Notions of "over-excited" nerves had replaced ones of humoral imbalances. Many factors, including individual temperament, heredity, and external circumstances, were connected to the faulty working of nerves.[155] Alongside somatic explanations that stressed the activities of the nervous system, a Lockean notion arose that psychiatric conditions resulted from faulty associations of ideas. Such psychological explanations vied with physiological ones.[156]

Despite the growing interest in nervous conditions, classifications of mental disorders remained rudimentary. At the end of the 18th century, Cullen notwithstanding, most categorizations remained very general. Amorphous notions of nervous disorders that entailed an extraordinarily capacious range of symptoms dominated. While the medical model was beginning to gain supremacy, religious, moral, and supernatural conceptions of madness remained popular, especially among laypersons. Before the 19th century, patients were more likely to take their psychological complaints to an array of healers including midwives, herbalists, and alchemists, than to physicians. A dedicated psychiatric profession had not yet emerged.[157] However, locating mental conditions in the nervous system paved the way for a widespread medicalization and specification of mental disorders during the following century. This development allowed medical professionals to gain a control over madness that persists to the present.

## Notes

1. While Hippocrates is probably the most acclaimed physician in history, nothing is known about the man himself, and it is not even clear that he actually existed. Neither of the two major figures in Greco/Roman medicine, Hippocrates and Galen, produced extensive accounts of any mental illness. More in-depth discussions of psychological disorders are found in the works of physicians practicing in Rome during the first two centuries AD, including Celsus, Aretaeus of Cappadocia, and Soranus of Ephesus.
2. Roccatagliata, 1986; Porter, 2002. Despite the centrality of madness in Ancient Greek culture, care for the mad was more a familial than a public responsibility. "If a man is mad, he shall not be at large in the city, but his family shall keep him in any way they

can," Plato wrote in the *Laws* (XI, 934). Their relatives often isolated and placed physical restraints on madmen who engaged in harm to persons or possessions (Rosen, 1968, 86).

3. Xenophon, quoted in Rosen, 1968, 94.
4. Herodotus, 1996, 160–69.
5. Quoted in Whitebook, 2017, 292.
6. Hippocrates, 2016, 43.
7. Simon, 1978, 22.
8. Aristotle, 2009,127.
9. http://www.philosophybasics.com/philosophers_diogenes.html.
10. Evans et al., 325.
11. http://www.brainyquote.com/quotes/authors/h/hippocrates.html.
12. Simon, 1978, 235. Greek physicians also developed a third condition of *phrenitis*, which was a type of mania that sometimes occurred during infections and fevers. Finally, they recognized hysteria and epilepsy as having broad resemblances to core conditions of madness. In contrast, little evidence suggests that the Greeks recognized conditions that resemble what we now call "schizophrenia" (Evans, McGrath & Milns, 2003).
13. Hippocrates, 2016, 56.
14. Simon, 1978, 152.
15. Rosen, 1968, 98.
16. Quoted in Rosen, 1968, 98–99.
17. Roccatagliata, 1986, 229.
18. The following material is adapted from Horwitz & Wakefield, 2007.
19. Jackson, 1986, ix.
20. Hippocrates, 1923–31, vol. 1, 263.
21. Hippocrates, 1923–31, vol. 4, 185.
22. Roccatagliata, 1986, 232.
23. Homer, 1998, 24: 53–56.
24. Rosen, 1968, 63, 80.
25. Quoted in Jackson, 1986, 32.
26. Jackson, 1986.
27. Roccatagliata, 1986, 232.
28. Roccatagliata, 1986, 52.
29. Boudon-Millot, 2013, 143.
30. Roccatagliata, 1986, 199.
31. King, 2013, 267–68.
32. Rosen, 1968, 96.
33. Quoted in Porter, 2002, 45.
34. Jackson, 1986, 29.
35. Quoted in Porter, 2002, 37.
36. Hippocrates, 2016, 57.
37. Porter, 1997, 55–62.
38. Simon, 1978, 234.
39. Rosen, 1968, 132.
40. Roccatagliata, 1986, 88.

41. Holmes, 2013.
42. Simon, 1978, 220–1.
43. Jouanna, 2013, 104; Roccatagliata, 1986, 41.
44. Simon, 1978, 71.
45. Plato, 2005, 26–38, 246a–256e.
46. Vogt, 2013, 191.
47. Roccatagliata, 1986, 60.
48. Plato, 1892. II, 50. The ancient Hebrews used the same word for behaving like a prophet and the ravings of a madman (Rosen, 1968, 57).
49. Plato, 2005, 244A.
50. Simon, 1978, 185.
51. Plato, 2005, 265A. See Dodds, 1951, 64; Guven, 2005, 24.
52. Homer, 1999, 23: 11–15, 455–56.
53. Indeed, Freud named his best-known concept, the Oedipus complex, after a character in Sophocles' play.
54. Sophocles, 2008.
55. Aeschylus, 1984.
56. Euripides, 1999, 2, 45.
57. Dodds, 1951, 65.
58. Herodotus, 1996, 348.
59. Rosen, 1968, 77, 83, 86.
60. Dodds, 1951, 68.
61. Rosen, 1968, 86.
62. Holmes, 2013.
63. Roccatagliata, 1986, 201.
64. Holmes, 2013.
65. Scull, 2015, 69.
66. Porter, 1997, 2002 and Scull, 2015 provide excellent overviews of mental illness in this period.
67. St. Augustine, 2009, 8.29.
68. Porter, 1997; Scull, 2015, 48–85.
69. Porter, 1997, 106–12.
70. Levack, 2013.
71. Avicenna, 2000, 77.
72. Quoted in Porter, 2002, 52.
73. Quoted in Shorter, 2015, 71.
74. As early as 1526, the Swiss physician Paracelsus questioned the Hippocratic/Galenic tradition, proclaiming that "proofs derive from my own experience and my own reasoning and not from reference to authorities" (Makari, 2015, 42).
75. Shorter, 1992, 15; Glas, 2003, 1.
76. Burton, 1621/2001.
77. Burton, 1621/2001, 143–44.
78. Burton, 1621/2001, 145–46.
79. Burton, 1621/2001, 137.

80. Burton, 1948, 331.
81. Burton, 1621/2001, 261.
82. Burton, 1621/2001, 131.
83. Burton, 1621/2001, 357–8.
84. Burton, 1621/2001, 358–59.
85. Burton, 1621/2001, III: 142, 148.
86. Burton, 1621/2001, II: 243.
87. Porter, 1997.
88. MacDonald, 1981.
89. MacDonald, 1981, 105.
90. MacDonald, 1981, 159.
91. MacDonald, 1981, 158.
92. MacDonald, 1981, 159.
93. MacDonald, 1981, 159.
94. MacDonald, 1981, 149.
95. MacDonald, 1981, 78.
96. MacDonald, 1981, 67, 76, 107.
97. MacDonald, 1981, 88–89, 99, 41.
98. MacDonald,1981, 187.
99. MacDonald, 1981, 197.
100. Shakespeare, 1988: *Hamlet*, 1.2, 136–38.
101. Hamlet 1.4 73–74.
102. Hamlet 1:2 133–34.
103. Hamlet 1.2 87–95.
104. Hamlet 1.5 197–98.
105. Hamlet 3.4 143, 151–53.
106. Hamlet 2.2 378–79; 3.4 194–95; 5.2 234–35; 2.2 205–6.
107. Hamlet 4.5 76–77.
108. King Lear 1:1 298–99; 1 3: 19.
109. King Lear 4 7: 58–61.
110. King Lear 3 4: 12–14. Gloucester, too, attributes his condition where "grief has crazed my wits" to his son's betrayal 3 4: 254–58.
111. King Lear 4 7: 15.
112. Macbeth 1 3:84–85.
113. Macbeth 2.2.
114. Macbeth 5.1 58–74.
115. Macbeth 5.3.
116. Reiss (2008, 81) observes: "In the first three decades of American asylum medicine, no figure was cited as an authority on insanity and mental functioning more frequently than Shakespeare." Indeed, during the first two decades after its founding in 1844 the *American Journal of Insanity* (the predecessor to the *American Journal of Psychiatry*) published no fewer than thirteen articles on Shakespeare (Whooley, 2019, 47).
117. Hieronymous Bosch, Ship of Fools, c. 1490–1500. Musee du Louvre, Paris.
118. See especially Makari, 2015.

119. Porter, 1997, 242–43; Jackson, 1986, 21.
120. Makari, 2015, 33.
121. Quoted in Robinson, 1995, 237.
122. Quoted in Scull, 2009, 167.
123. Quoted in Porter, 2002, 126.
124. Porter, 1987, 52.
125. Quoted in Porter, 1987, 46.
126. Quoted in Jackson, 1986, 297.
127. "The cause of madness," Rush asserted, "is seated in the blood vessels of the brain." Quoted in Shorter, 1997, 27.
128. Jackson, 1986, 126, 299.
129. Makari, 2015, 190.
130. Makari, 2015, 63.
131. Bichet quoted in Jackson, 1986, 132.
132. Scull, 2009, 154.
133. Cheyne, 1733/1991.
134. Porter, 1987, 108.
135. Quoted in Porter, 1987, 55.
136. Scull, 2009, 57.
137. Porter, 1991, xxxiii; Cheyne, 1733/1991, 194, 333.
138. Cheyne, 1733/1991, 343. See also 260–61.
139. Cheyne, 1733/1991, ii.
140. Shorter, 1992, 24.
141. Trotter, 1807, xvii.
142. Quoted in Porter, 1991, 182.
143. Porter, 1991, xxxii.
144. Porter, 2002, 86.
145. Shorter, 1997, 22–23; Micale, 2008, 23.
146. Porter, 1997, 194, 269.
147. Locke, 1690/1996, book II, Ch. xxxii, 5.
148. Quoted in Makari, 2015, 121.
149. Porter, 1997, 260.
150. Jackson, 1986, 24–5; Porter, 2002, 128.
151. Battie, 1758/1962, 35–36, 90, 34.
152. Quoted in Porter, 1997, 243.
153. Porter, 1987, 83.
154. Cheyne, 1733/1991, 52; Porter, 1991, xli.
155. Scull, 2009, 26; Shorter, 1992, 21.
156. Porter, 2002, 128–30.
157. Shorter, 1997, 22.

# 3

# A Biological Century

> Psychiatry, with its modern classification of ailments, methods of diagnosis, and treatment—compared with what it used to be it was a gigantic achievement.
>
> —Anton Chekhov, *Ward 6* (1892)

The 19th century brought about huge changes in explanations of and responses to mental illness. When the century began, many general physicians were still employing some variant of Hippocratic notions that diseases resulted from humoral imbalances.[1] They often combined these millennia-old views with various theological, folkloric, and social frameworks. Most diagnostic schemes only incorporated a few broad categories such as mania and melancholia that had nonspecific origins and impacts. Common treatments such as purges, emetics, and bleeding were not specific to mental illness but were applied to most diseases. The mad were often subject to harsh physical restraints and confined in jails, almshouses, or, less often, mental institutions. Lay people were more likely to consider insanity as a religious, moral, or legal issue than as a medical problem.

As the century progressed, the multiple meanings of madness coalesced into a single medically oriented framework.[2] Research on mental disorders, which adopted a biomedical paradigm that assumed mental illnesses were brain based and aspired to classify specific mental illnesses, flourished in French and German academic settings.[3] Psychiatrist Theodor Meynert (1833–1892) captured the essence of this view in the forward to his popular 1884 textbook: "The reader will find no other definition of 'Psychiatry' in this book but the one given on the title-page: 'Clinical Treatise on Diseases of the Fore-Brain.' The historical term psychiatry, i.e. 'treatment of the soul,' implies more than we accomplish, and transcends the bounds of accurate scientific investigation."[4] Initially, most researchers optimistically assumed that this somatic model would lead to discoveries that could cure insanity.

Around the beginning of the century a new profession of psychiatry arose that was associated with the widespread emergence of specialized inpatient institutions for the insane.[5] Several decades later, medically oriented outpatient therapies for nervous conditions greatly expanded. These changes had lasting impacts on ideas about the similarities and differences between mental and physical illnesses and about the causes of mental illness that persist at present. However, they had little influence over the division between normal from abnormal conditions, where the traditional without cause criteria remained prevalent in lay and professional conceptions alike.

## Medicalizing Inpatient Treatment

At the dawn of the 19th century an eclectic, pluralist, and divided array of medical, religious, and magical healers responded to the mentally ill.[6] Few establishments focused on dealing with the mad, who were usually confined in jails, prisons, almshouses, and other facilities that contained a variety of deviant populations. As the century unfolded, the first specialized profession of psychiatrists (initially called "alienists") arose that was closely tied to the emergence of mental hospitals. This group consolidated under medical authority the previous potpourri of explanations and treatments of madness and nerves.[7] The facilities dedicated to responding to mental disorder that emerged around the beginning of the century, however, were based on a decidedly nonmedical model grounded in environmental and psychic assumptions.

### Moral Treatment

In thoroughgoing contrast to their later manifestations, the institutions for the insane that developed around the turn of the century in England, Europe, and the United States were part of a utopian movement called "moral treatment," which emphasized how madness could be relieved when treated through humane principles of care. Locke's philosophy that insanity resulted from misconceived ideas was a major influence on the character of these hospitals. English physician William Battie wrote the first substantial book about asylums, *Treatise on Madness* (1758), which used Locke's ideas about the malleability of human behavior as a blueprint.[8] If, as Locke claimed, faulty associations of thoughts caused insanity, then it seemed to follow that controlling the environments in which these defective perceptions arose could cure it. Battie, therefore, proposed removing patients from their pathogenic surroundings as the antidote to madness. His treatments disdained both harsh methods and drugs, instead relying on therapeutic approaches that involved personal contact between patients and healers.

Drawing on Locke's associationist theory, the designers of asylums carefully structured the architecture, activities, and routines of patients in order to bring about mental changes. And, following Battie's therapeutic optimism, moral treatment assumed that rational and humane treatments could cure the insane. It was most likely to succeed when it was practiced in small hospitals, usually located in rural areas, with highly structured routines. Typical activities in these asylums encompassed walking in nature; attending concerts, theatrical performances, and lectures; and engaging in therapeutic talks with staff members.[9] Moral treatment was pragmatic and tailored to the needs of the individual. Its basic tenets overthrew centuries of horrific responses to the mentally ill, which often involved beatings, starvation, and various forms of torture.

Insanity, according to the advocates of moral treatment, need not be a chronic condition. Because deficient social environments precipitated insanity, new, therapeutically oriented environments could presumably cure it. For example, French physician Phillipe Pinel (1745–1826), who is often considered to be the founder of modern psychiatry, replaced external coercion with forms of care such as special diets, long walks, and strenuous physical activity for his patients.[10] One French asylum superintendent, who influenced Pinel, noted how

> My experience has shown, and shows daily, that to further the cure of these unfortunates one must treat them with as much kindness as possible, dominate them without mistreatment, gain their confidence, fight the cause of their illness, and make them envision a happier future. I have always fought this illness by psychological means and thus known the happiness of some favorable results.[11]

Conversely, physical restraints and punishments were only used as last resorts. In England, philanthropist William Tuke (1732–1822) and his descendants established the influential York Retreat, which minimized the use of coercion and emphasized humane, psychologically based treatments.

By the beginning of the 19th century, mental institutions using the principles of moral treatment had spread throughout Europe.

Most of these institutions were built in bucolic, rural settings because of the belief that removal from fast-paced urban environments that presumably led to insanity could help to cure it. Dorothea Dix (1802–1887), who founded or enlarged more than thirty mental hospitals, was the leading American advocate of moral treatment.[12] Like her European predecessors, she was confident that appropriate asylum care could cure madness. In 1843 Dix proclaimed: "I come to present the strong claims of suffering humanity. I come as the advocate of helpless, forgotten, insane and idiotic men and women; of beings, sunk to a condition from which the most unconcerned would start with real horror."[13] Before the mid-19th century, however, asylums held only small numbers of predominately middle- and upper-class patients.

## The Transformation of Mental Hospitals

The emergence of mental institutions largely stemmed from developments surrounding changing family structures. For millennia kin were accountable for containing the actions of their disruptive and embarrassing relations. Mental illness was rarely regarded as a communal problem, so governmental bodies had little to do with its control. Except for the most extreme and violent cases, relatives were responsible for insulating communities from the behaviors of the mad.[14] As the 19th century unfolded, overarching social trends including population growth, urbanization, increasing geographic mobility, and immigration rendered families less able to regulate their deviant members. A new, more individualistic social order developed that involved greater personal freedom from traditional restraints. At the same time as the strength of family control declined, more people relocated from rural to urban areas. This made the insane more visible and heightened public concern over safety issues. Moreover, many newcomers to cities were single males who lacked family ties.

The decline of familial tolerance for and capacities to deal with strange behaviors caused asylum populations to explode. For example, before 1810 in the entire United States, only 500 patients were kept in one public and a few private mental hospitals. By 1860, in addition to many private asylums, 28 of the 33 states had public institutions that contained the insane.[15] Between 1800 and 1880, commitments increased eight-fold from 1 for every

6,000 persons to 1 for every 750 persons.[16] Rates of inpatient confinement in Europe rose in similar fashion. In England the average asylum population grew from 116 patients in 1827 to 1,072 patients in 1910.[17]

An eclectic and diffuse set of reasons led to confinement in mental institutions. Some commitments arose after triggers such as grief, failures of love relationships, business losses, overwork, intemperate lifestyles, and the like.[18] Gerald Grob's study of over 2,000 cases admitted to the Worcester State Hospital in Massachusetts between 1833 and 1845 found that seven causes accounted for 60% of the total: "intemperance (278), ill health (38), religious excitement (191), masturbation (145), domestic affliction (219), loss of property and fear of poverty (131), and some form of personal disappointment (101)."[19] Men commonly entered asylums after financial problems, intemperance, or masturbation and women after childbirth, domestic troubles, or spiritual difficulties. In France, Pinel emphasized broad factors such as overambition, religious fanaticism, and failed love affairs as reasons for entry into mental institutions.[20]

As the century progressed, not just the number but also the nature of asylum residents changed radically. Marginal, poor, isolated, and often foreign-born patients replaced the earlier generally high status, well-off clients. At the same time, the proportion of inpatients with incurable and difficult to treat organic diseases such as neurosyphilis, arteriosclerosis, senility, and alcoholic psychoses rose.[21] As their occupants increasingly had incurable conditions, the character of asylums drastically changed from their initial therapeutic orientations to custodial approaches that functioned to segregate occupants from their communities. By the century's end, long stays, overcrowding, brutality, malnourishment, and high death rates marked these facilities. Theories grounded in concepts of heredity and degeneration that saw mental illness as incurable replaced the buoyant attitudes toward treatment that characterized earlier decades. The asylums that initially generated so much hope had become custodial warehouses.[22]

Even after mental institutions emerged and expanded, however, families remained the critical decision-makers over whether to hospitalize members they deemed as mad. Asylums did not actively seek patients but responded to whatever cases were referred to them. One study indicated: "In almost all cases, the decision to commit someone to a state asylum in 19th century New York was made by the patient's family."[23] Another examination of commitments in San Francisco found that relatives initiated 57% of commitments, physicians 21%, and the police 8%.[24] The relational context of most commitments to mental hospitals thus involved parents committing children, husbands committing wives or, more rarely, wives committing husbands, adult children committing elderly parents, or adult siblings committing each other. Most commitments initiated by police involved individuals without families who behaved oddly in public places. In effect, families or, occasionally, local police, not hospital personnel, decided who should be institutionalized.

## Psychiatry Emerges

The profession of psychiatry itself arose in tandem with mental hospitals. The superintendents of the initial asylums were generally physicians who adhered to the principles of moral treatment. They were somewhat of an outlier among doctors. For much of the 19th century, medicine had little cultural legitimacy. Physicians did not need to attend a medical school in order to practice, and few standards existed over who could enter the profession.[25] Unlike the vast majority of physicians in private practice, who were typically poorly paid and of low status, asylum doctors held secure, high-salaried, and prestigious positions.[26]

A major problem the first asylum superintendents confronted was that there was nothing specifically *medical* about their practices. They faced the dilemma that their institutions emerged within a framework that emphasized psychological, social, and religious, not biological, forces. None of the major aspects of moral treatment—creating highly structured therapeutic environments, personalized relationships between therapists and patients, uplifting cultural activities, and the like—had any connection to the specific expertise of medical personnel. Real doctors treated bodily, not mental, afflictions. Anyone, not just physicians, could possess the sympathy to the insane that moral treatment entailed. Only a model that showed how mental disorders had a physical basis could validate placing asylums within medical jurisdiction.[27] From mid-century onward, academic neurologists and psychiatrists provided this justification.

### Brain-Based Research Emerges

The growing influence of research-oriented psychiatrists and neurologists within academic medical institutions, especially in Germany, had enduring impacts on views of mental disorder.[28] This group used a biological model of mental illness that reduced psychic life to the operation of physical forces in the brain and nervous system. Their psychiatry had no psychology; it assumed that mental illnesses were in essence no different from bodily maladies. German Wilhelm Griesinger's pronouncement that "The etiology of mental illnesses in general is none other than the etiology of all other brain and nerve diseases" was emblematic of the era.[29]

Early in the century, German neuroanatomist Franz Josef Gall (1758–1828) established the enormously popular school of phrenology, which was, according to historian George Makari, "the single most influential theory of the mind over the next four decades."[30] Gall rooted all human behavior, including madness, in the material qualities of brains. He viewed brains as the organ of minds and taught that different mental faculties were localized in distinct brain regions. Gall postulated the existence of twenty-seven different areas, each responsible for a unique psychic function.[31] Because the shapes and sizes of skulls provided valid markers of brain differences among individuals, the careful study of crania could reveal different types of defective mental functioning and hence could be used to diagnose distinct mental disorders. An American popularizer of phrenology similarly insisted that the mind was "as capable of being sick as the body" so that mental diseases consisted "solely in functional derangement caused by organic disease."[32] French physician Paul Broca's (1824–1880) demonstration that aphasia resulted from lesions in the frontal lobe of the cortex (now called "Broca's area") provided the first proof of Gall's theorem that a mental disorder could stem from a localized brain defect. At the time, however, Gall, Broca, and others had no way to directly examine living human brains but instead relied on autopsies or observations of symptoms among brain-damaged individu als.[33],[34]

In contrast to Lockean models that influenced moral treatment, brain-based models held that ideas, thoughts, and motives were the effects of material influences, not their causes. The biological frame emphasized how, although madness could result from diverse sources, it was ultimately a product of malfunctioning brains. Beginning in the mid-19th century there was a "long period when biological accounts of madness ruled virtually unchallenged."[35] The biological conception had great appeal in an era when science was replacing religion as the most authoritative source of explanations and treatments of madness. The presumed somatic basis of insanity also entailed the primacy of medical practitioners, as opposed to religious or other healers, as the appropriate responders to it.

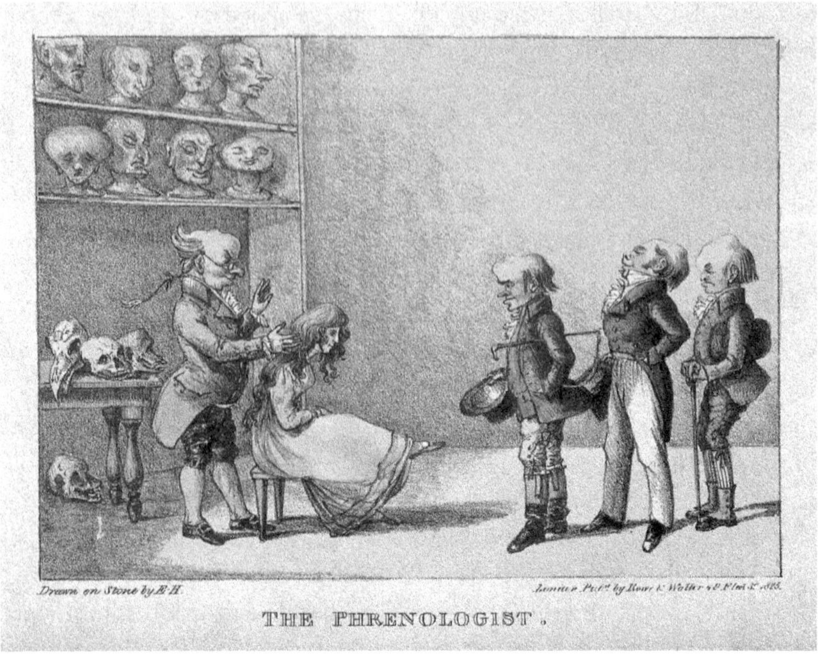


**Figure 3.1** Nineteenth-century phrenologists followed the teachings of neuroanatomist Franz Josef Gall, who believed that the close scrutiny of skulls could reveal the brain areas associated with mental illness and other psychological faculties.

Gall's influential phrenology delivered the validation that asylum keepers were dealing with brain disorders rather than the treatment of souls. If insanity was a somatic disease, then doctors were the most legitimate responders to it.

By mid-century, psychiatrists who identified as physicians had replaced moral treatment with understandings that insanity was rooted in defective brains and nervous systems. "Again and again," historian Normal Dain summarizes, "asylum superintendents, trying to convince the public of the value of mental hospitals, pointed out that insanity was subject to medical treatment and cure because it was a physical disease of the brain."[36] In 1844, the leaders of mental institutions in the United States institutionalized the medical control of asylums when they founded the Association of Medical Superintendents of American Institutions for the Insane. For many decades to come, the vast majority of psychiatrists found employment in mental hospitals.[37] The medical profession had gained control over treating madness, an authority that persists at present.

## Specificity Develops

Alongside attempts to localize mental disorders in brains, psychiatrists tried to delineate disorders from one another. The grouping of large numbers of the insane within institutions made apparent that they did not suffer from a single condition. This situation led psychiatrists to develop many classifications of mental illness that strove to specify what particular disorder patients displayed. Until the end of the century, however, none were theoretically grounded or widely accepted. Fierce debates over taxonomies persisted, concerning not just the adequacy of any particular classification for specifying etiology and/or prognosis but also about the very possibility of any rigorous systematization of mental diseases.

At one extreme were those who classified virtually any symptom presentation as a separate disorder, leading to the proliferations of diseases. American psychiatrist Pliny Earle enumerated sixty-one different physical and twenty-three different moral causes of insanity.[38] Eminent French psychiatrist John-Etienne-Dominique Esquirol's (1772–1840) *Atlas of 1838* contained twenty-seven diagnostic categories, each accompanied by a line drawing of a patient with that diagnosis.[39] At the other extreme were those who considered all mental disorders to be variants of a single disorder. The first U.S. census survey to ask about mental disorder in 1840 reflected this approach, containing just one category of mental disorder, "insanity." Pinel limited his classification to the four categories of melancholia, mania, idiocy (stupor), and dementia (lack of capacity for orderly thought).[40] Overall, Gerald Grob observes, "nosologies tended to be general and fluid, and judgments about individual patients represented pragmatic choices that had few practical consequences."[41]

While doctors strove to identify specific types of insanity and to localize them in the brain, their understandings of what factors distinguished madness from sanity remained the same: the "without cause" criteria that emerged with the Hippocratics. For example, Johann Reil (1759–1813), a German physician who coined the term "psychiatry" in 1808, described the heterogeneous collection of patients within asylums: "The madhouse has its usurpers, tyrants, slaves, criminals, and defenseless martyrs, fools who laugh without cause, and fools who torture themselves without cause."[42] Eminent American psychiatrist Isaac Ray (1807–1881) asserted: "Madness is not indicated so much by any particular extravagance of thought or feeling, as by a well-marked change of character, or departure from the ordinary habits of thinking feeling and acting, without any adequate external cause."[43]

The basic question of how to separate madness from sanity continued to rest on commonsensical notions. "[Psychiatrists'] definition of insanity reflected popular beliefs about right behavior so widely held that they needed little confirmation. . . . [They] had no more scientific means to identify insanity than the intelligent lay person," historian Nancy Tomes observes.[44] English psychiatrist John Charles Bucknill (1817–1897) posed the essential dilemma that the absence of a clear definition of mental disorder presented: "To the pathologist the substance of the brain is as yet practically structureless. Although the microscope reveals cells and tubes and intervening stroma, up to the present time it is unable to indicate when these are in a normal or abnormal state."[45] This problem, which current neuroscientists still have not been able to resolve, bedeviled all diagnosticians. Even the most prominent classifier of madness, Emil Kraepelin, maintained the without cause notion to define insanity itself.

## Emil Kraepelin

German psychiatrist Emil Kraepelin is consensually considered to be the most influential classifier of mental illness. "It is Kraepelin, not Freud, who is the central figure in the history of psychiatry," historian Edward Shorter writes.[46] For most of his career Kraepelin was a physician in a Munich asylum, using descriptions of inpatient cases to develop his classifications. He followed the tradition in physical medicine started by the 17th-century English physician Thomas Sydenham and developed by the 19th-century German pathologist Rudolph Virchow that considered mental disorders to be true diseases manifesting some brain pathology. This view had been highly successful in helping to distinguish physical diseases from each other, especially as knowledge of infectious agents and other physical pathogens rapidly grew over the course of the 19th century.

### Scientific Classifications Develop

When Kraepelin began his research, medical scientists were making striking advances in discovering specific organic causes for diseases including cholera, tuberculosis, tetanus, diphtheria, and typhoid fever. In contrast, Kraepelin confronted a field of psychiatric classification that was in intellectual chaos, with no consensual diagnostic system. Everyone since Greek times had employed symptoms to individuate disorders. But without any commonly shared principle for how to divide up the varied symptomatic presentations, the use of symptoms allowed physicians and psychiatrists to develop many different classifications.

Although the bacterial cause of syphilis—one of the most dreaded diseases at the time—was not discovered until 1905, the rapidly growing knowledge about this condition in the late 19th century provided Kraepelin with a model for psychiatric diagnoses. Syphilis indicated that mental disorders, like physical ones, could stem from an underlying physical pathology with no psychic causal component and thus fit directly within traditional medical diagnostic theory. In addition, syphilis showed how the same underlying disorder could be present although its symptoms changed over time, differing markedly at different stages of the disease. The moral of this stage model seemed clear: it is not just symptoms at any particular time but symptoms over the entire progression of a disease that served to identify any illness.

"I soon realized," Kraepelin explained about his initial effects to classify patients, "that the abnormalities at the beginning of the disease had no decisive importance compared to the course of the illness leading to the particular final state of the disease, just as happened with the various forms of paralysis (syphilis)."[47] Syphilis became Kraepelin's exemplar for how conditions identified with insanity could have underlying biological causes despite the changing manifestations of their symptoms. Accordingly, his primary classificatory principle was not based on symptoms but on the course and outcome of any illness.

Although Kraepelin generally assumed that mental disorders stemmed from disease processes in the brain, his methods were inductive and derived from his clinical experience of observing patients and listening to their self-reports. He came to emphasize two major categories of psychotic illness, rigidly separating cases of dementia praecox from those of manic depression. Dementia praecox (what is now called "schizophrenia") had several subtypes including the abnormal movements of the catatonic, the delusions of the paranoid, and the disorganization of the hebephrenic. Each typically arose among young people and usually had a progressively deteriorating course that ended in dementia.[48]

In contrast, manic-depressive psychoses were mood disorders marked by fluctuating paths that often improved over time. Kraepelin showed how a great variety of symptomatic presentations of affective disorder—for example, slowness of thinking, hopelessness, inner torment, inhibited activity, and inability to feel pleasure, as well as physical symptoms, such as sleep and appetite disorders and fatigue—in fact represented a single underlying pathology. Based on this hypothetical underlying unity of various symptom presentations, he classified even individuals who were only depressed and had no manic symptoms as having manic-depressive disorder. "In the course of the years," Kraepelin emphasized, "I have become more and more convinced that all [melancholic] states only represent manifestations of a *single morbid process*."[49] He also described dementias that developed in later life and were progressively degenerative as well as a condition where paranoid features dominated.[50]

Although he is commonly portrayed as the major influence on current psychiatric diagnostic manuals because of his insistence that each mental illness was a discrete, symptom-based entity, in fact, Kraepelin rejected diagnoses based on static depictions of symptoms at any particular point in time and instead focused on the course, prognoses, and, if possible, cause to differentiate various conditions.[51] He opposed the sole use of symptomatic criteria to infer which disorder was present. Although diagnosticians have to use symptoms as a main resource, Kraepelin did this in a way that was intended to transcend overt appearances to get at underlying pathology. Most important, he seemed to show how mental disorders could be described and distinguished in ways that were identical to observations of physical diseases.

Kraepelin's system was revolutionary because it deposed the longstanding focus on symptoms to instead concentrate on course and outcomes. His major accomplishment was to insist that each specific mental illness had a distinctive trajectory that unfolded across the patient's life course. He expected that the identification of anatomical lesions would eventually confirm his classifications.

## The Organic Causes of Mental Disorders

Hereditarian theories that reigned among both physicians and laypersons at the time strongly shaped Kraepelin's explanations for the emergence of insanity. Gall, for example, had argued that insanity was an inherited brain disease and downplayed the causal importance of environments. French surgeon Ulysse Trelat succinctly summarized this view in 1856: "There is one great cause of insanity, a primordial cause, the cause of causes, *heredity*, which fixes the disease in families and makes it transmissible from generation to generation."[52]

During the last half of the century theories that explained madness as the product of degeneration prevailed. French psychiatrist Benedict Morel (1809–1873) wrote an influential book, *Traits of Physical, Intellectual, and Moral Degeneration among the Human Species* (1857), that posited as mental illness passed through successive generations it became worse and worse. Neurasthenics might have children who were hysterical or hypochondriacal and their offspring, in turn, might develop dementia praecox. Degeneration theory, which was popular in both medical and lay circles, foreshadowed the later rise of social Darwinism as a dominant ideology in professional and general cultures alike.[53]

Kraepelin's causal theories lay squarely within the dominant 19th-century emphasis on heredity. He maintained that most disorders stemmed from inherited predispositions so that "insanity may be to an astonishing degree *independent of external influences.*"[54] Even many cases that seemed to arise normally from outside factors such as deaths, quarrels, unrequited love, infidelity, or financial difficulties actually were manifestations of disorders that stemmed from innate dispositions. "The real cause," Kraepelin wrote, "of the malady must be sought in *permanent internal changes,* which at least very often, perhaps always, are innate."[55] Consequently, "[T]he fact should not be underestimated that in the care of the insane in institutions we possess the only means by which it is possible to eliminate the most frequent cause of insanity, namely heredity."[56]

## What Is a Mental Disorder?

While Kraepelin is widely known for his distinctions of schizophrenic and bipolar conditions, commentators have ignored how he separated conditions that indicated madness from those that were sane. It is true that Kraepelin's central concern was to specify what particular disorders inpatients had, but he also embraced the without cause doctrine that had dominated distinctions between normality and pathology for thousands of years. He identified conditions as disorders when they were without "sufficient cause" in light of their circumstances or, when they initially seemed to be with cause, they became independent of circumstances and continued even after the provoking circumstances changed:

> Morbid emotions are distinguished from healthy emotions chiefly through the lack of a sufficient cause, as well as by their intensity and persistence. . . . Even in normal life moods come and go in an unaccountable way, but we are always able to control and dispel them, while morbid moods defy all attempts at control. Again, morbid emotions sometimes attach themselves to some certain external occasions, but they do not vanish with the cause like normal feelings, and they acquire a certain independence.[57]

Such cases include states that were initially disorders, as well as ones that began as normal responses but subsequently became morbid. These conditions could be distinguished from normal distress by telltale evidence such as inexplicable recurrence, psychotic ideation, or duration well beyond the cessation of any triggering event.

For example, Kraepelin addressed the issue of differentiating between manic-depressive disorder and normal sadness in some of his case presentations:

> I will first place before you a farmer, aged fifty-nine, who was admitted to the hospital a year ago. . . . On being questioned about his illness, he breaks into lamentations, saying that he did not tell the whole truth on his admission, but concealed the fact that he had fallen into sin in his youth and practiced uncleanness with himself; everything he did was wrong. "I am so apprehensive, so wretched; I cannot lie still for anxiety. O God, if I had only not transgressed so grievously!" . . . The illness began gradually seven or eight months before his admission, without any assignable cause. Loss of appetite and dyspepsia appeared first, and then ideas of sin. . . . The most striking feature of this clinical picture is the *apprehensive depression.* At first sight, it resembles the anxieties of a healthy person, and the patient says that he was always rather apprehensive, and has only grown worse. But there is not the least external cause for the apprehension, and yet it has lasted for months, with increasing severity. This is the diagnostic sign of its morbidity.[58]

Here, Kraepelin notes that even the extreme emotional and physiological symptoms of this patient were consistent with intense normal sadness, especially in a person with a dispositional tendency toward the melancholic side. But, he observed, the patient's symptoms started "without any assignable cause." Moreover, in addition to the fact that "there is not the least external cause for the apprehension," the condition had lasted months (and thus has a prolonged and seemingly inordinate duration) and had not, as normal sadness episodes do, displayed a trajectory of decreasing symptoms—far from it, it has shown "increasing severity" over time even though nothing new occurred in the circumstances to warrant such changes. This disconnection of the patient's condition from external events, and especially the lack of a trajectory showing normal coping and mastery, "is the diagnostic sign of its morbidity."

What is critical in Kraepelin's discussion is that, after reciting the duration and the symptoms, he noted that "at first sight, it resembles the anxieties of a healthy person," especially one with somewhat melancholic (but normal range) temperament. That is, Kraepelin recognized that symptoms of this duration and severity can be a normal response to events. It is not the duration or symptoms in themselves but their lack of proportional relation to any plausible external cause that allowed him to see that this condition was a disorder.

In another passage in which he reiterated the without cause criterion as central to diagnosis, Kraepelin made it clear that, even in his day when more severe cases were the rule among inpatient psychiatric patients, there was a real possibility of misclassifying a normal person as disordered because the symptoms could be identical:

> Under certain circumstances it may become very difficult to distinguish an attack of manic-depressive insanity from a *psychogenic* state of depression. Several times patients have been brought to me, whose deep dejection, poverty of expression, and anxious tension tempt to the assumption of a circular depression, while it came out afterwards, that they were cases of moodiness, which had for their cause serious delinquencies and threatened legal proceedings. As the slighter depressions of manic-depressive insanity, as far as we are able to make a survey, may wholly resemble the well-founded moodiness of health, with the essential difference that they arise without occasion, it will sometimes not be possible straightway to

> arrive at a correct interpretation without knowledge of the previous history in cases of the kind mentioned.[59]

Here, Kraepelin used the term "psychogenic" to refer to normal sadness states with sufficient external cause. The crucial point was that "the slighter depressions of manic-depressive insanity, as far as we are able to make a survey, may wholly resemble the well-founded moodiness of health, with the essential difference that they arise without occasion." Kraepelin understood that the symptomatic presentation of normal and disordered cases could be the same, so that the context in which they arose was the essential differentiating criterion.

In sum, while adhering to 19th-century aspiration to specify distinct diseases and to uncover their organic foundations, Kraepelin maintained the traditional distinction between conditions that were "with" or "without" cause. Not symptoms in themselves, but symptoms that became detached from their contexts and took on a life of their own, indicated disorder. Kraepelin offered symptoms as evidence to infer a diagnosis but he never attempted to define disorders solely in terms of necessary and sufficient symptoms. He clearly recognized normal episodes "with cause" that were proportionate to their triggers and that subsided after the stressor subsided, and he actively grappled with how to distinguish normal from disordered symptoms when their overt appearances were similar. What conditions were "with" or "without" cause did not stem from professional knowledge but arose from Kraepelin's commonsensical understandings, which echoed Hippocratic distinctions that had persisted for millennia.

## Nervous Conditions

The appeal of the biological model was strong enough to influence views of not just psychotic conditions found among inpatients but also the common psychogenic complaints that outpatients voiced. By midcentury, patients with a variety of nervous conditions flocked to medical practices. Echoing Cheyne's remarks about 18th-century England, many observers characterized the 19th century as a time that featured a remarkably high rate of nervousness.[60] Such moods were increasingly coming to be viewed as products of defective brains and nervous systems. Sufferers of physical and moral ailments such as headaches, phobias, insomnia, fatigue, lack of purpose, and the like were clearly not insane, but no distinct diagnostic term captured their complaints. Because asylums were associated with severe mental disorder, patients and their families increasingly became attracted to labels such as "nerves" that avoided the dire implications of terms such as "madness" or "insanity."

In 1869 American neurologist George Beard (1839–1883) developed what became the wildly popular notion of "neurasthenia."[61] Beard was committed to developing a scientific basis for his specialty, but he faced the dilemma of having to deal with outpatients who expressed an extremely diffuse array of woes that encompassed nonspecific distress, sexual problems, fatigue, depression, and anxiety. The dominant theories he confronted indicated that such problems were signs of hysteria or hypochondriasis and so were products of patients' minds with no somatic basis. Yet, these theories were not congruent with what had become standard medical thinking at the time: real diseases must be distinct entities with organic sources.

Beard developed the neurasthenia diagnosis to indicate that the condition he described was a physical, not a mental, state. It combined a protean mixture of various physiological and psychological conditions. According to Beard, symptoms of neurasthenia encompassed "dyspepsia, headaches, paralysis, insomnia, anaesthesia, neuralgia, rheumatic gout."[62] A weak nervous system underlay both the mental and somatic symptoms that fell into the neurasthenic category. "Nervousness," Beard declared, "is a physical not a mental state, and its phenomena do not come from emotional excess or excitability or from organic disease but from nervous debility and irritability."[63]

The grounding of neurasthenia in the nervous system helped account for the immense popularity the diagnosis enjoyed: it removed the stigma from people who suffered from what they and their physicians believed was a genuine disease. Accordingly, Beard did not focus on psychological therapies but emphasized treatments that employed electrical stimulation, drugs, surgical injections, and the like. Physicians could treat psychic conditions while proclaiming they were somatic complaints stemming from flaws in the central nervous system.

The neurasthenic diagnosis also illustrates how social accounts of mental disorder persisted alongside brain-based explanations and treatments. Beard echoed Cheyne's earlier notion that social forces created mental distress and disorder: the rapid increase of nervous conditions arose from the stresses of modern life.[64] Beard connected neurasthenia to the wholesale changes of industrialization, urbanization, and mechanized transportation that had emerged during the 19th century. In particular, "steam power, the telegraph, the periodical press, the sciences, and the mental activity of women," accounted for the high prevalence of the disease.[65] Also like Cheyne, Beard believed that the affluent and refined classes, especially those who worked with their minds, were more sensitive and thus more prone to nervous exhaustion.[66] Therefore, he associated neurasthenia with social status, culture, and distinction. Men, too, were particularly vulnerable to neurasthenia because of their greater participation in civilization. In contrast, women were more prone to develop the psychogenic condition of hysteria.

For Beard, neurasthenia could only have arisen as a product of 19th-century American conditions, which led people to become overly tense, worked up, on edge, and overwhelmed by stress. Too much strain overtaxed people's coping abilities and depleted their nervous energy. He noted: "No age, no country, and no form of civilization, not Greece, nor Rome, nor Spain, nor the Netherlands, in the days of their glory, possessed such maladies."[67] Although Beard claimed neurasthenia was a uniquely American affliction, it also became an extremely popular diagnosis in Europe. "The name of neurasthenia is on everybody's lips; it is the fashionable new disease," Paul Dubois, a French physician, observed.[68]

Like Cheyne, as well as many psychiatrists and physicians at the time, Beard did not see brain-based and environmental causes as mutually exclusive. Although he believed that social factors led to the ubiquity of neurasthenia, he also emphasized how heredity determined which particular people succumbed to the pressures of civilization. Individual lifestyles and experiences provoked inherited susceptibilities to nervous weakness.[69] The biologically predisposed could give in to nervous collapse not just from the hectic demands and constant stimuli of modern life but also through decadent activities such as illicit sex, masturbation, or gambling.[70]

The diffuse diagnosis of neurasthenia imposed an artificial coherence and precision on what had been seen as a disparate array of symptoms. It provided patients with a somatic-sounding label for symptoms that fit no existing category of organic disease.[71] Both patients and doctors also preferred a diagnosis of neurasthenia over one of hysteria or hypochondria because it protected patients from beliefs that their problems were purely psychic. It also insulated them from charges that they were malingerers; they received assurance that their ailments were genuine and not imaginary. Perhaps most importantly, neurasthenia was not associated with madness. "Patients," Edward Shorter writes, "found the notion of suffering from a physical disorder of the nerves far more reassuring than learning that their problem was insanity."[72]

The most important contribution of neurasthenia was to direct medical attention to distressed but not institutionalized patients. Neurasthenia was, in Beard's term, a "disease of the street," although this street was likely to be found in the well-to-do section of town.[73] The diagnosis greatly broadened conceptions of what sorts of conditions were legitimately treated as diseases in outpatient practices. As one 19th-century French psychiatrist proclaimed: "Everyone submits more or less to the influence of this morbid nervous excitability."[74] Moreover, it provided a nonstigmatizing label that provided respite for those who were overcome by the pressures of modern civilization. "Within a decade of Beard's death in 1883," historian Charles Rosenberg notes, "the diagnosis of nervous exhaustion had become part of the office furniture of most physicians."[75]

Neurasthenia focused a previously scattered diagnostic scene into a single category that dominated views of nervous conditions for the remainder of the century among both physicians and the general educated public. In retrospect, this capacious diagnosis did not enhance the amorphous classificatory scene that existed at the time. It was, in essence, scientifically useless. As prominent English physician T. Clifford Allbutt (1836–1925) noted at the end of the century, "[T]he so-called diseases of the nervous system [were] a vast, vague, and most heterogeneous body, two-thirds of which may not primarily consist of diseases of nervous matter at all."[76] Nevertheless, neurasthenia became the major diagnoses of nervous afflictions in the late 19th and early 20th centuries because it satisfied the needs of patients to receive a physical label for their diffuse ailments and of doctors for a distinct diagnosis of nervous complaints. The best source of legitimacy for mental afflictions at the time was belief in their organic nature. The immense popularity of neurasthenia helped to shape the early writings of Sigmund Freud, who was to become by far the most prominent psychiatrist for the next half-century.

## The Biomedical Freud

The pull of the biological view during the late 19th century was strong enough that even Viennese neurologist and psychoanalyst Sigmund Freud (1856–1939), who would make a clean break with this tradition at the turn of the century, accepted the prevailing materialist view of mental illness. Freud's initial training occurred within the dominant biomedical model. For the first twenty years of his career, he was a neurologist and an anatomist, studying such topics as the nervous systems of primitive fish. Freud was taught within medical and academic settings that, much like the present, emphasized how brains and nerves could unlock the secrets of human behavior, including mental diseases. Psychological processes, as well, could be fully explained through physiology without reference to subjectivity.

### Freud as Diagnostician

Initially, Freud held the view that, ultimately, the human sciences would be explained through reference to the natural sciences. Indeed, his early writings had the goal of transforming psychology into a biological and physiological discipline.[77] The opening sentence of his unpublished *Project for a Scientific Psychology* (1895) read: "The intention is to furnish a psychology that shall be a natural science: that is to represent psychical processes as quantitatively determinate states of specifiable material particles."[78] Moreover, much like Kraepelin, Freud strove at the onset of his career to differentiate various types of mental disorder.

Freud's pioneering writings in the 1890s, many co-authored with physician Josef Breuer (1842–1925), were aligned with then-current trends to view neurotic complaints as disguised manifestations of sexual impulses.[79] At this point, Freud's goal was to use physiology, particularly as it applied to sexual instincts, to explain psychology. This was an especially challenging endeavor because he believed that the most fundamental processes involved in producing neuroses were unconscious and thus not easily accessible to either patients or their clinicians. Freud's views about the neuroses thoroughly changed both within his early writings and between them and his later works. Yet, two forces remained constant for generating psychological problems: the importance of sexual forces and the formative influences of experiences during infancy and early childhood.

Despite the radically mentalistic turn his work would come to take, Freud's early writings, no less than Kraepelin's, remained within the traditional 19th-century biomedical framework. Freud's original view of the neuroses was more neurological than psychological, reflecting the emphasis on delineating specific conditions that dominated medical thinking at the time. Over the course of the 1890s he defined a variety of neurotic conditions, many of which—generalized anxiety, phobias, obsessions, and panic—have largely persisted until the present. "Obsessions and phobias are separate neuroses, with a special mechanism and aetiology," Freud presented as examples.[80] Indeed, his most lasting achievement from this period was to precisely separate and describe the various nervous conditions that were thrown together in the then dominant neurasthenic diagnosis.

Much in current DSM style, Freud used symptoms to identify distinct conditions. At the most general level, he divided the neuroses into two categories. The first were the actual (or current) neuroses that stemmed from somatic processes related to the present sexual life of patients.[81] These included neurasthenia and anxiety neuroses. The second were the psychoneuroses, including hysteria and obsessional neuroses, which stemmed from repressed psychic memories of sexual traumas in childhood.[82]

Freud not only severed organically based anxiety conditions from neurasthenia and hysteria but also delineated several partially distinct anxiety conditions through closely observing their particular symptoms. "I call this syndrome 'Anxiety-Neurosis,'" Freud wrote in 1894, "because all its component elements can be grouped round the central symptom of 'morbid anxiety' and because individually they each have a definite connection with this."[83] He described the specific psychic components of threat, irritability, and inability to concentrate and the somatic components of heart palpitations, breathing problems, tremors, sweating, and gastrointestinal disturbances that remain the central components of anxiety disorders.[84] Indeed, Freud's description of anxiety attacks contained ten of the twelve symptoms listed in the DSM-III (1980) diagnostic criteria for panic disorders.

Likewise, Freud's description of anxious expectation, a free-floating state of nervousness and apprehensiveness, closely resembles the diagnosis of generalized anxiety disorder that emerged in the DSM-III:[85]

> A woman who suffers from anxious expectation will imagine every time her husband coughs, when he has a cold, that he is going to have influenzal pneumonia, and will at once see his funeral in her mind's eye. If when she is coming towards the house she sees two people standing by her front door, she cannot avoid the thought that one of her children has fallen out the window; if the bell rings, then someone is bringing news of a death, and so on.[86]

Anxious expectation, Freud noted, could attach itself to any suitable ideational content. Numerous commentators from St. Augustine through Kierkegaard had described similar anxiety conditions that were omnipresent aspects of the human condition. Freud's conception, however, moved generalized anxiety from the religious and philosophical spheres to the domain of medicine. It was a significant step toward the medicalization of anxiety that had no specific object.

Freud not only distinguished the symptoms of the major neuroses from one another but also provided an etiological basis for separating these conditions from each other and from other types of neuroses. Following the dominant neurological approach at the time, he assumed that many psychic experiences reflected physical and chemical processes. Freud applied the laws of the conservation and transformation of energy developed in physics to explain the various neuroses.[87] They stemmed from varying levels of sexual excitation, which were

> something which is capable of increase, decrease, displacement and discharge, and which extends itself over the memory-traces of an idea like an electric charge over the surface of the body. We can apply this hypothesis . . . in the same sense as the physicist employs the conception of a fluid electric current. For the present it is justified by its utility in correlating and explaining diverse psychical conditions.[88]

Sexual energy could be transformed into a variety of psychic, behavioral, and somatic manifestations that were representations of the same underlying force.

The actual neuroses resulted from current sexual disturbances and reflected problems in expressing somatic energy.[89] Anxiety neuroses were thus especially common among sexually frustrated practitioners of *coitus interruptus*, sexual abstinence, and the like.[90] "Forced abstinence, frustrated sexual excitement (not gratified by sexual intercourse), incomplete or interrupted coitus (not attaining gratification)" resulted in excessive sexual tension and consequent anxiety.[91] Indeed, Freud suggested that ultimately anxiety neuroses would be removed completely from consideration as neurotic disturbances because, at bottom, they reflected metabolic disturbances of sexuality.[92] Neurasthenia, which resulted from the unsatisfactory satisfaction of sexual needs, was another actual neurosis. While anxiety neuroses stemmed from the accumulation of more sexual excitation than people could psychically assimilate, neurasthenia resulted from inadequate forms of sexual release. Like other physicians at the end of the 19th century, Freud viewed masturbation, in particular, as a major source of neurasthenia.

Freud distinguished the anxiety neuroses from hysteria. Unlike anxiety conditions, which stemmed from current deficiencies in sexual gratification that are "not derived from any psychical source," hysterical conditions were rooted in psychological repression of earlier real or imagined traumatic memories. Hysterical patients, Freud and Breuer memorably wrote, "suffer principally from reminiscences."[93] Psychic repression converted their memories of early sexual traumas into more acceptable somatic expressions. Freud's view thus grounded hysterical suffering in the psychological realm and anxious conditions in somatic processes: "anxiety-neurosis is actually the somatic counterpart of hysteria."[94]

The genius of Freud's early works on nervous conditions did not reside in his etiological theory but instead in his differentiation of different types of neuroses. They unified under the neurotic umbrella a variety of previously distinct syndromes or symptoms associated with other conditions. The creation of a distinct class of anxiety disorders; the split of anxiety from neurasthenia, hysteria, and other states; and the separation of distinct anxiety states both culminated the trend toward growing specificity that marked 19th-century medicine and anticipated the DSM-III diagnostic revolution in 1980. Indeed, the anxiety conditions that Freud delineated continue to lie at the heart of these diagnoses in current manuals. Freud's subsequent writings developed different views, placing far less emphasis on specific diagnoses and the organic etiology of the neuroses. Yet, ironically, his initial conceptions of the neuroses have more in common with the recent symptom-based DSM classifications than does Kraepelin's focus on the prognosis and outcomes of various disorders.

Freud's writings about hysteria contained in embryonic form the notion that mental disorders could be purely psychic. Nevertheless, the general thrust of his early work—no less than Kraepelin's and other prominent 19th-century diagnosticians—fell very much in the dominant tradition that emphasized disease specificity. Nineteenth-century psychiatrists disagreed about many things including the causes, prognoses, and treatments of mental disorders, but almost all of them, including Freud, embraced a medical model. Like Kraepelin, Freud strove to distinguish distinct conditions from one another, but, unlike his German competitor, Freud used presumed etiology rather than course and prognosis to make his divisions.

## Freud's Initial Causal Theories

In contrast to his adherence to conventional 19th-century views on the need to specify distinct mental disorders, Freud's early views about causation granted more weight both to purely psychic and to external forces than the dominant model. In a marked split with reigning views, Freud discarded a hereditarian focus to emphasize how everyone, not just the biologically predisposed, could suffer from the various neurotic conditions. For example, he stated that "neurasthenia is one of those affections which anyone may easily acquire without having any hereditary taint."[95]

Freud's initial causal theories of the neuroses featured a situational orientation. Each "*betray as their common source, the sexual life of the person concerned, either a disturbance of his present sexual life or important events in his past life*."[96] In this first, short-lived stage, Freud posited that neurotic disturbances resulted from traumatic events in childhood that were repressed but reappeared in distorted form as hysterical symptoms. What became known as the "seduction theory" posited that sexual abuse in infancy or early childhood led to adult pathology. Childhood traumas caused—not just triggered—neurotic symptoms in later life. "External events," Freud and Breuer declared in 1893, "determine the pathology of hysteria to a far greater extent than is known and recognized."[97]

Three years later, Freud proposed a radical new thesis: sexual traumas that befell children accounted for every case of hysteria:

> The event, the unconscious image of which the patient has retained, is a premature sexual experience with actual stimulation of the genitalia, the result of sexual abuse practiced by another person, and the period of life in

> which this fateful event occurs is early childhood, up to the age of eight to ten, before the child has attained sexual maturity. *A passive sexual experience before puberty*: this is the specific aetiology of hysteria.[98]

Freud insisted that: "whatever case and whatever symptom we take as our starting point, *in the end we infallibly come to the realm of sexual experience* as the origin of hysteria."[99] He emphasized: "*These sexual traumas must . . . occur in early childhood (before puberty) and they must consist in actual excitation of the genital organs (coitus-like processes)*."[100] These experiences had little impact at the time they occurred; their pathogenic effects only became manifest during puberty when some later event reactivated memories of the earlier trauma.[101] Patients repressed these traumas, although they remained present in an unconscious state. The task of their physicians was to bring repressed memories to consciousness, a process that Breuer and Freud called "abreaction."

Breuer, too, emphasized the traumatic impact of sexual events, although he focused on the time of marriage: "I do not think I am exaggerating when I assert that *the great majority of severe neuroses in women have their origin in the marriage bed*."[102] Freud, however, rejected Breuer's focus on the traumas of recently married women. He believed that erotically charged events such as sexual abuse, witnessing parental intercourse, or experiencing guilt over masturbation in early childhood or even infancy provoked hysterical neurosis.

Freud concluded *Studies on Hysteria* with a query from a patient: "Why, you tell me yourself that my illness is probably connected with my circumstances and the events of my life. You cannot alter these in any way. How do you propose to help me, then?" Freud famously replied: "But you will be able to convince yourself that much will be gained if we succeed in transforming your hysterical misery into common unhappiness."[103]

Freud's initial emphasis on how traumatic sexual events in childhood had profound psychic consequences in later life did not last. By 1897, he had already begun to doubt that repressed early traumas were the source of later neurotic symptoms. In what Freud considered to be the cornerstone in the discovery of psychoanalysis, he abandoned the seduction theory, repudiating his initial view that caretakers and relatives seduced his female—and sometimes male—patients into having incestuous relations when they were very young:

> Under the influence of the technical procedure which I used at that time, the majority of my patients reproduced from their childhood scenes in which they were sexually seduced by some grown-up person. With female patients the part of seducer was almost always assigned to their father. I believed these stories and consequently supposed that I had discovered the roots of the subsequent neurosis in these experiences of sexual seduction in childhood. . . . If the reader feels inclined to shake his head at my credulity, I cannot altogether blame him. . . . When I had pulled myself together, I was able to draw the right conclusions from my discovery: namely that the neurotic symptoms were not related directly to actual events but to wishful phantasies, and that as far as the neurosis was concerned psychical reality was of more importance than material reality.[104]

Freud subsequently asserted that these accounts stemmed not from actual seductions but from illusions derived from universal oedipal sexual longings that very young children have for their opposite sex parents. "I had in fact," Freud stated, "stumbled for the first time upon the Oedipus complex, which was later to assume such an overwhelming importance, but which I did not recognize as yet in its disguise of phantasy."[105] Patients' later recollections of molestations were less likely to reflect actual events than their own unconscious desires. Freud thus turned his attention from the environment toward unconscious mental life. This move changed the origin of later psychic disturbances from traumas perpetuated by adults to the child's own innate erotic feelings.

Freud changed his views to deemphasize experiences themselves in favor of internal psychic responses: "It is not the experience itself which acts traumatically, but the memory of it when this is re-animated after the subject has entered upon sexual maturity," he wrote in 1896. Most stories of sexual seductions that patients told him were, in fact, "phantasies" that emerged as defenses against their own childhood sexual activities such as masturbation. Hysterics now suffered from "fictitious" traumas.[106] "I have since learned to unravel many a phantasy of seduction and found it to be an attempt at defence against the memory of sexual activities practiced by the child himself," Freud acknowledged. He goes on to say: "The important thing, therefore, was evidently not the sexual stimulation that the person had experienced during childhood; what mattered was, above all, how he had reacted to these experiences, whether he had responded to them with 'repression' or not."[107] This change marked a "decisive turning point in psychoanalysis" because it moved Freud to study unconscious intrapsychic dynamics and their ambiguous relationship to tangible experiences.[108] The outer material world and the inner psychic one were thoroughly intertwined because all external "reality" was formed through unconscious and conscious interpretations. Memories were not grounded in some objective reality but were constantly revised in light of successive life experiences.

After Freud rejected the environmental position that the seduction theory entailed, he came to focus on the power of innate, sexual instincts and their consequent psychic repression as propelling neuroses. The symptoms of all neuroses were products of sexual activity, whether current or past, real or fantasized. Other forces such as heredity or temperament might be predisposing, contributing, or supplementary, but only sexuality had the power to cause a neurosis in the absence of any other factor. Ultimately, the fundamental cause of neuroses lay in the conflict between universal sexual instincts and the restrictions that human civilization placed on these natural proclivities.[109]

In contrast to Freud's initial classificatory system of the neuroses, which largely persists in current psychiatric manuals, history has not been kind to his etiological theory that was based on sexual energy. The failure of Freud's early causal theories seems largely due to the degree to which they reflected his particular cultural circumstances rather than universal somatic and psychic processes. Viennese culture was obsessed with thoughts of sex, which often became manifest through repressed and distorted expressions. It featured an especially powerful fear of masturbation; many physicians emphasized its horrific consequences, among which included insanity, premature death, and vulnerability to many diseases.[110] The primacy Freud gave to repressed sexual energy might have accurately reflected the situation of a particular stratum of his own society but was thoroughly inadequate as the basis for a general theory of the neuroses.

## What Is Normal and What Is Neurotic?

In contrast to organically oriented psychiatrists at the time who usually, if tacitly, adopted the traditional without cause distinction between normal and pathological behaviors, Freud proposed a new kind of separation. During his first, biologically oriented stage he used a hydraulic model of sexual energy to separate normal from disordered conditions.

He single-mindedly pressed for the sexual basis of all mental disorders: the compelling power of frustrated sexual energy served as the basis for distinguishing morbid from healthy behaviors. Psychological well-being depended on an adequate discharge of sexual excitation. Normally, this energy stayed in equilibrium through everyday mental and physical, as well as erotic, activities. Disturbances in either their past or present sexual life, however, could lead people to have either too much or too little sexual energy.[111]

Some disorders were abnormal because they stemmed from excessive levels of undischarged sexual energy.[112] Hysteria, Freud's central concern during the 1890s, exemplified this dynamic. It was a broad label that encompassed a wide variety of symptoms such as paralysis, fainting, convulsions, seizures, and choking that arose from repressed memories of sexual activities in early childhood. Erotic feelings led hysterics to experience disproportionately high amounts of sexual excitation that exceeded their adaptive capacities. In opposition to normal memory processes, which could be actively recollected and expressed, the erogenous memories of hysterics did not disappear but were repressed, remaining present in an unconscious state.

In Freud's hydraulic model, drives that were blocked in one way would find expression through some other way. Pathological memories of early eroticism persisted with their original intensity because they were not voiced.[113] Repressed memories sought expression through inappropriate somatic pathways and consequent hysterical symptoms. Unlike normal memories, which gradually dissipated with the passage of time, Freud and Breuer observed about neurotic memories: "At first sight it seems extraordinary that events experienced so long ago should continue to operate so intensely—that their recollection should not be liable to the wearing away process to which, after all, we see all our memories succumb."[114]

The hydraulic model also dictated a second source of abnormal conditions that arose from inadequate levels of sexual discharge. In contrast to hysteria, which was a psychic disturbance with roots in childhood, anxiety neuroses were somatic conditions stemming from current failures to achieve appropriate levels of sexual release. Freud emphasized processes such as abstinence, *coitus interruptus*, and masturbation that prevented people from achieving equilibrium in their sexual lives. For example, he claimed that "Neurasthenia arises whenever a less adequate relief activity takes the place of the adequate one, thus, when masturbation or spontaneous emission replaces normal coitus."[115] He was particularly concerned with preventing masturbation: "The prevention of masturbation in both sexes is a task that deserves more attention than it has received up to the present time."[116] The physician's task in restoring normality "consists in bringing the patient to abandon all injurious varieties of sexual intercourse and to enter into normal sexual relations."[117] Once individuals have regular heterosexual intercourse resulting in orgasm their neurosis should disappear. "No neurosis," Freud concluded, "is possible with a normal *vita sexualis*."[118]

Freud's grounding of pathology in levels of sexual energy did not last. Like his causal theories, his views of normal and abnormal sexual discharge were permeated with the ruling values of his era. Freud's view of normality reflected a time and place that negatively sanctioned all but the most limited forms of sexuality. Culturally relative ideas about sexual behavior, not universal biological or psychic processes, set Freud's dividing line between normal and pathological conditions. Once moral standards changed regarding what were "injurious varieties of sexual intercourse," Freud's conceptions of how to divide unnatural from natural conditions became anachronistic. In contrast, the without cause criterion continues to influence the division of sanity from madness through the present.

## Conclusion

At the outset of the 19th century, mental illnesses were few in number, loosely defined, explained through many diverse and often competing theories, and treated by a wide variety of healers. As the century progressed, theological views faded as understandings coalesced around a medical model that understood mental disturbances as comparable to organic diseases. They were brain malfunctions that were often transmitted through hereditarian processes and that should be specified and distinguished from each other through their etiology, course, and outcome.

The most seriously ill often entered mental institutions that were run by superintendents affiliated with the new medical specialty of psychiatry. The more numerous classes of nervous patients came to seek help from somatically oriented doctors. Although explanations of mental disturbances still encompassed both internal and external factors, the balance had tilted sharply toward the former. At the beginning of the 20th century a revolutionary new conception of mental illness emerged that overturned many of the developments over the past 100 years that equated mental and physical disorders.

## Acknowledgments

The section "Emil Kraepelin" is adapted from Horwitz & Wakefield, 2007, 75–82. The section "The Biomedical Freud" is adapted from Horwitz, 2013.

## Notes

1. Porter, 1997, 493.
2. Scull, 2015, 210.
3. In contrast, 19th-century American psychiatry was, in Edward Shorter's (2015, 20) colorful description, "a small tail wagged by a very large European dog."
4. Theodore Meynert, 1884, https://en.wikipedia.org/wiki/Theodor_Meynert. The first American psychiatric textbook, Benjamin Rush's *Medical Inquiries and Observations Upon the Disease of Mind* (1812), likewise presented a thoroughgoing somatic view of the nature and treatment of mental disorder.
5. These institutions were successively called "madhouses," "lunatic asylums," and "psychiatric hospitals" (Porter, 1997, 494).
6. Porter, 1997.
7. At the time, there was little difference in the subject matter of psychiatry and neurology. Instead, the major distinction between the two was that most neurologists had outpatient practices in contrast to asylum-based psychiatrists (Scull, 2015, 277).
8. Battie, 1758/1969.
9. Reiss, 2008.
10. Goldstein, 1987, 212.
11. Jean Baptiste Pussin quoted in Davidson, 2013, 209.
12. Grob, 1994, 47.
13. Quoted in Grob, 1994, 1.
14. The family-centered system of social control, however, seldom involved providing loving and sympathetic care to disturbed kin. Instead, families routinely subjected members they perceived as mad to coercive control. Shorter (1997, 2) describes a representative case around the turn of the century in Germany: "A youth of sixteen, who for years had lain in a pigpen in the hut of his father, a shepherd, had so lost the use of his limbs and his mind that he would lap the food from his bowl with his mouth just like an animal." This case resembles the famous confinement of Mr. Rochester's first wife, Bertha Mason, in *Jane Eyre* (331): "In the deep shade, at the farther end of

the room, a figure ran backwards and forwards. What it was, whether beast or human being, one could not, at first sight tell: it groveled, seemingly, on all fours; it snatched and growled like some strange wild animal; but it was covered with clothing, and a quantity of dark, grizzled hair, wild as a mane, hid its head and face."

15. Grob, 1973.
16. Tomes, 1984.
17. Shorter, 1997, 47.
18. Tomes, 1984, 93–95.
19. Grob, 1994, 61.
20. Vandermeersch, 1994, 223.
21. Shorter 1997, 53–60; Grob, 1991a.
22. The tenets of moral treatment persisted through the 20th century in small, private institutions such as the Menninger Clinic that catered to wealthy clients. See Friedman, 1990.
23. Dwyer, 1987, 87.
24. Fox, 1978, 84.
25. Starr, 1982.
26. Grob, 1994, 76.
27. Scull, 2015, 211–17. Porter, 1997, 498, 509.
28. In the 19th century, American psychiatry was centered in mental hospitals, not in academic settings.
29. Quoted in Shorter, 1992, 209. Although modern accounts equate Griesinger with a biological approach, in fact, he acknowledged the critical importance of social, cultural, and historical influences on mental disease. See Arens, 2013.
30. Makari, 2015, 474. Phrenology was also highly influential in the United States during the three decades before the Civil War.
31. Makari, 2015, 461.
32. Orson S. Fowler, quoted in Dain, 1964, 163.
33. The most famous example was a railroad worker, Phineas Gage, who experienced dramatic personality changes after a stake went through his frontal lobe (Damasio, 1994).
34. Franz Joseph Gall examining the head of a pretty young girl. Attributed to Edward Hull, 1825. Wellcome Collection, London.
35. Scull, 2015, 217.
36. Dain, 1964, 66.
37. Wallace, 1994.
38. Tomes, 1984, 81.
39. Grob, 1991.
40. Just three years before Pinel developed his four-fold classification, he wrote a psychiatric textbook that contained hundreds of different species of mental disease (Menninger, 1963, 20).
41. Grob, 1991, 422.
42. Quoted in Makari, 2015, 443.
43. Quoted in Dain, 1964, 72.
44. Tomes, 1984, 87, 122.

45. Whooley, 2019, 39.
46. Shorter, 1997, 100.
47. Quoted in Healy 2008, 71.
48. Although Kraepelin's emphasis on the steady deterioration of most schizophrenic patients persists in much psychiatric thought, longitudinal studies indicate that a substantial proportion of persons with this condition remain stable, improve, or recover over time (Bleuler, 1978; Davidson, 2013).
49. Kraepelin, 1921/1976, 1, italics in original.
50. Kraepelin's work underwent many changes between the publication of the first edition of his textbook in 1883 and its posthumous ninth edition in 1927.
51. E.g. McNally, 2011, 184.
52. Quoted in Porter, 2018, 105.
53. See especially Porter, 2018.
54. Kraepelin, 1921/1976, 181, italics in original.
55. Kraepelin, 1921/1976, 180, italics in original.
56. Kraepelin quoted in Porter, 2018, 207.
57. Kraepelin, 1907/1915, 68.
58. Kraepelin, 1904/1917.
59. Kraepelin, 1904/1917, 199–200.
60. Gay, 2002, 132–33.
61. Although Beard first used this term in an 1869 article, his renowned book did not appear until 1881.
62. Quoted in Shorter, 1992, 221.
63. Beard, 1881, 17. While 18th-century physicians viewed the nerves as strong and energetic forces, by the time that Beard wrote they were associated with exhaustion.
64. Madness, too, could stem from social pressures. "Insanity is," famed American physician and statistician Edward Jarvis wrote, "a part of the price we pay for civilization. The causes of the one increase with the developments and results of the other" (quoted in Grob, 1994, 61).
65. Rosenberg, 1997, 105.
66. Lutz, 1991.
67. Beard, 1881, vii–viii.
68. Quoted in Shorter, 1992, 221.
69. Another example stems from the work of the most prominent 19th- theorist of sexual deviance, Richard Krafft-Ebing. He postulated that any form of nonprocreative sexuality, such as homosexuality, arose from a combination of inherited predispositions to psychopathology and particular life experiences (Bayer, 1981, 20).
70. Rosenberg, 2007, 83. See also Lutz, 1991.
71. Scull, 2009, 97; Micale, 2008, 156–57; Shorter, 1994, 132.
72. Shorter, 1997, 113.
73. Quoted in Glas, 2003, 7.
74. Goldstein, 1987, 225.
75. Rosenberg, 1997, 108.
76. Quoted in Oppenheim, 1991, 9.
77. Jones, 1953, 272.

78. Freud, 1895/1976, 295. Freud retroactively characterized this work as "delirium, babbling, or gibberish" (Roudinesco, 2016, 57).
79. Sulloway, 1979.
80. Freud, 1895/1959, 128.
81. Freud scholar and critic Frederick Crews (2017, 400) notes that "current" neuroses is a more accurate translation than "actual" neuroses.
82. Ellenberger, 1970, 487.
83. Freud, 1894b/1959, 77.
84. See, for example, Freud, 1916–17/1989, 15–16; Freud, 1933/1965.
85. American Psychiatric Association, 1980, 233.
86. Freud, 1894b/1959, 79.
87. Ellenberger, 1970, 534–46.
88. Freud, 1894a/1959, 75.
89. Freud, 1898/1959, 242.
90. Freud, 1894b/1959, 97. See also Makari, 2008, 88.
91. Freud, 1896/1959, 147.
92. Freud, 1905, 282.
93. Freud, 1893/1959, 29; Freud, 1894b, 96.
94. Freud, 1894b, 105.
95. Freud, 1898/1959.
96. Freud, 1896b/1959, 145, italics in original.
97. Freud, 1893/1959, 54.
98. Freud, 1896b/1959, 149, italics in original.
99. Freud, 1896c/1959, 193, italics in original.
100. Freud, 1896a/1959, 156, italics in original.
101. Freud, 1898/1959, 243.
102. Freud & Breuer, 1895/1974, 328, italics in original.
103. Freud & Breuer, 1895/1974, 393.
104. Freud, 1925/1959, 33. Female caretakers of boys were responsible for about a third of the cases of seduction that Freud recounted.
105. Freud, 1925/1959, 34. While Freud believed that most accounts of sexual abuse were products of unconscious fantasies, he never denied that many cases were real. In his *Introductory Lectures* of 1916–17 (370) he states: "You must not suppose . . . that sexual abuse of a child by its nearest male relatives belongs entirely to the realm of phantasy. Most analysts will have treated cases in which such events were real and could be unimpeachably established."
106. Freud, 1896c/1959, 157; Freud, 1905/1959, 276; Freud, 1914/1959, 300.
107. Freud, 1905, 276, 279.
108. Ellenberger, 1970, 488.
109. Freud, 1905, 276.
110. Gay, 2002, 46, 147–51.
111. Freud, 1896b/1959, 145.
112. Many reasons could account for why some people had "excessive" amounts of sexual excitation—actual experiences, psychological predilections, or biological predispositions (Freud & Breuer, 1895/1974, 320–1).

113. Freud & Breuer, 1895/1974, 58, 62.
114. Freud & Breuer, 1895/1974, 58.
115. Freud, 1894a/1959, 98.
116. Freud, 1898/1959, 239.
117. Freud, 1898/1959, 237.
118. Freud, 1905/1959, 276.

# 4
# Freud's Transformation of Normality

To us he is no more a person
Now but a whole climate of opinion
—W. H. Auden, "In Memory of Sigmund Freud" (1940)

At the beginning of the 20th century, the Austrian neurologist-turned-psychoanalyst Sigmund Freud and his disciples developed a revolutionary approach to the study of psychological phenomena. Although Freud's interest in mental illness never wavered, his major works in the new century—*The Interpretation of Dreams* (1900), *The Psychopathology of Everyday Life* (1901), *Three Essays on the Theory of Sexuality* (1905), and *Jokes and Their Relation to the Unconscious* (1905)—turned sharply away from studying specific mental disorders toward interpreting conventional aspects of human behavior such as dreams, jokes, mistakes, and sexuality. These writings posited that, at the heart of sane, no less than neurotic, behaviors lay deeper cruel, murderous, and incestuous instincts that originated in infancy. "Through abnormality," German novelist Thomas Mann wrote, Freud would "succeed in penetrating most deeply into the darkness of human nature."[1]

Freud's revised views placed less emphasis on examining the particular neurotic conditions—hysteria, anxiety, neurasthenia, phobias, obsessions—that were his chief concern during the 1890s. Instead, it focused on identifying the presumably unconscious dynamics that underlay both normal and pathological phenomena. Freud's attention turned from diagnosing the specific disorders patients had toward understanding the symbolic meanings involved in their dreams, fantasies, and erotic activities. The psychodynamics behind these universal aspects of human nature were not discrete from, but were highly overlapping with, the processes that brought about mental disorders. From 1900 onward, Freud thoroughly blurred the boundaries between normal and disordered occurrences. Mental illnesses became embedded in general aspects of human behavior rather than in specific pathologies.

Freud's efforts between the beginning of the century and the onset of World War I turned attention from the pathological to the normal, from biology to psychology, and from specific mental disorders to broad psychic dynamics, especially those that arose from the inherent conflict between human instincts and the institutions that strove to curb them. The inescapable struggle between human nature and social regulations led Freud to emphasize the process of repression. He developed a more psychological view of all the neuroses, especially emphasizing how memory generally worked to suppress unpalatable thoughts. In thoroughgoing contrast to the dominant strain of psychiatric thought, including his own work in the previous century, at the heart of Freud's new psychoanalytic approach was the effort to understand pathological conditions in terms of purely psychic processes. Freud focused on repressed desires, psychological conflicts, and universal mental processes, all of which had little to do with hereditary predispositions, organic etiologies, or any other direct physical cause.

The onset of World War I led Freud's work to take yet another divergent turn. The final stage of his career, which lasted until his death in 1939, was marked by a greater concentration than previously on contemporaneous rather than childhood dynamics, external rather than internal reality, and the efforts of individuals to reconcile their instinctual demands with social constraints. Both phases of Freud's 20th-century writings helped revolutionize the study of human behavior, moving pathological phenomena from its fringes to its core. Conversely, Freud questioned the idea of normality, calling it an "ideal fiction."[2]

## Blurring the Boundaries Between Normal Emotions and Mental Disorders

During his first biologically oriented stage, Freud relied on a broad hydraulic theory of sexual energy combined with culturally infused definitions of healthy sexuality to claim that the diagnoses he studied—hysteria, anxiety, neurasthenia, and so on—were pathological and not normal. In the period from the turn of the century to his death in 1939, however, Freud's work generally abandoned the attempt to divide mental disorders from normality instead emphasizing their fuzzy boundaries and shared overlap.

*The Interpretation of Dreams* inaugurated Freud's second stage of thought.[3] This book "has generally been considered Freud's single most important work, nothing short of magnificent in its psychological achievement."[4] Toward the end of his career Freud himself still considered that "It contains, even according to my present-day judgment, the most valuable of all the discoveries it has been my good fortune to make. Insight such as this falls to one's lot but once in a lifetime."[5] Freud's lengthy self-analysis over the previous decade led him to explore dreams because they provided "the royal road to a knowledge of the unconscious activities of the mind."[6] All dreams, regardless of their manifest content, represented symbolic fulfillments of wishes that dated back to early childhood or even infancy. They seemed uniquely suited to uncover the bedrock sexual instincts behind both normal and pathological processes.

More than any other previous thinker about mental illness, Freud highlighted the extensive commonalities between normal and abnormal mental states. In his autobiography (1925) Freud reflected on *The Interpretation of Dreams*: "But when it came to dreams, it was no longer dealing with a pathological symptom, but with a phenomenon of normal mental life which might occur in any healthy person." He goes on to say that because this is true "then psycho-analysis was no longer an auxiliary science in the field of psychopathology it was rather the starting-point of a new and deeper science of the mind which would be equally indispensable for the understanding of the normal."[7] Likewise, the final sentence of his *Three Essays on Sexuality* concluded that his explicit goal was to construct "a theory adequate to the understanding alike of normal and pathological conditions."[8] Or the *parapraxes* (Freudian slips) that were the focus of *Psychopathology of Everyday Life* typified normal as much as neurotic individuals. "The borderline between the normal and the abnormal in nervous matters is a fluid one," Freud summarized.[9]

When Freud focused on pathology it was in order to better understand normality.[10] He provided a general psychology that used a single framework to understand all human behavior, not just neurotic behavior. His goal was to show how both ordinary and neurotic behavior stemmed from the same roots in universal childhood experiences such as unresolved oedipal complexes, toilet training practices, infants sucking at their mother's breasts, and parental giving and withholding of love and hostility. The genesis of the highest virtues and great scientific and artistic achievements, as well as the worst perversions, stemmed from the same psychic bases. Repression of normal sexual development might in one case lead to sexual perversion but in another case to great artistic achievement.

The central works of Freud's second stage, *The Interpretation of Dreams* and *Psychopathology of Everyday Life*, focused on examples based on nonclinical cases. The mechanisms that underlay dreams or slips of the tongue among presumably sane people were analogous to the mechanisms that brought about the symptoms of neurotics. Dynamic psychiatry used the same techniques to uncover the unconscious meanings that fairy tales, jokes, dreams, or neurotic symptoms expressed. The major writings of the last part of Freud's career, *The Future of an Illusion* (1928) and *Civilization and Its Discontents* (1930), applied dynamic theories to major social institutions. The use of common explanations for sane as well as for pathological behaviors thoroughly blurred the boundary between normality and abnormality and linked both ordinary and neurotic behaviors to the same principles of human development. This served at the same time to make pathological, nonpsychotic behavior commonplace and to pathologize ordinary, everyday behavior.[11]

Sexual desires and experiences provide a good illustration of the highly overlapping nature of normality and abnormality in Freud's 20th-century texts. He believed that all humans were naturally polymorphous at birth; fixed objects of desire did not arise until later stages of the life course. Social norms and life experiences, not innate desires, typically drove sexual instincts into particular channels. Although most people confined their sexuality to heterosexual intercourse as they matured, those who do not were stimulated by forces that were present among everyone at earlier points of life. Sexual "perverts" displayed behaviors that manifested the same drives that most people had repressed: "The conclusion now presents itself to us that there is indeed something innate lying behind the perversions but that it is something innate in everyone though as a disposition it may vary in its intensity and may be increased by the influences of actual life."[12]

Most remarkably, at the core of infantile desires was what Freud later came to call the Oedipus complex. As early as 1897, Freud had stated in a letter to his friend Wilhelm Fliess: "A single idea of general value dawned on me. I have found in my own case too, [the phenomenon of] being in love with my mother and jealous of my father, and I now consider it a universal event in early childhood."[13] The psychic role of incestuous urges and their social prohibition that he came to develop in *The Interpretation of Dreams* remained at the core of Freud's psychoanalytic theory for the rest of his career. Crucially,

Freud viewed oedipal desires as a window to view normal, as much as pathological, sexuality. They were not unique to neurotics but "No human individual is spared such experiences."[14]

Freud believed that sexual desires arose very early in life. Infants typically attached their erotic energy to their parents, although siblings and other caregivers were occasionally the objects of their longings. "Psychoanalytic investigations have shown beyond the possibility of doubt that an *incestuous love choice* is in fact the first and regular one," he insisted. His analysis of dreams indicated that boys wished to kill their fathers and sleep with their mothers while girls sexually desired their fathers and so wanted to eliminate their mothers. "Children's sexual wishes," Freud wrote, "awaken very early and . . . a girl's first affection is for her father and a boy's first childish desires are for his mother."[15] Children viewed their same-sex parents as competitors for the parent for whom they lusted after. Yet, Freud maintained, society is only possible if social norms force humans to repress these innate inclinations, especially their sexual inclinations. For him, the incest taboo that prohibited sexual behaviors between children and their parents perfectly illustrated this dynamic because it led people to suppress their most basic impulses.

In *The Interpretation of Dreams* Freud famously highlighted the ancient Greek playwright Sophocles' myth of King Oedipus to justify his conception of incestuous desire. Indeed, Freud believed that his discovery of the Oedipus complex where children desired to kill their same-sex parent and marry their opposite-sex parent was the towering accomplishment of his career: "If psychoanalysis could boast of no other achievement than the discovery of the repressed Oedipus Complex," Freud wrote toward the end of his life, "that alone would give it a claim to be included among the precious new acquisitions of mankind."[16]

The Oedipus complex illustrated a fundamental and universal human dilemma. Oedipus was the son of Laius, the King of Thebes, and his wife Jocasta. He was abandoned as an infant because an oracle had warned Laius that his son would grow up to murder his father. Oedipus was rescued and brought up in a different court. When he was an adult he met Laius—who was unknown to him—on the road, quarreled with him, and then killed him. Oedipus eventually became the King of Thebes and married Laius' widow, Jocasta. They conceived four children together without realizing they were mother and son. Eventually, Oedipus discovered that he had murdered his father and slept with his mother and pulled out his eyes to punish himself for these abominations.

For Freud, the tragedy of Oedipus was so moving because it revealed a universal human conflict: "King Oedipus, who slew his father Laius and married his mother Jocasta, merely shows us the fulfillment of our own childhood wishes . . . Like Oedipus, we live in ignorance of these wishes, repugnant to morality, which have been forced upon us by Nature."[17] Freud acknowledged the general disgust incest aroused but insisted that it stemmed from cultural prohibitions that harshly repressed the expression of natural passions. To him, the oedipal legend showed that because innate incestuous desires are so opposed to moral norms, they must remain unconscious and evoke vigorous denials of their existence.

His interpretation of the Oedipus complex also led Freud to posit a fundamental conflict that all humans, the normal as well as the neurotic, faced: they must adhere to social strictures that unavoidably repress their most forceful desires. Severe cultural prohibitions were necessary to counteract children's universal sexual longings and to insure they were never acted upon. Because these urges were so powerful, their cultural repression created lasting psychic injuries. "The prohibition against incestuous object choice," Freud asserted in one of his last works, "is perhaps the most maiming wound ever inflicted throughout the ages on the erotic life of man."[18]

The Oedipus complex illustrates the extensive overlap between the normal and the pathological. Freud concluded that incestuous desires among children had nothing to do with mental disorder in particular but were natural and universal. He wrote of Oedipus:

> His destiny moves us only because it might have been ours—because the oracle laid the same curse upon us before our birth as upon him. It is the fate of all of us, perhaps, to direct our first sexual impulse towards our mother and our first hatred and our first murderous wish against our father. Our dreams convince us that that is so. King Oedipus, who slew his father Laius and married his mother Jocasta, merely shows us the fulfilment of our own childhood wishes.[19]

Neurotics, Freud claimed, "are only distinguished by exhibiting on magnified scale feelings of love and hatred to their parents which occur less obviously and less intensely in the minds of most children."[20]

## Distinguishing Normality from Pathology

Given their extensive overlap, how did Freud distinguish normal from neurotic phenomena? The first way was the point in the life cycle when they appeared; the second was the age-old criteria of the context in which they emerged. The oedipal complex illustrates the importance of developmental timing in separating natural and pathological states. The incestuous and murderous wishes that this process entails were only natural during a particular stage of the life course, generally from around the ages of four through six. At this point, most children successfully repressed their sexual feelings for parents and then transferred them to opposite-sex peers during adolescence. Only neurotics maintained intense, although repressed, erotic longings for parents into adulthood. The incestuous conflicts of neurotics, however, were only more obvious and severe than those everyone else experienced.[21]

Anxiety neuroses provide another example. Various anxiety conditions could be normal or pathological depending on the life stage when they developed. Each period in human development had prototypical, normal sources of anxiety. The primary source of anxiousness among infants was responding to the absence of their mothers. During the next stage, fears of being left alone, darkness, and strangers were the most common manifestations of anxiety. During the subsequent phallic phase, object loss was the chief source of anxiety among girls while boys especially feared the castrating power of their fathers. Older children were afflicted by anxiety that stemmed from fears of punishment by the moralistic superego. Typically, all of these characteristic fears dissipated as the child matured.[22]

Normal people, in Freud's view, outgrew their childhood fears while neurotics still behaved as if old danger situations still existed: "The phobias of very young children, fears of being alone or in the dark or with strangers—phobias which can almost be called normal—usually pass off later on; the child 'grows out of them.'"[23] Disordered states of anxiety were regressions to earlier stages of development:

> We consider it entirely normal that a little girl should weep bitterly at the age of four if her doll is broken, at the age of six if her teacher reprimands her, at the age of sixteen if her sweetheart neglects her, at the age of twenty-five, perhaps, if she buries her child. . . . We should be rather surprised, in fact, if this girl, after she had become a wife and mother, should weep over some knickknack getting broken. Yet, this is how neurotics behave.[24]

Neurotics reverted to states of anxious helplessness even though the sources of danger no longer existed. They "remain infantile in their attitude to danger and have not surmounted obsolete determinants of anxiety."[25]

The fear of castration illustrates this dynamic. Freud considered castration fears among boys during the oedipal period to be realistic because "what is decisive is that the danger is one that threatens from outside and that the child believes in it." Indeed, for males, the "fear of castration is one of the commonest and strongest motives for repression and thus for the formation of neuroses."[26] Fears of castration were normal when experienced during the appropriate stage of childhood but could become neurotic when they persisted well beyond this period and became unconscious sources of symptoms later in life.

One of Freud's most famous cases, Little Hans (1909) illustrates the interplay between normal and neurotic processes.[27] Hans was a five-year-old boy who developed a serious case of a horse phobia. Shortly after seeing a horse fall to the ground, the boy became intensely anxious, clung to his mother, and was terrified of going onto the street because of a fear that a horse would bite him. Freud interpreted Hans' symptoms as repressed castration anxiety that typified the oedipal period and viewed his phobic symptoms as mechanisms that served to relieve the anxiety and tension that his unconscious conflict with his father created. Repression forced Hans' unacceptable thoughts into his unconscious, blocked their overt discharge, and transformed them into the substitutive symptoms of a horse phobia.

According to Freud, Hans wanted to kill his father in order to sleep with his mother but instead transferred his murderous instincts toward his father onto horses. This displacement allowed Hans to remain unaware of his rage toward his father and so to maintain his love for him. Moreover, the horse phobia let Hans avoid what he feared by simply not going outdoors. The true object of his fear, his father, could not be so easily avoided. Repression had displaced undischarged sexual energy as the cause of Hans' phobia.[28]

The case of Little Hans illustrates Freud's distinction between neurotic versus realistic anxiety. On the surface, it might seem that a boy's anxiety that his father would castrate him would be classifiable as a neurotic expression of anxiety. However, Freud believed that all five-year-old boys normally love and want to possess their mothers and, as a result, fear their fathers' jealousy and retribution, which they interpret as a threat of castration. "If 'Little Hans,'" Freud wrote, "being in love with his mother, had shown fear of his father, we should have no right to say that he had a neurosis or a phobia. His emotional reaction would have been entirely comprehensible. What made it a neurosis was one thing alone: the replacement of his father by a horse." Normal Oedipal fears became neurotic only when they were displaced from the real fear of castration and "directed to a different object and expressed in a distorted form, so that the patient is afraid, not of being castrated by his father, but of being bitten by a horse or devoured by a wolf."[29] The unconscious displacement of the original offensive idea allowed Hans, who simultaneously hated and loved his father, to recognize only his loving feelings while he displaced his hatred of the father onto the bad horse. Fathers were natural sources of fears of castration; horses were not.

In addition to the stage of the life cycle when some condition emerges, the context in which symptoms arose separated normal from disordered states. Freud's writings on anxiety and depression illustrate this distinction. In his later works, Freud came to separate normal and neurotic forms of anxiety on the basis of the classic Hippocratic contextual criteria. The symptoms of anxiety reactions were not peculiar in themselves; what made them neurotic is that they appeared in contexts that seemed to be either inadequate or inappropriate causes of the response. He distinguished realistic fears ("realistic anxiety"), which were normal, naturally programmed responses to external dangers, from anxiety neuroses. Normal fear arose as a reaction to actual dangers, signaling the ego about some threatening situation and so had the "indispensable biological function to fulfill as a reaction to a state of danger."[30]

Freud developed the unique idea that—in contrast to realistic fears that were products of external situations—anxiety neuroses stemmed from some inner threat, such as feeling overwhelmed by some instinctual drive that one had to suppress. Neurotic anxiety involved a disproportion of internal emotions and external threats. In neurotic disorders, fear mechanisms that were constructed to alert people to the presence of actual dangerous situations were activated in the face of unconscious dangers from internal sources. Individuals then mistakenly treated unconscious mental threats as coming from some outside stimulus and therefore reacted as if nonthreatening situations or objects were potentially dangerous. Thus, the dangers seemed as absurd and irrational to the person experiencing them as they did to others.

Freud's distinction between natural and neurotic anxiety did not lie in the nature of the anxiety itself, because normal and pathological anxiety shared common manifestations. Nor did it lie in how much distress or social impairment the condition caused, because responses to external threats could be just as distressing and impairing as internal ones. Even extreme fears were normal when they arose and persisted in contextually appropriate situations. "A person suffering from anxiety," Freud emphasized, "is not for that reason necessarily suffering from anxiety neurosis."[31]

Instead, the central difference was that, unlike fear that arises in response to realistic dangers, neurotic anxiety is a contextually inappropriate response to the degree of danger in the actual circumstances and so seems enigmatic and pointless. In neurotic disorders fear mechanisms that were constructed to alert people to the presence of actual dangerous situations were transformed to activate in the face of unconscious and, therefore, unknown internal dangers.[32] Fear involved a proportionate response to an external stimulus; morbid anxiety represented a disproportionate internal response. Fears of extreme severity could be normal when they were appropriate responses to traumatic external threats. Conversely, mild symptoms could be neurotic when they were not related to realistic contexts. Freud could not precisely specify what were "realistic" fears or "unrealistic" anxiety disorders; he stressed the loose boundaries and common mechanisms between normal and abnormal concerns. Both normal fears and anxiety disorders had gradients of severity, but they did not fall on different ends of the same continuum.

Depression provides a second example of Freud's use of contextual criteria to separate normal from pathological conditions. Analytic attempts to explain depression were based on traditional assumptions about the differences between depressive conditions that arose with and without expectable environmental causes. In his central work on depression, "Mourning and Melancholia," (1917) Freud explicitly compared the normal phenomenon of mourning to the pathological phenomenon of melancholy. Both states featured profound dejection, loss of interest in the outside world, an inability to feel pleasure, and an inhibition of activity. The distinction between mourning and melancholia did not stem from their symptoms so much as from the fact that the former was a normal reaction to loss whereas the latter was pathological.

Freud began his essay by noting the differences between the normality of grief and the disorder of melancholia, explaining that

> Although grief involves grave departures from the normal attitude to life, it never occurs to us to regard it as a morbid condition and hand the mourner

> over to medical treatment. We rest assured that after a lapse of time it will be overcome, and we look upon any interference with it as inadvisable or even harmful.[33]

Symptoms associated with mourning are intense and are "grave departures from the normal," in the sense that grief is greatly different from usual functioning. Nevertheless, grief is not a "morbid" condition; that is, it is not a medical disorder that represents the breakdown of a biologically normal response. Thus it does not require medical treatment; indeed, Freud emphasized that it would "never occur to us" to provide medical treatment to the bereaved. In addition, he stressed that grief is naturally self-healing, so that with time the mourner would return to a normal psychological state. Medical intervention, he suggested, could actually harm the grieving person through interfering with this natural process.

Freud's version of the Hippocratic distinction between depressions with cause (mourning) and without cause (melancholia) allowed him to elucidate the different psychodynamics that underlay the two conditions. For mourners, the world came to feel empty and without meaning due to conscious losses, whereas melancholics experienced the ego as impoverished due to unconscious losses. The self-reproaches of melancholics pathologically redirected their internalized hostility from earlier love objects onto the self. Therapy, therefore, should teach them to express their inward anger toward the objects that are its actual targets. In contrast, people who experience normal sadness are going through a natural and necessary process that was "inadvisable or even harmful" to disrupt with medical treatment. Freud and other psychoanalysts largely accepted as self-evident the traditional distinction between normal intense sadness resulting from loss and symptomatically quite similar pathological depression disproportionate to loss.

Freud thus used context, as well as life cycle stage, as criteria for separating normal from disordered conditions. He did not, however, adhere to the dimensional view of this distinction that is often attributed to him. Many analysts who followed Freud—especially in the post–World War II period—did view the neuroses as continuous with normality.[34] This conception, however, misconstrues Freud's own portrayal of the difference between the normal and the pathological. It is the case that neurotic and normal individuals shared many traits, so there was no sharp break between them. Yet, this relationship does not resemble a continuous distribution where normal phenomena are at one end, severe neuroses are at the other end, and mild and moderate conditions fall between these two extremes. Instead, Freud felt that all people had the capacity to become neurotic; those who actually did usually faced various life contingencies that brought out a potential that exists in everyone. His view of the relationship between normality and abnormality better fits a highly overlapping Venn diagram than a continuous distribution.

Freud was able to focus on the intersection between the normal and the neurotic because he split both from madness. For thousands of years, the essential distinction between normality and mental disorder lay between sanity and the clearly incomprehensible symptoms of madness. While many cases were difficult to place on one side or the other of this boundary, psychotic behaviors were usually readily distinguished from normality; they differ qualitatively, not just quantitatively, from normal behavior. Freud, however, concentrated on neurotic and normal actions and rarely wrote about or analyzed people with psychoses who he believed were "unsuitable" prospects for psychoanalysis.[35] He discouraged his followers from treating psychotics because of their inability to develop transference relationships, which were at the heart of psychoanalytic treatment: "I am skeptical," he wrote, "about the effectiveness of analysis for the therapy of psychoses."[36]

The essence of psychoanalytic classification was to abolish the boundary between neurosis and normality, not the boundary between psychosis and other behavior. Through bracketing the psychoses, Freud enabled a view that thoroughly intertwined neuroses with normality. In contrast to the fairly obvious distinctions between the mad and others, the boundaries between neurotics and normals were vague, blurry, and hard to define. This indistinctness allowed psychiatrists to expand their practices to encompass a large range of people who were neither psychotic nor totally normal. The resulting isolation of the psychoses facilitated the emergence of a new clientele for psychiatry who were decoupled from the residents of mental asylums and connected to widespread psychosocial afflictions.

Freud's views about the relationship between mental illness and normality sharply broke from previous thinking. He did not view neuroses such as sexual perversions, hysteria, obsessions, compulsions, phobias, and anxiety as much forms of illness as related to normal behavioral functions. His assumption of a near universality of psychopathology both made the abnormal less strange and heightened the strangeness of the normal. Perhaps more than any other figure, Freud revealed how the bizarre disguises the ordinary while the familiar camouflages the strange.

## Mental and Physical Illnesses Are Distinct

Freud soon repudiated the association he made during the first stage of his career between psychic problems and the kinds of organic states that were of concern to physicians. Indeed, he retrospectively stated in his *Autobiographical Study* that even during his early years, medicine never had an interest for him: "Neither at that time nor in later life, did I feel any particular predilection for the career of a doctor. I was moved, rather, by a sort of curiosity, which was, however, directed more towards human concerns than towards natural objects."[37] In contrast to his initial desire to show how "human concerns" reflected "natural" forces, Freud came to disconnect the two and emphasize the unique aspects of psychological conditions.

Although Freud never stopped insisting that psychoanalysis was a science, in fact, its interpretative methods made it far more akin to fields such as literature, history, and philosophy than to medicine, biology, and chemistry.[38] Core analytic concepts such as repression and the unconscious could not be measured, quantified, or subject to experimental study and so intrinsically stood apart from the essential features of scientific medicine. Moreover, the reliance of analysts on ingenious interpretations of case studies meant that the field could not incorporate a basic aspect of the scientific method: the replicability of results by independent observers.

For Freud mental illnesses were distinct from physical ones in every significant way. "We have found it necessary," he wrote in 1913, "to hold aloof from biological considerations during our psychoanalytic work and to refrain from using them for heuristic purposes so that we may not be misled in our impartial judgment of the psychoanalytic facts before us."[39] These "psychoanalytic facts" consisted of symbols in need of interpretation, not organic diseases in need of accurate diagnoses. Freud's goal was to reveal the deeper levels of meanings of psychic reality that were reducible neither to brain functioning nor to conscious thought.[40] They emerged from the unconscious, which was a purely psychological realm with no organic substrate.

The study of dreams reinforced Freud's developing ideas that many mental systems were rooted in meaningful symbols, not in physical processes. His earlier work had already shown how the physical-seeming symptoms of hysteria had no association with either anatomical or neurological defects. Hysteria was a purely psychological phenomenon that arose from biographical, not brain-based, factors. Despite the fact that they often presented themselves as distortions of physical functioning, hysterical symptoms were divorced from biological grounding: "The lesion in hysterical paralysis must be completely independent of the nervous system, since in its paralyses and other manifestations hysteria behaves as though anatomy did not exist or as though it had no knowledge of it."[41] For example, a young woman whose lover has just rejected her might unconsciously recall having waved goodbye to him and subsequently develop a paralyzed arm.[42] Analysts would need to understand the symbolic significance of the symptom before they could cure it, but they would not require anatomical knowledge of how arms work.

Freud came to hold the view that not just hysteria but all mental illnesses must be understood on their own terms, which were distinct from the principles that governed organic processes. Clinicians had to understand the hidden meanings behind symptoms, rather than uncover some presumed biological cause. The case of Little Hans illustrates how Freud had almost completely lost his original interest in identifying distinct disorders on the basis of the symptoms they displayed. Instead, overt manifestations such as a horse phobia were not diagnostic indicators that distinguished various conditions from each other but were symbolic representations that required interpretation. Symptoms were the starting point in a complicated search to find the core dynamics that lay beneath them.

The centrality of memory is another factor that helped divorce psychoanalysis from medicine. Memories of the distant past, especially of infantile and early childhood experiences, were especially important components of analysis. In contrast, such memories are generally irrelevant to the mechanisms that produce physical disease. The dividing of cancerous cells, the inflammation of airways in bronchial asthma, or the elevated levels of blood sugar that characterize diabetes do not depend on the recollections of their carriers. In contrast, psychologically grounded illnesses are intrinsically connected to how people think about their origins. Freud came to believe that there was no "there" for accurate memories to uncover. "There are no indications," he wrote, "of reality in the unconscious, so that one cannot distinguish between truth and fiction that has been cathected with affect."[43]

Freud came to assert that mental disorders were thoroughly intertwined with the ways they were remembered rather than with objective indicators. Memories were not facts but were continually reconstructed in light of later experiences. Their meanings stemmed from how they related to other memories rather than to either real events or brain processes. They constantly shifted their nature in order to preserve psychological equilibrium in the present. This meant that the truths that psychoanalysis uncovered were not objective facts but aspects of narratives that involved the stories that patients tell to themselves.[44] This process could hardly be more distinct from the methods of physical medicine.

Freud discarded his goal not just of deriving an organic substructure for mental disturbances but also of discovering appropriate diagnoses for them. Psychoanalysts faced the problem that, unlike medicine, overt symptoms did not indicate some underlying disease. "The hysterical symptom," Freud wrote, "does not carry [any particular] meaning with it, but the meaning is lent to it, soldered to it, as it were; and in every instance the meaning can be a different one, according to the nature of the suppressed thoughts which are struggling for expression."[45] Instead, life histories, which varied from individual to individual, determined what specific appearances the general symptoms of neuroses would take. Freud's later theories were thus far less diagnostically specific than his initial discussions of the neuroses.

The particular manifestations of symptoms were not as important as the general mechanisms, anxiety, in particular, that underlay them. A capacious concept of anxiety emerged in the second stage of Freud's work that in certain respects turned psychiatric nosology back to the more general views that dominated its classifications before the mid-19th century. Anxiety itself, rather than its various expressions, became "the fundamental phenomenon and main problem of neurosis."[46] The chief difference with prior conceptions was that anxiety, rather than melancholia, nerves, or neurasthenia, became the unifying process that explained the variety of diverse expressions of the neuroses.

Freud became far more concerned with uncovering deep truths of human nature than with differentiating what particular condition bedeviled patients. This stance split the subject of psychoanalysis from the prevailing concerns of physical medicine toward a far greater emphasis on subjectivity. Ultimately, the neglect of diagnostic specificity in Freud's later works, which contradicted the dominant trends of 20th-century medicine, helped facilitate the subsequent downfall of psychoanalysis itself.

Psychoanalytic treatments were another factor that sharply divided analysis from physical medicine. The analyst's task was to interpret the meanings of symptoms, not to use them to define a particular ailment and to specify its management. Dreams became especially important vehicles in this endeavor. Clinicians had to go beyond their manifest content to uncover the disguised meanings that dreams contained. Interpretation involved specifically mental and interpersonal processes. Unlike the handling of a physical condition, psychoanalysis was a "talking cure"; therapy consisted of linguistic exchanges between patients and their clinicians. Words, not brains, nerves, or any other physical process, were the medium Freud used to explore the puzzles of human nature.[47] Free association, where patients say whatever comes into their heads, became the focus of the analytic session. Nothing could be further from the drugs, surgeries, and other physical procedures that doctors use to treat bodily conditions.

In addition to free association, Freud came to stress the importance of transference where patients would relive past relationships, traumas, and experiences in their encounters with therapists. This emphasis on the curative power of healers themselves foregrounded the bonds that developed between patients and therapists as aspects of the therapeutic procedure.[48] Because patients transferred their old feelings, especially toward parents, within their present sessions with clinicians, therapists could confront, interpret, and make fully conscious these typically unconscious past ordeals. This placed the specific relationship that unfolded between patients and their analysts at the core of treatment. Freud explained: "We overcome the transference by pointing out to the patient that his feelings do not arise from the present situation and do not apply to the person of the doctor, but that they are repeating something that happened to him earlier. In this way we oblige him to transform his repetition into a memory."[49] The active utilization of the rapport between the analyst and the patient distinguished psychoanalysis from treatments in other branches of medicine.

## Analysis and Medicine

Freud himself recognized the essentially nonmedical nature of psychoanalysis and did not believe that analysts must be physicians. However, the purely psychic thrust that dominated the field since the early part of the 20th century posed a particular dilemma for American analysts, who vociferously insisted that only medical school graduates could enter analytic training institutes. The vastly influential Flexner Report of 1910 had critiqued the deplorable state of medical education at the time and insisted on the necessity of implementing higher standards grounded in rigorous scientific training, especially in anatomy and physiology.[50] Yet, the biochemical curriculum that newly infused medical schools was based on the very paradigm that Freud had rejected. At the time few medical schools offered courses in psychiatry because they perceived it as a basically unscientific field.

The splitting of mental from organic conditions that made psychiatry a purely psychological endeavor placed psychiatrists in a vulnerable professional position. There was no reason that medically trained practitioners would have any special expertise in treating neurotic conditions.[51] The emerging outpatient sector of psychiatry faced a variety of professional competitors. Many people with psychic afflictions still consulted general physicians. Many others sought consolation from religious healers.[52] A potpourri of spa doctors, fortune tellers, and counselors of various stripes also offered remedies. Later, social workers and clinical psychologists would enter the therapeutic marketplace. Psychoanalysts would have to establish their legitimacy without possessing any medically based understandings or treatments.

## Suggestion

Freud's emphasis on the psychic nature of phenomena such as dreams, unconscious fantasies, and repressed desires raised a fundamental issue about the nature of psychoanalysis that remains controversial at present. The mental and interpersonal aspects of psychoanalytic explanations and therapies made them vulnerable to serious problems that are rarely present in the management of physical illnesses. These processes depend on unconscious memories that patients are initially unable to recall. They are only brought into consciousness after strenuous efforts their therapists make to uncover and interpret them. Freud provided the example of deciding that one of his patients had neurasthenia. After making this diagnosis, the clinician

> may fearlessly require the patient's confirmation of one's surmises. Denial at the beginning should not mislead the physician; every resistance is finally overcome by firmly insisting on what has been inferred, and by emphasizing the unshakable nature of one's convictions. In this manner one learns all kinds of things about the sexual life of men and women.[53]

Especially given the largely unknowable aspects of long-past memories that were more likely to represent desires than actual events, analysts' interpretations can reflect their own theoretical predilections more than the "sexual life of men and women." Freud's famous patient Dora illustrates this process: when she emphatically declared she had no interest in the sexual advances of a family friend, Freud claimed that "in such a case 'No' signifies the desired 'Yes.'"[54]

The centrality of patients' emotional ties to their analysts made them especially susceptible to suggestive influences. The Oedipus complex provides an example. In fact, patients rarely spontaneously recounted scenes of parental seduction, whether real or imagined, to Freud. Instead, he admitted: "One only succeeds in awakening the psychical trace of a precocious sexual event under the most energetic pressure of the analytic procedure, and against enormous resistance."[55] Freud went on to describe how patients had no memories whatever of these seduction scenes but "only the strongest compulsion of the treatment" evoked them. Most of his patients knew nothing about their supposed sexual desires for relatives before they entered treatment, and, indeed, they often vigorously resisted Freud's interpretations. But, for Freud, denials of incestuous longings provided evidence of their presence. Indeed, he thought that the more strongly people protested, the more intense their incestuous wishes were likely to be.

Because psychoanalytic treatments rely on the interpersonal relationship that develops between patients and therapists, they are highly vulnerable to iatrogenic processes resulting from suggestion. Freud seems to have first developed his theory regarding the instinctual basis of children's sexual longings for their parents and then induced his patients to confirm his view. "It rather appears that [patients] retrospectively confirmed his theoretical hypotheses, only *after* he had suggested the latter by insistent questions, encouragements, admonishments and the reframing of reality," Freud scholars Mikkel Borch-Jacobsen and Sonu Shamdasani conclude.[56]

## Inner and Outer Sources of Pathology

As the last chapter indicated, from an early point in his career, Freud rejected hereditarian explanations and insisted that repressed mental conflicts explained the development of many psychoneuroses. For a short period of time between 1895 and 1897, Freud proposed that the sexual seduction of infants and young children by adults led to hysterical symptoms in later life. In the latter year, Freud realized this theory was "an error." The claimed seductions "had never taken place," but patients' later memories of molestations were fantasies that stemmed from their own unconscious desires and fantasies. These yearnings typically reflected universal, unconscious desires, such as the Oedipus complex, which were so strong that they overpowered the impact of actual events.[57] Children were no longer innocent victims of adult abusers but were themselves highly sexual creatures by nature. In his autobiography Freud stated: "If the reader feels inclined to shake his head at my credulity, I cannot altogether blame him."[58] Freud's self-repudiation of the seduction theory turned his views away from real events in the outer world toward intrapsychic processes. From about 1900 until the First World War, Freud became far more interested in patients' subjective desires than in their external experiences.

World War I led to another turning point in Freud's thoughts about the external and internal causes of psychic disturbances. This war not only involved unprecedented carnage but also created a gigantic rift between the optimistic outlook of the prewar decades and the darkness of the time that accompanied and followed it. Freud could not ignore the massive slaughter of the war and the deep social and cultural shifts that arose in the postwar period. A focus on inner experience was unsustainable in light of the huge changes that Freud's world was undergoing.

Although Freud never shed his emphasis on the unconscious repression of sexually charged desires from early childhood, the onset and aftermath of World War I led his theories to evolve in major new directions. The war posed a basic challenge to his conception of neuroses: childhood sexual traumas and desires, whether real or imagined, seemingly could not account

**Figure 4.1** Freud's trip to Clark University in 1909 was the only time he visited the United States (from left (front): Sigmund Freud, G. Stanley Hall, and Carl Jung; (back): Abraham A. Brill, Ernest Jones, and Sandor Ferenczi). Photograph first published in September 1909. Library of Congress Digital ID: cph 3b41123.

for why so many soldiers mentally succumbed to the terrors of the battlefield. Thousands of combatants developed "shell-shock," which typically involved the same sorts of paralyses of limbs; hysterical loss of functions of sight, hearing, or speech; and various tics and fits that characterized hysterical female patients who were not in current peril. Soldiers had to worry about real, present dangers, not repressed memories from the earliest periods of their lives.

World War I forced Freud to turn his attention from the psychoneuroses that were products of infancy and childhood to neuroses that resulted from contemporaneous traumas.[59] Freud had to take into account the powerful impact of external, nonsexual, and current traumas on the development of mental disturbances. These widespread shocks challenged his body of work over the past two decades and paved the way for a fundamental revision of his theory. His later writings emphasized the differences between traumatic and hysterical neuroses. In contrast to hysterical conditions that originated in childhood experiences and desires, traumatic neuroses arose from unexpected external shocks.[60] Wartime traumas stemmed from actual events, not from intrapsychic, sexual conflicts. They also afflicted males and not the distressed females who dominated analytic outpatient practices.

In *Beyond the Pleasure Principle* (1920) Freud elaborated on the distinction between neuroses of peace where people defended themselves from internal threats and war neuroses that stemmed from external dangers. He developed the idea that traumas were marked by powerful, unexpected mental excitation that broke through the protective shields that individuals used to maintain their psychic equilibrium.[61] People who anticipated a traumatic experience were unlikely to develop neuroses. Those who were unprepared for the violent event and so were surprised by it, however, were flooded with more frightening stimuli than they were able to master and accommodate. Traumatic neuroses thus resulted from unanticipated, unpleasant, and overwhelming experiences in the present, not in the past.

Freud also wrote about the distinct form that dreams of traumatized soldiers took:

> The traumatic neuroses give a clear indication that a fixation to the traumatic accident lives at their root. These patients regularly repeat the traumatic situation in their dreams; where hysteriform attacks occur that admit of an analysis, we find that the attack corresponds to a complete transplanting of the patient into the traumatic situation. It is as though

> these patients had not yet finished with the traumatic situation, as though they were still faced by it as an immediate task which has not been dealt with; and we take this view quite seriously.[62]

The recurrent aspect of traumatic memories starkly opposed Freud's earlier interpretations of dreams as functioning to symbolically fulfill repressed wishes.

Freud's observations of wartime traumas led him to thoroughly revise his theory of dreams and the unconscious. In contrast to the principle that dreams represented the fulfillment of wishes, he stressed that battlefield traumas overwhelmed the capacities of people to cope with them; they remained present through dreams that involved the compulsion to repeat, and therefore to master, the traumatic experience. He observed that, in contrast to their waking lives when people tried to avoid thinking about traumas, the dreams of disturbed soldiers repeatedly took them back to the traumatic situation.[63] His prior theory that dreams were forms of wish fulfillment could not explain this phenomenon: "Now dreams occurring in traumatic neuroses have the characteristic of repeatedly bringing the patient back into the situation of his accident, a situation from which he wakes up in another fright. This astonishes people far too little."[64]

Freud concluded that soldiers continued to experience traumas such as exploding shells in their dreams as attempts to retrospectively master and stabilize their inner lives, not as ways to fulfill wishes or obtain pleasure.[65] Psychic repetitions of the event while dreaming represented attempts to anticipate, react to, and assimilate the trauma so that healing could occur.

In response to the war, Freud revised his conception of the neuroses. Far more than any previous thinker, he made anxiety the central aspect of neurotic disorders. Anxiety had been at the heart of Freud's notions during the initial formulation of his theories, when he posited that it resulted from a transformation of repressed sexual energy in the past, the present, or both. After the war, his thinking about anxiety changed in significant ways as Freud came to believe that anxiety was a fundamental cause of repression rather than vice versa: "It was anxiety which produced repression and not, as I formerly believed, repression which produced anxiety."[66]

Freud not only put anxiety at the heart of his theory but also reversed the traditional hierarchy of psychiatric classification, relegating depression (or melancholia) to a secondary status. Anxiety, in his view, lay behind most forms of neurotic behavior including not only such direct manifestations as phobias, obsessions, panic, and general anxiety but also in hysteria, sexual dysfunctions, psychosomatic problems, and depression, among others. "Anxiety," Freud wrote in *Civilization and Its Discontents* (1929), "is always present somewhere or other behind every symptom."[67]

During the 1920s Freud also turned away from his earlier focus on libidinous energy and the unconscious toward how individuals negotiated their relationship with the external world. Freudian theory in general changed its emphasis from the study of drives and instincts to the study of the ego and its defenses. In a major revision of his thinking, Freud deemphasized the distinction between unconscious and conscious mental phenomena. His new structural theory divided the mind into three parts, the ego, the id, and the superego, which roughly represented the processes of reason, passion, and conscience. Freud's central insight, containing Platonic echoes, was that these three elements were in perpetual conflict with each other. *The Ego and the Id* (1923) placed the ego at the center of psychological processes. Its major tasks were to mediate between the difficulties of the external world on the one hand and the demands of the instincts on the other. "As a consequence of these new theories," psychiatrist and historian Henri Ellenberger asserts, "the ego was now in the limelight of psychoanalysis, especially as the site of anxiety: reality anxiety, that is, fear caused by reality, drive anxiety from pressures from the id and guilt anxiety resulting from the pressures of the superego."[68]

*Inhibitions, Symptoms, and Anxiety* (1926) summarized Freud's later writings on anxiety. In it, he moved the key source of anxiety from somatic and intrapsychic factors to the ways in which the ego reacted to dangerous situations, especially separations from mothers.[69] Freud's conception of anxiety also broadened from its intimate connection with sexual energy toward a more general unpleasant response. He came to reject the hydraulic model of sexual energy that underlay his initial conceptions of neuroses, replacing it with the idea that anxiety served as a danger signal that arose as a response to trauma. Moreover, anxiety could stem from not just present dangers but also the anticipation of future ones. Freud also placed a greater emphasis on indefinite anxiety (*Angst*) that lacked any object. Anxiety was no longer an exclusive product of sexual repression but could emerge whenever a potent external stimulus overwhelmed people. Its quantity varied according to the strength of external demands people faced. Any danger situation that was more powerful than the resources people had to protect themselves could lead them to feel helpless and, consequently, anxious.[70]

In Freud's revised view anxiety served as a signal of some threat, warning of impending danger and activating defense mechanisms. He described how the ego judges that some situation that is not yet traumatic had the potential to become so:

> The conclusion we have come to, then, is this. Anxiety is a reaction to a situation of danger. It is obviated by the ego's doing something to avoid that situation or to withdraw from it. It might be said that symptoms are created so as to avoid the generating of anxiety. But this does not so deep enough. It would be truer to say that symptoms are created so as to avoid a *danger-situation* whose presence has been signalled by the generation of anxiety.[71]

While real anxiety stemmed from some known external danger, neurotic anxiety was a reaction to some unconscious threat. Such internal threats included sexual ones but could also stem from helplessness related to separations or threats to self-preservation. The idea of anxiety as arising from some inner danger was unique and had lasting influence among dynamic psychiatrists.[72]

In the final stages of his career, Freud's writings emphasized the interactive nature between individual inclinations and external pressures from parents, culture, and society that led to the repression of natural desires. All cultures depended on finding ways to insure that people renounced their most fundamental predispositions. The struggle between intraindividual and social forces was inevitable; analysts had to take into account both internal and external factors and, especially, their interrelationships. Psychoanalytic treatments turned away from recovering repressed unconscious memories toward strengthening the ego as the embattled mediator of instinctual energies and the moralistic superego. Their goal was to allow the ego to better maneuver between external reality, the superego, and the inner drives of the id in order to maximize the inherently limited realm of individual choice. Freud did not tell people what goals they should pursue but strove to increase their capacity to make informed decisions.

By the end of his life Freud had turned away from his initial intrapsychic orientation toward a view that increasingly employed the conflict between internal, inherent, and universal desires with external reality as the central psychic dynamic. As his thinking matured, Freud became more concerned with the role of social institutions—religion, law, morality—and not just intrafamilial dynamics as powerful external sources of repression. He viewed the nefarious impact of sexual repression on resultant psychopathology not only as an inner process but also as a social necessity. Suppression of instinctual desires was the price that individuals paid so that civilization itself could function.[73] Because personal gratifications and social demands stood in fundamental opposition to each other, selves were inevitably sites of constant tension, strife, and ambivalence.

## Conclusion

Freud's career underwent three major stages. The first phase, discussed in Chapter 3, remained within the boundaries of traditional 19th-century biologically and diagnostically based medicine, although it began to show some sharp divergences from this model. The next period that arose at the beginning of the 20th century concentrated on elucidating the common processes—repression, the unconscious, childhood sexuality, and the libido—that gave rise to both normal and neurotic phenomena. World War I initiated the final stage of his thinking, which focused on external traumas and the role of the ego in mediating between the conflicting demands of the id on one side and the superego and civilization on the other. Until the end of his life, Freud focused on general rather than specific psychic disturbances, on interpersonal and external sources of danger more than somatic processes, and on the intrinsic conflict between environmental repression and universal drives.

Freud's lasting contribution to thinking about mental illness was to successfully expand the range of disorders well beyond conditions thought to have an organic basis or the psychoses. Neuroses resulted from interactions between individuals and their human environments: no physical defect needed to be present. Psychiatrists came to have legitimate claims to treat distressing states that had previously been viewed as purely psychogenic in nature and therefore outside of the medical or neurological realm. Moreover, his blurring of the boundaries between the neuroses and normality created a zone of ambiguity that mental health professionals came to exploit. They became the legitimate responders to a vast expanse of common occurrences including interpersonal conflicts, sexual difficulties, and the psychic results of traumas. At the same time, the seriously mentally ill could be isolated in mental institutions and so not taint the new outpatient clientele.

Psychoanalytic thought was especially influential in the United States. After Freud's death, a wholesale psychosocial approach emerged in the country that explained virtually all sorts of human behavior. Beginning in the 1920s, dynamic views became influential in a diverse array of child guidance clinics, agencies dealing with juvenile delinquents, marriage counseling, and media reporting about psychic disturbance.[74] During the decades that followed World War II, analytic thinking powerfully influenced not only conceptions of mental illness and the practice of psychotherapy but also literature, film, advertising, and explanations of human action in general. In 1940, the poet W. H. Auden provided perhaps the best summary of Freud's influence:

> To us he is no more a person
> Now but a whole climate of opinion.[75]

## Acknowledgments

Portions of this chapter are adapted from Horwitz, 2013; Horwitz, 2018; Horwitz & Wakefield, 2007; Horwitz & Wakefield, 2012.

## Notes

1. Quoted in Whitebook, 2017, 101.
2. Freud, 1937, 235.
3. Perhaps symbolically, the book was finished in 1899 but received a publication date of 1900.
4. Sulloway, 1991, 320.
5. Freud, 1900/1965, xxxii.
6. Freud, 1900/1965, 647.
7. Freud, 1925/1959, 47.
8. Freud, 1905/1962, 109.
9. Freud, 1901/1990, 278.
10. Philip Rieff (1959, 47) noted that for Freud: "The understanding of the normal character through the neurotic character, of health through sickness, is indeed his master trope."
11. Roazen, 1992.
12. Freud. 1905/1962, 171.

13. http://www.freudfile.org/psychoanalysis/arcvhive_4.html. Freud and many of his followers date his discovery of the Oedipus complex to the late 1890s, but others, notably Frederick Crews (2017, 506–18), assert that he retroactively claimed to uncover the universality of oedipal processes in order to justify his later thinking.
14. Freud, 1940, 184.
15. Freud, 1900/1965, 291.
16. Freud, 1940/1989, 192–93.
17. Freud, 1900/1965, 296–97.
18. Freud, 1920/1953, 220–21; Freud, 1930/1962.
19. Freud, 1900/1965, 296–97.
20. Freud, 1900/1965, 294.
21. Freud, 1900/1965, 291.
22. Freud, 1936/1963, 83–84; 88–89.
23. Freud, 1926/1989, 79.
24. Freud, 1936/1963, 89.
25. Freud 1933/1989, 779.
26. Freud, 1936/1963, 86–87. The female counterpart to castration was feeling insufficient due to the lack of a penis.
27. Freud, 1909/1955. In fact, Freud only saw Hans himself just once. Hans' father, a disciple of Freud's, conducted the analysis.
28. Nemiah, 1985, 885.
29. Freud, 1926/1959, 25, 53.
30. Freud, 1926/1989, 63.
31. Freud, 1910/1989, 354.
32. Freud, 1926/1989, 103.
33. Freud, 1917/1957.
34. See especially Menninger, 1963.
35. Malcolm, 1981, 131.
36. Quoted in Roazen, 1992, 142. For example, Freud dissuaded Karl Menninger from using psychoanalysis among his hospitalized patients because he "never had success with psychoanalysis on severely mentally ill patients" (Friedman, 1990, 109).
37. Freud, 1925, 8.
38. Robinson, 1993, 185.
39. Freud, 1913/1955, 181–82.
40. Whitebook, 2017, 225.
41. Freud, 1893/1955, 169.
42. Simon, 1978, 24.
43. Quoted in Zaretsky, 2004, 34.
44. E.g. Sacks, 2017.
45. Freud, 1905/1963, 57.
46. Freud, 1926, 75.
47. Phillips, 2014.

48. Freud's emphasis on transference contains echoes of Franz Anton Mesmer's wildly popular earlier procedure of mesmerism, a technique similar to inducing hypnotic trances (Ellenberger, 1970).
49. Freud, 1916–1917/1989, 552.
50. See Starr, 1982.
51. Shorter, 1997, 146; Hale, 1995, 292.
52. Caplan, 2001.
53. Freud, 1898/1959, 228.
54. Freud, 1905/1989. 203. Earlier, French physician Hippolyte Bernheim charged that Charcot's use of hypnosis actually created the hysterical symptoms that he claimed to discover.
55. Freud, 1896b/1959, 153.
56. Borch-Jacobsen & Shamdasani, 2012, 156, italics in original. See also McNally, 2003, 111–12; 162–68; Ofshe & Watters, 1994, 293.
57. Sulloway, 1991, 247.
58. Freud, 1925/1959, 36.
59. Ellenberger, 1970, 837–38.
60. Freud, 1920/1989. "When war conditions ceased to operate the greater number of the neurotic disturbances brought about by the war simultaneously vanished," Freud wrote (1919/1955, 207).
61. Freud, 1920/1989, 33.
62. Freud, 1916/1989, 274–75.
63. Freud, 1933/1965, 28.
64. Freud, 1920/1989, 11.
65. Makari, 2008, 317.
66. Freud, 1926/1989, 32.
67. Freud, 1930/1962, 82.
68. Ellenberger, 1970, 516.
69. Freud, 1936/1963, 20; Freud, 1926/1989, 63.
70. Freud, 1936/1963, 113–14.
71. Freud, 1926/1989, 57.
72. Glas, 1996. Freud did not completely abandon his initial theory of anxiety. While he placed more emphasis on external traumas, he also stressed how actual traumatic events in themselves rarely led to neuroses, maintaining that "it would seem highly improbable that a neurosis could come into being merely because of the objective presence of danger, without any participation of the deeper levels of the mental apparatus" (Freud 1926/1989, 58). People who experienced traumas and consequently developed neuroses also had deeper, unconscious factors that led to their symptoms. Traumas could evoke conflicts that were already present so latent predispositions and external events jointly created disorders.
73. Sulloway, 1979, 374.
74. Burnham, 2014, 1, 26.
75. Auden, 1940/1994.

# 5

# Mental Illness Becomes Ubiquitous

> Most people have some degree of mental illness at some time, and many of them have a degree of mental illness most of the time.
>
> —Karl Menninger, *The Vital Balance* (1963)

A growing number of conditions seen as indicating mental illness and needing professional mental health care marked the quarter century that ran roughly from Freud's death in 1939 through the mid-1960s. A variety of factors contributed to this expansion of pathology in the United States. Some of these involved developments within psychiatry, whose mandate enlarged to the extent that, as a president of the New York Psychoanalytic Institute reflected: "Scarcely any human problem admits of solution other than psychoanalysis."[1] Another source of the growing range of disorders was the radical reshaping of concepts of normality and abnormality that emerged from the experiences of military psychiatrists during World War II. After the war, a newly activist federal government turned its attention to the mental health of entire populations, not just identified patients. At the same time, the transformation of the primary locus of psychiatric treatment from inpatient institutions to outpatient practices mandated a sweeping revision of psychiatry's system of classifying mental illnesses to encompass many more conditions. Psychoactive drug treatments, too, expanded to attract a huge proportion of Americans; the current templates for anxiolytic, antidepressant, and antipsychotic drugs all arose in the 1950s. The result of these changes was a wholesale rethinking of who was likely to succumb to mental illness. Another feature of this period—a dimensional view of mental illnesses as running from mild to moderate to severe—was repudiated after 1980 but has reemerged among contemporary researchers.

Other key aspects of this era did not last. A strong environmental emphasis characterized the postwar period as psychiatric attention turned to such influences as life events, changing familial roles, workplace difficulties, and economic burdens. At the same time, interest in the possible biological underpinnings of mental disorders diminished as did Kraepelinian approaches that sought the distinct indicators, causes, and prognoses associated with specific mental illnesses. By 1980 the biomedical portrayals that were preeminent before the turn of the 19th century had regained a dominance that continues to the present.

## Psychiatric Thought

Before World War II, the vast majority of psychiatrists held positions within inpatient institutions. Beginning in the 1920s they developed a number of somatic treatments, including organ removal, lobotomy, malaria fever treatment, hydrotherapy, insulin coma therapy, electroconvulsive therapy, and sedative drugs.[2] Like their European counterparts, most American psychiatrists held degenerative hereditarian theories of mental disorder, and many advocated for sterilizing institutionalized patients to prevent future generations of the mentally ill from being born. The vast majority had little use for, or interest in, Freudian ideas. Because they practiced in isolated institutions, aside from the occasional scandal over the deplorable conditions hospitalized residents faced, inpatient psychiatry had little presence or esteem in the general culture at the time. The emergence of the mental hygiene movement, the importation of psychoanalysis to the United States, and, especially, World War II would thoroughly transform and elevate the profession into a powerful social and institutional force.[3]

### Biopsychology and Mental Hygiene

Despite the fact that psychiatry was mostly inpatient based, by the 1920s the United States had already become a congenial national context for psychoanalysis.[4] A number of leading academics including Adolf Meyer at Johns Hopkins, James Jackson Putnam at Harvard, and G. Stanley Hall at Clark advocated Freudian theory and methods.[5] When the flood of European émigré psychiatrists arrived in the United States during the following decade, Meyer (1866–1950) was the most dominant American psychiatrist and "for more than thirty years, he was an almost overwhelmingly influential force in American psychiatry."[6] A German immigrant to the United States, he held the important institutional positions of, first, director of the Pathological Institute of the New York State Hospital and, then, psychiatrist in chief at Johns Hopkins Hospital from 1910 to 1941.[7] In these roles, he trained many prominent American psychiatrists.[8]

Like Freud's, Meyer's early work lay squarely in the biomedical tradition. However, also like Freud, Meyer came to downplay the organic aspects of mental illness, instead focusing on its psychosocial features. He particularly objected to Kraepelin's emphasis on defining specific psychiatric diseases. Instead, he developed an eclectic psychobiological model that emphasized how mental illnesses were holistic responses resulting from individuals' maladaptive reactions to their environments: "We do not think of disease entities but of processes, that is, miscarriages and deviations of functions."[9] Each person's condition stemmed from his or her unique combinations of biology, family history, personal experiences, life problems, and social positions. Because these factors were distinctive for any given individual, Meyer's approach was not conducive to producing generalizations. Nevertheless, his work was the primary influence on the first *Diagnostic and Statistical Manual of Mental Disorders* (DSM) that developed after the Second World War.

Meyer was not dogmatic, loosely connecting biological and, especially, psychological and social factors in an all-encompassing but amorphous synthesis of how people reacted to life events. One of his important legacies was as the cofounder (with former mental patient Clifford Beers) of the National Committee on Mental Hygiene in 1909. The mental hygiene movement marked a major expansion of efforts that strove to identify and treat mental illnesses at early stages before they became severe. It focused on preventing full-scale mental disorders from emerging through educating parents about how to recognize the initial signs of disturbance in their children and teachers on spotting potentially troubled students in their classrooms. It also developed clinics that targeted children who were thought to be at high risk for becoming mentally ill. The mental hygiene movement was an early aspect of psychiatry's turn away from mental institutions toward the vast terrain of distressed and maladjusted but not yet mentally disordered populations.

Meyer's acceptance of a wide range of causes exemplified the trend to promote social, external, and interpersonal influences as coequal with individual, innate, and intrapsychic ones. He also advocated for the use of a combination of psychological, social, and drug treatments. His open-minded views about the nature of mental illnesses and their treatment facilitated the entry of European émigrés into the American psychiatric scene.

## The Freudians Come to America

Freud's changing understandings of mental illness were a precursor to the vast expansion of perceived mental pathology and its psychosocial causes. His thrust in the 1920s had already shifted the focus of analysis from the study of drives and instincts toward the ego and its defenses. From this period through his death in 1939, Freud concentrated on the question of how people—neurotics and normals alike—dealt with the conflict between internal forces and the external demands of civilization. The boundaries of the various neuroses were unmistakably distinct from psychoses but were highly overlapping with normality. A consequence of his work was to encompass conditions that were clearly not psychotic into the legitimate realm of mental illness.

After Freud's death, analysis fractured into many distinct schools, none of which was preeminent. The rise of Nazism in Europe during the 1930s produced a mass emigration of European analysts to the United States, moving the centers of analytic thought from Vienna, Berlin, and Budapest to New York, Chicago, and Los Angeles. Once its major figures relocated to the United States, dynamic thought became more socially oriented. Although many analysts maintained the classic mode of multilayered interpretations of inner pathological symptoms, other new, more psychosocially oriented schools flourished. Their concerns gravitated toward the social and behavioral sciences, and various psychosocial approaches dominated the theory and practice of American psychiatry from the 1940s through the mid-1960s, when dynamic psychiatry began a steep decline from which it never recovered.

### Ego Psychology and Personality Disorders

After Freud's death, the theory and practice of dynamic psychiatry turned toward ego psychology and the study of personality disorders, which dominated mainstream American analysis for the following three decades. Ego psychology focused on problems that characterized the total personality of patients and their responses to the environment. These conditions were distinct from the classic neuroses because they involved stable and organized patterns of behavior that characterized how individuals acted, thought, and felt. The movement toward examining personality disorders freed psychiatrists "from the abnormal and bizarre and brought them up against what was normal."[10]

In 1945, one of Freud's disciples, Otto Fenichel (1897–1946), an Austrian émigré to the United States, summarized analytic thinking at the time in what became the standard textbook of dynamic psychiatry, *The Psychoanalytic Theory of Neurosis*. Historian Nathan Hale argues that this text "completed the transition from the early biologism of psychoanalytic character theory to a far more socially determined system, one in which the environment played a key role. This very environmentalist emphasis would make psychoanalysis more congenial to Americans."[11] Fenichel's work pushed analysis further in the more psychosocial direction that Freud adopted in the latter stages of his career. "There is no 'psychology of man,' in a general sense, in a vacuum, as it were," he explained, "but only a psychology of man in a certain concrete society and in a certain social place within this concrete society."[12] Fenichel observed how character types were highly responsive to social contexts: "The same persons may be extroverts under certain circumstances, introverts under others."[13]

Fenichel exemplified the trend to view modern character traits as communally determined, resulting from a society that featured less repression than Freud's cases had encountered several decades earlier. He noted how the problems patients presented were quite distinct from the classic neuroses that the field featured during its formative period. Midcentury clients were much more likely to display troublesome character structures than the neurotic fixations found in earlier decades. First, their problems were aspects of whole personalities rather than collections of symptoms. Second, their dilemmas stemmed less from intrapsychic states than from stable behavioral patterns that marked how they interacted with the external world. Third, in contrast to the traditional neuroses, many individuals with personality disturbances did not feel ill and were not bothered by their character structures. Instead, other people often urged them to enter clinical treatment because of the difficulties they encountered in dealing with such problematic personalities.

German émigré analyst Heinz Hartmann (1894–1970) was a major figure in the turn toward ego psychology. "Hartmann," according to psychiatrist and historian George Makari, "set the theoretical agenda for American ego psychology for the following three decades [after 1945]."[14] Echoing Freud, Hartmann replaced the power of instinctual drives with the ability of the ego to master the external world. Analysts were becoming more oriented to interpersonal processes that characterized social interaction in general, turning their attention away from the id toward individuals' relationships with other people, social norms, and cultural demands. Their clients increasingly sought help for diffuse relational difficulties, identity problems, and general life dissatisfactions.

The work of two other émigrés, the Austrians Heinz Kohut (1913–1981) and Otto Kernberg (b. 1928), helped turn the focus of American analysts "from Oedipus to Narcissus" and bring narcissistic personality disorders to popular attention.[15] Both Kohut and Kernberg saw narcissism as a personality type, not as feelings, traits, or behaviors that were isolated from a holistic character structure. They differed from each other, however, on the value they placed on narcissistic conditions.

Kohut challenged Freud's assumption that narcissism among adults was a pathological sign of immaturity representing a regression to an infantile state of being.[16] Instead, he insisted that narcissism was usually a healthy aspect of mature personalities. His views were attuned to the individualistic culture of the 1960s that celebrated the primacy of the self over the tyranny of social norms. Kohut also allied his work with the counterculture's turn toward celebrating inner experience, whether through drugs, music, or Eastern philosophy.[17] Kernberg, in contrast, asserted that most narcissists seethed with rage and aggression below their charming and seductive veneer. "Narcissism has always been," historian Elizabeth Lunbeck observes, "simultaneously pathological and normal."[18]

Social forces, especially the trend toward smaller families, more intense mother–child relationships, and easing sexual repression led to an upsurge, or at least the perception of one, in the number of narcissistic personalities in the post–World War II period. "Narcissism is as ubiquitous as anxiety in human motivation or psychic outcomes," prominent analyst Leo Rangell observed.[19] Awareness of conditions such as narcissism escaped the confines of the psychiatric theory and practice to become firmly embedded in mainstream culture. Social critics such as Christopher Lasch tied the ubiquity of self-absorption to the material prosperity and consequent selfishness of the era.[20] Regardless of whether narcissism had actually become more common, this notion became thoroughly entrenched in portrayals of typical personalities at the time.

The focus on personality structures broadened the range of experiences that were considered as "mental illnesses" to maladaptive patterns of behavior, character, and personal problems. The blurry boundaries between mental health and illness created a vast new territory for mental health professionals to colonize. The clients of dynamic clinicians came to be people who were dissatisfied with themselves, their relationships, their careers, and their lives in general. Psychiatry had become a discipline that was divorced from concerns about madness but intensely involved with phenomena that were hard to distinguish from normality.[21] "Postwar psychiatric thinking," historian Gerald Grob points out, "reflected an extraordinary broadening of psychiatric boundaries and a rejection of the traditional distinction between mental health and mental abnormality."[22] The flourishing of socially oriented explanations reached its peak in the popular works of the neo-Freudians.

## The Neo-Freudians

The rise of the Neo-Freudian School, which united psychoanalysis with the social sciences, was an influential source propelling the social origins of mental health and illness into the cultural spotlight. Among the most prominent neo-Freudians was the German-born psychiatrist Karen Horney. Horney (1885–1952) was one of a number of female analysts who entered the profession during the 1920s and 1930s and whose primary interests lay in exploring and transforming women's life experiences. Most important, Horney rejected the dynamic notion of the importance of inner drives and focused on the power of social and historical influences on human development.[23] She took a relativist conception that emphasized the historically and culturally specific nature of definitions of normality and abnormality:

> The conception of what is normal varies not only with the culture but also within the same culture, in the course of time. Today, for example, if a mature and independent woman were to consider herself "a fallen woman," "unworthy of the love of a decent man," because she had had sexual relationships, she would be suspected of a neurosis, at least in many circles of society. Some forty years ago this attitude of guilt would have been considered normal.[24]

Unlike Freud, Horney focused on the situation of women in modern societies. Mothers replaced Freud's oedipal fathers as the chief source of influence during early childhood. Horney also saw mother–daughter relationships as the primary generator of psychic conflict in later developmental stages. Women were in especially precarious positions because of social norms that placed numerous limitations on their opportunities for achievement. They required autonomy, power, and recognition that could only arise through social movements as opposed to psychological understandings. Her work was especially relevant to the situations of those postwar housewives who lacked meaningful social roles as they raised their children in isolated suburban communities while their husbands worked long hours away from home.

Horney's view of neurosis was suffused with external influences. In particular, except for "exceptional cases" she rejected the classical Freudian view that erotic forces underlay the neuroses.[25] In the short period since Freud developed his sexually grounded theory, cultural attitudes had changed to such an extent that sexual repression was no longer seen as consequential for producing psychic disturbances. Likewise, even such apparently universal phenomena as the Oedipus complex were not biologically innate but resulted from particular cultural conditions that generated hostility among family members.

Horney asserted that culture, not sexual or aggressive instincts, created neurotic personalities. Disturbed character structures were formed in childhood through parental incapacities to provide love and nurturance. Modern life and child-rearing practices produced people who had at the core of their characters a basic anxiety, which is "a feeling of being small, insignificant, helpless, deserted, endangered, in a world that is out to abuse, cheat, attack, humiliate, betray, envy."[26] Anxiousness drove characteristic individuals in the mid-20th century to be hostile, competitive, inferior, and emotionally isolated. Neurotics rarely confronted these feelings directly but instead projected them onto objects, situations, political events, or indistinct feelings of doom. Likewise, they employed a number of defenses such as the quest for power and prestige, material possessions, and affection from others. These conflicts were neither innate nor psychically based but stemmed from the cultural contradictions of American life. The struggles of neurotics were, as the title of Horney's best-known work indicated, "of our time."

Another popular neo-Freudian émigré from Germany, Erich Fromm (1900–1980), also conceived of psychic disturbances as fundamentally social. Like Horney, Fromm gave short shrift to libido and other sexual forces; instead he emphasized the historical shaping of psychological emotions and the need for social relatedness.[27] Fromm's major project was to merge Freud's insights with those of Karl Marx. His most popular work, *Escape from Freedom* (1941), viewed anxiety as the central problem of modern society. For Fromm, anxiety in the modern world was a product of unchecked capitalism, periodic structural unemployment, overpopulation, and a host of other social ills. Lacking the security of encompassing belief systems, individuals turned to totalitarian movements such as Nazism that protected them from the anxiety, isolation, and loneliness that freedom engendered.

Fromm rooted insecurities in the particular conditions, especially the economic systems, of each society rather than in biological or psychological universals. For example, he posited that the Oedipus complex was not universal but was a product of patriarchal societies, which placed sons in competition with fathers because they would inherit their property after their death. "Individual psychology," Fromm asserted, "is fundamentally social psychology."[28] The key problem people faced in this era was not their instincts or psychic repressions but understanding and overcoming an oppressive society.[29]

The neo-Freudians gained much attention and admiration in the general culture; Fromm and Horney's books were best-sellers. More mainstream analysts, however, did not appreciate their fame and expelled both from analytic institutes.[30] Their failure to establish their own institutional bases precluded the neo-Freudians from influencing psychiatric theory and practice. They did, however, garner tremendous popular acclaim from the late 1930s through the 1960s.

## World War II

Psychiatry's turn to biopsychology, mental hygiene, ego psychology, and personality disorders as well as the popularity of neo-Freudian views in the general culture was one factor serving to break down the boundaries between sanity and insanity and expand the range of pathology. A second influence that both broadened the realm of mental illness and turned it in a more environmental direction was the encounter of military psychiatrists with the mental health difficulties of draftees and soldiers during World War II.

The Second World War introduced a new type of alliance between psychiatry and the state. At the beginning of the war, the military joined with psychiatrists in a huge enterprise that screened millions of draftees for their mental fitness to serve.[31] It was psychiatry's first large-scale attempt to assess the mental health of the general population, in this case of draft-eligible men. The endeavor, which used broad criteria for what constituted mental illness, resulted in nearly two million of the fifteen million men screened being rejected as psychically unsuitable for military service. The major consequence of this massive screening effort was to legitimate the idea that psychiatrists could differentiate normality from pathology outside of any treatment setting and thus be the guardians of mental health among ordinary citizens.

The issue of which soldiers were normal and which were disordered also became the central concern of military psychiatrists during the war itself. They came to stress how psychically impaired soldiers responded normally to an abnormal environment. Psychiatrists Roy Grinker and John Spiegel's observation was emblematic: "It would seem to be a more rational question to ask why the soldier does *not* succumb to anxiety rather than why he does."[32] Grinker and others developed a social psychiatry based on their wartime experiences that focused on environmental and interpersonal situations. Their question shifted from explaining what kind of soldiers would break down to specifying what external conditions led to psychic disintegration.

Sociologist Gilbert Beebe and psychiatrist John Apple calculated that the average soldier would break down after about eighty-eight days of constant combat.[33] According to military psychiatrists Roy Swank and Walter Marchand: "One thing alone seems to be certain: Practically all infantry soldiers suffer from a neurotic reaction eventually if they are subjected to the stress of modern combat continuously and long enough." Indeed, Swank and Marchand considered as "aggressive psychopathic personalities" the less than 2% of soldiers who could withstand combat for an inordinate length of time.[34] The official American report on the war, *Combat Exhaustion*, indicated that almost all soldiers psychically broke down after three or four months of incessant combat. It concluded: "Each moment of combat imposes a strain so great that men will break down in direct relation to the intensity and duration of their exposure . . . psychiatric casualties are as inevitable as gunshot and shrapnel wounds in warfare."[35] Mental breakdowns under extreme stress were normal, not abnormal; they were likely to affect everyone, not just a predisposed minority.

The very high rates of psychic collapse during the war led to an abrupt shift in thinking about what factors led to mental disturbance. As the war progressed, most American psychiatrists "underwent a marked change of view, switching from their initial belief that 'a clear cut distinction [could] be made among men as between the "weak" and the "strong,"' to the view that 'every man has his breaking point.'"[36] Studies during the war showed that factors such as constitutional dispositions, heredity, or neurotic personality traits did not predict which soldiers became psychic casualties. Instead, by far the most important explanation of wartime breakdowns was the intensity and duration of combat experiences. Researchers came to discard psychological and biological accounts and to focus on factors such as varying levels of group cohesion, adequate training, and morale to explain the proportion of psychic casualties across different units.

By the end of the war, the environmental view that wartime traumas led normal men to develop psychiatric conditions thoroughly prevailed. Everyone, not just the biologically or psychologically vulnerable, and men as well as women, was susceptible to traumatic exposure. The experiences of mental health personnel during the Second World War dramatically expanded the social model of mental disorder in psychiatry.

The view of mental illnesses as resulting from situational stressors that afflicted huge numbers of previously normal people was not the only major change that World War II facilitated. The war also shaped a new way of viewing environmental factors as the major influence on the success of mental health treatments. Military psychiatrists rediscovered the "PIE" principles that had emerged in World War I: treatment of mentally traumatized soldiers should be *proximate* to their combat units rather than removed from the battlefield; it should be *immediate*, so that little time passed between the identification and treatment of psychiatric casualties; and psychically wounded soldiers should be *expected* to return to combat after a brief period of supportive care often accompanied by barbiturate drugs.[37]

A postwar consensus emerged that brief therapies near battlefields had been remarkably effective in restoring the mental health of traumatized combatants.[38] Psychiatrists and other policymakers inferred that this practice could be imported into peacetime to replace isolated mental institutions with outpatient practices in the community. A major reorientation toward the treatment of mental illness began that emphasized how psychic disturbances were unlikely to become chronic if they received rapid therapeutic responses. Policymakers also came to believe that early intervention for mental problems could prevent subsequent hospitalizations. Throughout the postwar decades the rhetoric of community care and treatment and, by implication, the obsolescence of mental hospitals shaped public debates and agendas.

The practice of military psychiatry in World War II both enhanced the status of the field and changed its thrust. The war led psychiatrists to view neuroses more as the product of current environmental stressors than of instinctual or early childhood experiences. It suggested that psychiatrists should be concerned with the mental health of normal men and women as well as those who suffered from serious mental illnesses. More than anything else, their experiences during World War II led psychiatrists to believe that purposeful social interventions could alter psychological outcomes. Military psychiatrists assumed prominent positions in the government and academia and shaped a socially oriented psychiatry that would have major influences on American life in subsequent decades. It would be directed not just toward the mentally ill but also toward the sane: "The greatest prerequisite for peace . . . must be sanity—sanity in its broadest sense, which permits clear thinking on the part of all citizens," President Harry Truman wrote to the American Psychiatric Association (APA) in 1948.[39]

Another lesson that psychiatrists learned from their wartime experiences was the thoroughgoing inadequacy of extant classifications of mental disorder. The diagnostic system at the time could not encompass the conditions of the vast majority of soldiers who succumbed to psychic traumas. An entirely new categorization of mental disorder soon emerged.

## The DSM-I and DSM-II

Before World War II most American psychiatrists practiced within mental institutions, where the vast majority of psychiatric patients resided. When the war began, about two-thirds of psychiatrists worked in these hospitals, using biological accounts of and somatic treatments for mental illnesses. In the postwar era, the traditional preoccupation with persons with severe and persistent mental illnesses gave way to a concern with the psychological difficulties of a far larger and more diverse population as well as with social problems more generally. Persuaded that there was a continuum from mental health to mental illness, psychiatrists increasingly shifted their activities away from the psychoses toward the less severe and more common neurotic conditions. This emphasis complemented the vast transformation of psychiatric practice that was occurring at the time. By 1956 only about 17% of the ten thousand or so members of the APA were employed in inpatient settings.[40]

Nowhere was the change in the nature of postwar psychiatry better illustrated than in the creation of a new classification of mental disorders. The first standardized diagnostic system in the United States, the *Statistical Manual for the Use of Hospitals for Mental Diseases*, had been intended for use within mental institutions, which employed it to categorize patients upon admission, assign them to various wards, and classify them at release.

First issued in 1918, this manual divided mental disorders into twenty-two principal groups, only one of which represented all the psychoneuroses.[41] The *Statistical Manual* guided psychiatric classification from its initial edition through its tenth edition in 1942.

Over the course of this period, however, the center of gravity in American psychiatry was shifting from state hospitals, which focused on psychotic cases, to psychodynamic therapies of outpatients with less severe pathology. The classifications of psychotic disorders that dominated the *Statistical Manual* were thus no longer relevant for the growing numbers of noninstitutionalized patients. In 1952, the APA produced the first edition of a new manual, the *Diagnostic and Statistical Manual of Mental Disorders* (DSM-I), which better reflected the nature of the psychiatric profession's changing patient population.[42] Meyer's work, along with Freud's, was the most important influence on the first major diagnostic manual that encompassed psychiatric conditions found outside of mental hospitals.

Few psychiatrists at the time believed in the specificity of mental disorders. Despite Kraepelin's efforts to apply the diagnostic model that prevailed in general medicine to psychiatric conditions, the amorphous terms "neuroses," "nervous disease," or "stress" characterized the heterogeneous range of psychosocial conditions in the United States during much of the 20th century. The views of Karl Menninger, a leading dynamic psychiatrist, were representative. Menninger (1893–1990) argued that most people moved back and forth between mental health and mental illness. He believed that separating individual mental disorders into discrete categories with unique symptom characteristics—scientific medicine's *modus operandi*—was a mistake: "Instead of putting so much emphasis on different kinds and clinical pictures of illness, we propose to think of all forms of mental illness as being essentially the same in quality and differing quantitatively." He summarized: "There is only one class of mental illness—namely mental illness."[43]

This insistence on the diffuseness of mental illness stemmed from the dynamic and Meyerian beliefs that the same underlying psychosocial process of individuals' failure to adequately adapt to their environments produced a wide variety of psychic consequences. Instead of being discrete diseases, mental illnesses ranged from the severe adaptive failures of the psychoses to the moderate and minor range failures of the neuroses.[44] Rather than treating the symptoms of mental disorder, Menninger, like Meyer, urged psychiatrists to explain how the individual's failure to adapt came about and its meaning to the patient. In other words, "What is *behind* the symptom?"[45]

His views widely diffused into popular culture through his regular column in the *Ladies Home Journal*.

The DSM-I (and second edition, DSM-II) did not adhere to Menninger's extreme position about the similarities of all mental disorders: after all, a diagnostic system intrinsically must make some distinctions among various conditions. It did, however, place more emphasis on the dynamics behind its various entities than on their symptomatic manifestations. Aside from the dwindling group that remained in asylum-based practices, psychiatrists at the time had little interest in brain diseases. The forces of most interest to the majority of clinicians in outpatient practices were far more psychological and social than biological.

The "basic division" in the DSM-I severed mental disorders that involved organic brain disturbances from those without such disturbances.[46] The first were cases in which a primary impairment of brain tissue function resulted in or precipitated the disturbed mental function. The manual separated these organic conditions from functional disorders that had psychogenic origins with no defined physical cause. The DSM-I paid considerably more attention to the functional group, which it divided into nonorganic psychoses, neuroses, and personality disorders. It particularly focused on the latter two groups. Following Meyer, it called nonorganic conditions "reactions" (e.g., "depressive reaction," "gross stress reaction," etc.) because, unlike brain disturbances, they presumably arose in response to problematic life histories and social circumstances.

At the heart of the DSM-I (and DSM-II) was the concept of "neurosis." It was the synthesizing rationale behind the psychoneurotic category of the manual, which itself was the class at the center of clinical practice. The term referred to the psychological conflicts that emerged as ways to deal with underlying conscious or unconscious anxiety. The first sentences of the summary description for the overall Psychoneurotic Disorders grouping were indicative:

> The chief characteristic of these disorders is "anxiety" which may be directly felt and expressed or which may be unconsciously and automatically controlled by the utilization of various psychological defense mechanisms (depression, conversion, displacement, etc.). . . . "Anxiety" in psychoneurotic disorders is a danger signal felt and perceived by the conscious portion of the personality. It is produced by a threat from within the personality (e.g., by supercharged repressed emotions, including such

> aggressive impulses as hostility and resentment), with or without stimulation from such external situations as loss of love, loss of prestige, or threat of injury.[47]

This definition indicates the extent to which psychodynamic assumptions infused the DSM-I classifications. Anxiety expressed defense mechanisms that were largely unconscious and that emerged from some inner threat or external situation. Moreover, the ways patients expressed anxiety, through such mechanisms as "depression, conversion, or displacement," were secondary to the fundamental process of anxiety that was behind each overt manifestation. The specific categories of neuroses (e.g., phobic reaction, obsessive compulsive reaction, depressive reaction, etc.) were divergent expressions of common underlying conflicts.

The DSM-I definition of depression (which it called "depressive reaction") typified the psychodynamic approach:

> The anxiety in this reaction is allayed, and hence partially relieved, by depression and self-depreciation. The reaction is precipitated by a current situation, frequently by some loss sustained by the patient, and is often associated with a feeling of guilt for past failures of deeds. The reaction in such cases is dependent upon the intensity of the patient's ambivalent feeling toward his loss (love, possession) as well as upon the realistic circumstances of the loss.[48]

This definition, like the others in the nonorganic categories of the DSM-I, focused on the dynamics (loss, guilt, ambivalence) that presumably led to depressive conditions. Some patients could become depressed because of "feelings of guilt" and others because of "realistic circumstances of the loss." Thus, the criteria recognized both intrapsychic and external causes of depressive neurosis but excluded organic origins. As psychiatry became more and more oriented toward brain research during the 1970s, the segregation of biological from psychosocial processes and the focus on the latter would become a serious problem.

The definitions of the other nonorganic categories in the manual—the psychotic, personality, and stress-related disorders—reflected the same principles. None provided any guidance for how one could identify or measure these conditions; most were infused by psychosocial assumptions of how they arose. Others, such as "transient personality reactions to acute or special stress," encompassed normal reactions that were, by definition, expected to disappear soon after the precipitating stressor ended.

Sixteen years later the APA issued the DSM-II in 1968. The second edition maintained the general psychodynamic orientation of the first DSM although it no longer used the term "reaction." Both the DSM-I and DSM-II focused on psychodynamic explanations that, following Meyer, directed attention to the total personality and life experiences of each individual patient. The DSM-II made few changes in the definitions of the various diagnoses and continued to describe each condition in perfunctory and theory-infused ways. Depression (now called "depressive neurosis"), for example, noted that this "disorder is manifested by an excessive reaction of depression due to an internal conflict or to an identifiable event such as the loss of a love object or cherished possession."[49] It stipulated that either psychological or social factors could lead to depression but did not mention what symptoms characterized this condition.

Some of the various definitions in the DSM-I and DSM-II attempted to separate disordered from normal conditions. For example, the DSM-II noted in regard to anxiety neurosis: "This disorder must be distinguished from normal apprehension or fear, which occurs in realistically dangerous situations."[50] Likewise, it indicated that depressive neurosis "is manifested by an excessive reaction of depression" so that proportionate depressive responses that were not "excessive" would not be disorders.[51] Or, cyclothymic personality disorders "are not readily attributable to external circumstances."[52] These distinctions, while vague and hard to apply in particular cases, at least suggested that contextually appropriate behaviors (fear in dangerous situations, nonexcessive sadness after loss, moods that change in response to environmental cues) were not mental disorders. Nevertheless, the manual's perfunctory definitions made the profession vulnerable to criticism that psychiatry was too subjective, medically unscientific, and overly ambitious in terms of its ability to define and explain the conditions it was treating.

In emphasizing the overlaps between different types of mental disturbances while downplaying their symptomatic presentations, the definitions in the DSM-I and DSM-II distinctly diverged from models of traditional biomedical diseases. The particular symptoms that were essential starting points in the Kraepelinian framework had virtually no role in the first two DSMs. Instead, as Menninger stressed, they strove to understand what was beneath superficial symptomatic appearances.[53] Dynamic psychiatry focused on investigating the particularities of individual lives.

This necessarily involved rejecting standard scientific research protocols that involved testing hypotheses through using clearly operational variables and control groups. The manuals were also far removed from what Charles Rosenberg has called the essence of classifications of biomedical diseases: separating the nature of the condition from the characteristics of the person who has the illness.[54]

The DSM-I and DSM-II reflected the character of American psychiatry in the postwar decades. They were developed for clinicians who needed to understand particular patients; they did not require standardized knowledge about how each patient compared to others. At the time, psychiatric researchers were few in number and had little professional prestige. Psychodynamically oriented psychiatrists dominated departments of psychiatry in medical schools, which were far more concerned with training clinicians than researchers. Gerald Grob observes that by 1960 "virtually every chairperson of a department of psychiatry stated unequivocally that the psychodynamic frame of reference (as contrasted with the descriptive or organic) was dominant" in training and education.[55] By the mid-1960s, psychoanalysts chaired about 60% of psychiatry departments.[56] During this era, case studies that demonstrated interpretative ingenuity provided standards of evidence: statistical analysis of aggregates of people was neither highly developed nor highly valued. Accordingly, neither the DSM-I nor DSM-II were central resources for psychiatric research.

The lack of specificity in the diagnostic manuals of the 1950s and 1960s was unsurprising. The DSM-I and DSM-II's nonexplicit definitions were not liabilities for mental health practitioners during this era. Particular diagnoses had little role to play in the explanations or practices of psychodynamic psychiatry. Dynamic therapies were also largely nonspecific so that distinct diagnoses were unnecessary for guiding treatment plans. Most outpatients at the time paid for their own therapy; no private or public third parties required diagnoses to reimburse clinicians. The resurgence of biologically oriented psychiatry and the attendant field of psychopharmacology in the 1970s sounded what would become the death-knell of the dynamic view.

## A New Federal Presence

The DSM-I went into effect at the same time as the role of the federal government in mental health issues was undergoing dramatic changes. Before

World War II, states and localities were responsible for programs concerning mental health and illness.[57] Their efforts focused almost exclusively on the upkeep of inpatient mental institutions; they had little else to do with the social response to mental illness. People without severe disturbances were not relevant to public policy efforts.

The dramatic social and economic changes that the New Deal brought to American life during the 1930s provided a congenial context for the turn to a socially oriented psychiatry. Its policies legitimated the welfare state and enlarged the scope of federal intervention in making social reforms. The ideological climate of the postwar period was conducive to sympathetic attitudes toward psychological injuries, the idea that stressful social environments could produce psychic disturbances in otherwise normal individuals, and a stronger role of psychiatrists in formulating national mental health guidelines.

## From Asylum to Community

Psychiatric experiences during World War II led to the emergence of a new template of mental health and illness that focused on the entire population, not only those who were treated for some mental illness. Just as all soldiers had been at risk for developing psychic disturbances, all citizens were in jeopardy of becoming mentally ill. The attention of policymakers turned away from serious, psychotic conditions toward the treatment of widespread neuroses and problems of living in society at large.

Concern with general psychological well-being became the center of major public programs. The views of the Group for the Advancement of Psychiatry (GAP), which was formed in 1946 by psychiatrists who had served in the war, were particularly important in expanding psychiatric influence over an extremely broad range of human experience. While the mental hygiene movement had already penetrated social institutions including the juvenile justice system, adult courts, and prisons in the 1920s, GAP strove for much broader effects.[58]

GAP promoted programs that would advance the mental health needs of whole populations. It advocated for community-based interventions for a vast array of individual, family, and social harms. Its first leader, Karl Menninger's brother William (1899–1966), urged psychiatrists to move beyond individual patients and "provide leadership and counsel to the family, the community, the state, welfare workers, educators, industrialists, religious leaders, and others."[59] The legitimate range of psychiatrically relevant conditions, which had already expanded from mental hospitals to outpatient treatment, was coming to encompass everyone's mental health. Proponents viewed their mission in these terms: "all social, psychological, and biological activity affecting the mental health of the populace is of interest to the community psychiatrist, including programs for fostering social change, resolution of social problems, political involvement, community organization planning, and clinical psychiatric practice."[60] Moreover, psychiatrists associated with GAP believed that appropriate social programs could prevent mental illnesses from arising at all. This group would come to have an especially powerful impact on governmental policy toward mental health and illness during the postwar period.

## The National Institute of Mental Health

The National Institute of Mental Health (NIMH) was created in 1949 to be the agency in charge of research, training, and services for mental health and illness. Reflecting the newly activist federal policy agenda, the agency's title included "health" but not "illness," "disease," or "disorder." Its first director, Robert Felix (1904–1990), was a former military psychiatrist who was closely allied with GAP, sharing its expansive definition of mental illnesses and its focus on the social roots of these conditions. The NIMH had little concern with people with serious mental illnesses who resided in state hospitals. Instead, its mission was to redirect resources and attention away from inpatient institutions toward the mental health of the entire population. It embraced the new socially oriented psychiatry that was concerned with "the whole social framework of contemporary living."[61]

One aspect of the agency's agenda involved a sweeping scope of psychiatric activism that emphasized how dire social environments produced pervasive amounts of mental illness. In 1953 Felix warned a congressional committee that "mental illness has reached epidemic proportions." He urged psychiatrists to "go out and find the people who need help and—that means, in their local communities."[62] Mental disorders were not confined to the small group with severe conditions but extended to unserved groups who faced neurotic breakdowns, general social maladjustments, and diffuse psychic disturbances. The agency would strive to promote mental health through detecting, treating, and preventing cases in the broader society. Conversely, it would let the various states keep control of and fund mental institutions.

Between 1950 and 1960 the NIMH's budget grew from $8.7 million to over $100 million.[63] During its early years, the agency emphasized community-oriented research; 60% of its grants were awarded to psychologists, sociologists, and other social scientists.[64] It also stimulated the growth of the psychiatric profession, devoting about 70% of its budget to promoting education. Consequently, between 1948 and 1976 the number of psychiatrists increased from just 4,700 to about 27,000.[65] In the 1960s over 10% of medical school graduates became psychiatrists. By 1970 they were treating about a million outpatients.[66]

At the core of NIMH policies was the development of hundreds of community mental health centers (CMHCs) whose mandate encompassed not just treating individuals with mental disorders but also helping those who were distressed, promoting broad social changes in the community and attempting to prevent mental illnesses from arising at all.[67] Between 1966 and 1979 the NIMH established 789 such centers. CMHCs became involved in efforts that included alleviating poverty, combating juvenile delinquency and violence, and promoting positive mental health. The NIMH also sponsored research on the mental health aspects of broad social problems such as racism, poor housing, divorce, and the like that presumably contributed to the development of psychological disturbances. It especially emphasized the role of prevention, reasoning that policies should aim to stop disorders from emerging or to confront them at early stages before they became chronic. The NIMH both emerged out of and contributed to the activist agenda that brought a huge range of phenomena into the realm of pathology during the postwar period.

## Stress Research

In addition to promoting community-based policies, another thrust of the NIMH during the 1950s and 1960s was to sponsor an influential body of research about stress-related illnesses. In 1956, physiologist Hans Selye (1907–1982) termed the many consequences of stress the "general adaptation syndrome," a condition that closely resembled the 19th-century malady that George Beard had termed "neurasthenia."[68] It encompassed a diffuse and multifaceted array of psychic, somatic, and interpersonal problems that often arose as responses to the strains of everyday life. The common psychological features of stress-related problems included a combination of symptoms involving nervousness, sadness, and malaise. Typical physical symptoms consisted of headaches, fatigue, back pain, gastrointestinal complaints, and sleep and appetite difficulties. Both the psychic and the bodily symptoms accompanied struggles with interpersonal, financial, and occupational concerns. These types of complaints accounted for a large proportion of disturbances found among patients in general medical and outpatient psychiatric treatment.[69]

Researchers applied the stress model in surveys of mental distress among untreated community populations. Naval psychiatrists Thomas Holmes and Richard Rahe developed a widely used scale for use in survey research that correlated the overall amount of social readjustment individuals underwent after experiencing common stressful life events with the resultant amounts of their distress. The scale encompassed not just widely occurring negative life events such as the death of a spouse or other close family member, divorce or marital separation, or losing a job but also such conventional activities as getting married, obtaining a mortgage, or going to a new school. Even mundane happenings—getting a traffic ticket, taking a vacation, or starting to go to a gym—could be a source of mental pathology.[70]

The Midtown Manhattan Study, an interdisciplinary effort between psychiatrists and social scientists that the NIMH partly funded, was probably the most important project in the stress tradition. Begun in the late 1950s, it used a sample of nearly two thousand randomly selected Manhattan residents to relate a range of sociodemographic data and indicators of social strains to the kinds of symptoms that Selye had emphasized. The study applied a continuous notion of mental illness, where even a single symptom indicated a "mild" form of mental illness. The survey found that psychiatric symptoms were extraordinarily widespread: only 18.5% of individuals were free of symptoms; 36.3% had mild symptoms; 21.8% had moderate symptoms; and 23.4% had severe symptoms. Respondents' socioeconomic status in large measure accounted for their location on this continuum.[71]

The Midtown Manhattan Study was especially notable because of its focus on a large, untreated sample, its use of common indicators of distress as signs of mental illness, its emphasis on social factors as causes of psychic disturbances, and its findings regarding the enormous amount of emotional disabilities in this urban setting. The study exemplified how far psychiatric interest had turned from mental hospitals and even from outpatient treatment toward the untreated suffering of entire populations. The borders between sanity and madness were becoming imperceptible. Not only was there

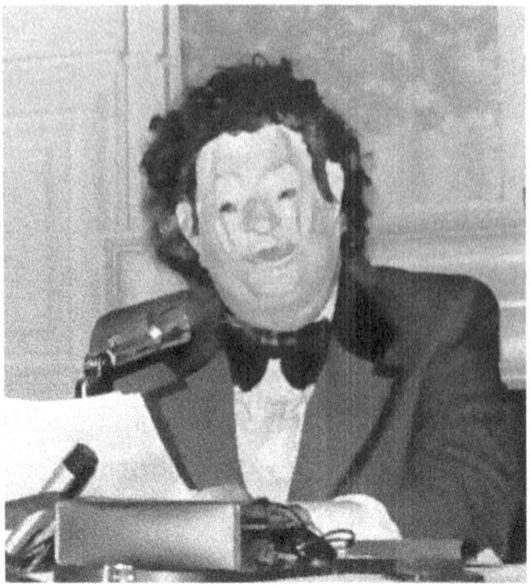

**Figure 5.1** Tranquilizer ads portrayed these drugs as ways to help their predominately female consumers calmly perform their daily activities. Meprospan ad, *JAMA*, July 16, 1960.

no sharp distinction between normality and pathology, but few people could claim to be "normal."

## Psychotropic Drugs in the 1950s and 1960s

A more socially focused dynamic psychiatry that marked postwar mental health activities facilitated the entry of environmental conceptions of mental illness into American culture. Paradoxically, another key route for the transmission of a social view of psychic disturbances was through the explosion of prescriptions for tranquilizing drugs during the 1950s and 1960s.[72] Unlike later medications that were marketed for the control of brain-based chemical imbalances, these wildly popular pills entered public consciousness as ways to deal with widespread life stressors that had no apparent biological grounding.

In 1955 the development of meprobamate (Miltown) dramatically changed the nature of treatment for the mental health problems of patients in outpatient psychiatric and, especially, in general medical practice. While earlier drug classes including the opioids, bromides, and barbiturates were widely used for psychic problems, none had become cultural sensations to the extent of the minor tranquilizers. Miltown was promoted far more for its ability to relieve ordinary problems of stress than as a remedy for specific psychiatric conditions. Advertisements, which were directed toward general medical practitioners as well as psychiatrists, touted an extremely broad array of indications for its use, including "tense, nervous" patients, "insomniacs," "anxious depression," "alcoholics," and "problem children."[73] It was, in short, the ideal treatment for ubiquitous everyday difficulties stemming from the "stress," "nerves," or "tension" that accompanied everyday life.

Miltown, historian Andrea Tone reports, created a climate where "it was okay to see doctors for drugs to make them feel better about the vagaries of life, not just to treat diseases."[74]Harking back to Beard, the tranquilizers treated conditions that stemmed from the pressures of modern life, especially those that women faced. For example, one ad encouraged physicians to prescribe Miltown so that "Pregnancy can be made a happier experience." Another featured the "battered parent syndrome" suffered by housewives whose children led them to be "physically and emotionally overworked, overwrought and—by the time you see her—probably overwhelmed."[75]

Within a year of its introduction, Miltown became the best-selling drug in American history. By 1965 physicians and psychiatrists had written 500 million prescriptions for it.[76] The spectacular success of the benzodiazepine Librium, which was introduced in 1960, displaced Miltown. Valium, in turn, succeeded Librium as the newest blockbuster tranquilizing drug. By 1971 Valium was the single most prescribed drug of any kind: 20% of all American women and 8% of men reported using a minor tranquilizer in that year.[77]

The promotion of the benzodiazepines echoed the earlier themes used for Miltown. Valium and Librium were touted for their ability to ease a huge range of conditions that heightened nervous tension. Advertisements in medical journals touted their ability to relieve distress associated with problems with spouses and children, overwork, career failures, social role transitions, housekeeping, and even traffic jams. One ad for Librium portrayed the circumstances of typical college students:

> A whole new world of anxiety. The new college student may be afflicted by a sense of lost identity in a strong environment. Today's changing morality and the possible consequences of her "new freedom" may provoke acute feelings of insecurity. She may be excessively concerned over competition, both male and female, for top grades. Unrealistic parental expectations may further increase emotional tension.[78]

An ad for Valium even explicitly advocated its use for people without any psychiatric symptoms: "For this kind of patient [an overworked doctoral candidate who was finishing a thesis and had indigestion and gastrointestinal problems]—with no demonstrable pathology—consider the usefulness of Valium."[79]

The popularity of the tranquilizing drugs was especially due to their usefulness in treating the diffuse kinds of psychosocial problems seen in general medical practice; primary care physicians wrote between 70% and –80% of prescriptions for Miltown, Librium, and Valium.[80] Studies in the 1950s and 1960s found that only about a third of these drugs were prescribed for specific mental disorders, while the rest were given as a response to more generalized complaints.[81] In 1963, the feminist author Betty Friedan provided perhaps the best characterization of what the tranquilizers treated—"the problem that has no name"—the ubiquitous malaise, tension, and anxiousness that results from the gap, which was especially wide among women, between the expectations of a fulfilling life and the realities of a stifling existence.[82] During the 1950s and 1960s, the tranquilizing drugs could be successfully marketed as remedies for general life stresses and protean conditions such as unsuitable spouses, overdemanding parents, taxing jobs, or financial stresses with little consideration about whether or not they treated specific disease states. By the 1970s, as the next chapter details, this situation had changed considerably.

In addition to the spread of the tranquilizers in the general population, a revolution in drug treatments for serious mental illnesses arose in the early 1950s. Major breakthroughs occurred as pharmacologists serendipitously discovered a number of new medications including chlorpromazine for schizophrenia, lithium for bipolar disorder, and imipramine and the monoamine oxidase inhibitors for depression.[83] Indeed, the effectiveness of these early drugs remains unsurpassed. Subsequent chapters consider these medications in more detail.

## A Psychotherapeutic Culture Emerges

The quarter century between World War II and the mid-1960s was marked by more socially oriented psychiatric theories, a new classification system geared toward outpatient conditions, population-based federal policies that encouraged community treatments, and new, widely prescribed drugs.

Another major development was the emergence of a therapeutic culture whose members held favorable views of mental health experts and who self-initiated outpatient therapy. Western societies had undergone a major fissure after World War I: the war's wholesale carnage brought about a widespread disillusionment with previous values. As shared meanings lost their power, many individuals came to focus less on their places in kin, communal, and religious groups and more on their inner lives and personal experiences. The disintegration of communities led selves to become the new center of concern. What sociologist Philip Rieff called "psychological man," whose highest values were personal autonomy and well-being, provided the basis for a new, more individualistic culture among intellectuals, artists, and other highly educated groups in Europe and the United States.[84]

The war shattered faith in traditional cultural and religious institutions, but psychoanalysis was poised to serve as a meaningful alternative in an age devoid of spiritual principles.[85] As people were detached from communities, personal identity became both a problem and a project for them. Analysis had the promise of teaching inhabitants of the modern world "how to live without belief."[86] In a society drained of compelling collective ideals, many people turned to psychotherapists to find meaning in their lives.

Initially, analysis promised to replace outdated chains of social and sexual repression with a journey of self-discovery that released personal freedom. At the time, prominent critic Malcolm Cowley noted: "Freudian psychology provided a philosophical justification and made it unfashionable to be repressed."[87] The new therapeutic mentality featured growing rates of voluntary help-seeking from analysts and other mental health professionals to discuss "diffuse anxiety, loss of identity, inability to create, unhappiness."[88]

Over the middle part of the 20th century, dynamic psychiatry created a language through which troubled people could interpret and seek relief for a wide variety of conditions. The "golden age" of analysis blossomed after World War II, running from 1945 through the mid-1960s.[89] Statistics, which became available starting in 1955, show nearly a tripling of psychiatric outpatients from 379,000 in that year to over a million ten years later and to almost two million by 1968.[90] At the same time, the problems people brought to treatment broadened from the neuroses that were common in Freud's time to more general maladaptive patterns of behavior and character and to the even more nebulous realm of personal problems. According to Rieff: "All experience is symptomatic now. People seek treatment because they sleep poorly, or have headaches, or feel apathetic toward loved ones, or because they are dissatisfied with their lives."[91] Confirming his assertion, a study appearing in the *American Journal of Psychiatry* in 1950 found that the most common complaints among psychiatric outpatients were nervousness, internal tension, feelings of vague unreality, and depression.[92]

During the era that psychosocial theories dominated the mental health professions, poor marriages, troubled children, failed ambitions, nervousness, diffuse anxiety, and general discontent with life became reasons for seeking psychiatric help.[93] Psychiatry was transformed from a profession concerned with insanity to one responding to types of distress that were hard to distinguish from normality. Sociologist Charles Kadushin aptly summarized the nature of the problems of psychiatric outpatients that he studied at the time: "The empirically derived clustering here presented covers a wide range of problems; indeed it is one way of grouping all of life."[94]

During the postwar period the significance of psychiatric conceptions extended far beyond just those who entered treatment. Discussions of mental health and illness—as framed from a psychodynamic perspective—were a prominent, and highly influential, presence in American culture as psychiatrists became the preeminent authorities regarding mental health as well as mental illness. Historian John Burnham notes: "At the high point of both the popularization and prestige of psychoanalysis in the United States in the 1940s–1960s, it was difficult to separate the core psychoanalytic movement from the pervasive cultural impact."[95] Women's magazines, in particular, were filled with articles about analysis, which they considered to be the exemplary mode of psychological therapy.[96]

Especially in the post–World War II period, dynamic ideas spread from highly educated intellectuals, permeating the general culture to a greater extent than any other body of psychiatric thought to date. Remarkably, a thousand-page edition of *The Basic Writings of Sigmund Freud* that appeared in 1947 sold over a quarter of a million copies.[97] Psychodynamic views also came to infuse literature on child-rearing, education, and the social sciences. "It hardly matters whether Freud's ideas are true or false. What matters is that they have impregnated our entire culture and the way in which we understand the world through films, art, comic books, and television," historian Michael Roth declared.[98]

The cinema was an especially influential purveyor of psychiatric thought. In 1944, leading director Joseph Mankiewicz stated that "I am convinced that the next period of years will bring psychiatry in general, and psychoanalysis in particular, into great prominence as a most important source of literary, dramatic and motion picture material."[99] His prophecy was fulfilled: particularly in the late 1950s and early 1960s, numerous films presented highly idealized portrayals of compassionate and insightful psychiatrists who routinely solved a wide variety of personal problems.[100]

Psychiatry's constituency had moved from the residents of mental institutions and outpatient practices to troubled individuals and the public at large. An opponent of analysis, behavioral psychologist O. Hobart Mowrer observed: "Anyone who reached adulthood prior to 1950 knows how perversely Freudian theory and practice dominated not only in the specific field of psychotherapy, but also education, jurisprudence, religion, child rearing, and art and literature, and social philosophy."[101] According to critic Janet Malcolm, psychoanalysis "detonated throughout the intellectual, social, artistic, and ordinary life of our century as no cultural force has (it may not be off the mark to say) since Christianity."[102] Psychiatrists spoke with authority about issues concerning not just mental disorders but also about how individuals ought to feel, families ought to function, and institutions ought to act. Their advice applied "to normal people and their normal problems in normal communities."[103] The first director general of the World Health Organization promoted the most expansive agenda: "With the other human sciences, psychiatry must now decide what is to be the immediate future of the human race. No one else can. And this is the prime responsibility of psychiatry."[104]

## Conclusion

None of the major schools that developed in the quarter century between Freud's death and the mid-1960s—the biopsychosocial model, ego psychology and the study of personality disorders, neo-Freudianism, and stress research—sharply separated neurotic from normal psychosocial phenomena. These views were congruent with the notion that the lines between sanity and madness were fluid. Karl Menninger's observation was exemplary: "most people have some degree of mental illness at some time, and many of them have a degree of mental illness most of the time."[105] In addition to the loose boundaries between the sane and the insane, psychiatrists came to emphasize how these processes were continuous, running from normality on the one end through mild and severe forms of neuroses through psychoses on the other end. The basic challenge that everyone—whether normal or mentally ill—faced was how to adapt to his or her environment.[106] The result was simultaneously to normalize the concept of mental illness and to neglect people with serious mental illnesses who most clearly fit this category. As an APA position paper stated:

> The concept of mental illness has been broadened from one in which psychiatry was concerned almost exclusively with psychotic patients (who represented a small percent of the population) to one in which the emphasis has become the commitment of the community's total mental health resources to serving and maintaining the mental health needs of the entire population.[107]

The vast expansion of the realm of mental illness was the most lasting impact of developments in the period between World War I and 1980. The psychosocial scope of psychiatric theory, military psychiatry's experiences with normal soldiers, the emergence of GAP and the NIMH, the growing focus on the mental health of the general population, and the prominence of the dimensional view in the postwar era resulted in a huge growth of the range of psychic pathology. Acclaimed studies such as the Midtown Manhattan research asserted that nearly everyone had at least a "mild" form of disorder. Psychiatrists and other mental health professionals became the legitimate responders to problems ranging from unhappy spouses and unruly children to diffuse unhappiness. The diagnostic manuals that followed the initial two DSMs would take advantage of the wide realm of pathology they classified, making sure to cast these problems within a biomedical framework that the earlier guides lacked.

Another prominent feature of postwar psychiatry lay in the primacy it gave to the social causes of mental illness. Freud's later works had already turned analytic attention toward the external world. Ego psychology, neo-Freudianism, the focus on traumatic conditions, and stress research all moved psychiatry toward the social world and away from biology. Even Meyer's psychobiological approach gave short shrift to the "bio" aspects of mental illness. The extraordinarily popular tranquilizing drugs that emerged in this period, too, were framed as remedies for problems in living rather than for defects of brain chemistry.

Following Freud and Meyer, postwar psychiatry—as the DSM-I and DSM-II illustrate—sharply differentiated mental from physical illnesses. Psychic conditions arose from a unique combination of psychological dispositions interacting with life events and broader social conditions. They were best treated through interpersonal explorations, perhaps supplemented with a prescription. None of the schools discussed in this chapter challenged the assumption that mental illnesses were distinct from organic diseases, and none strove to link their efforts to biomedical conceptions. All discarded the 19th-century efforts to specify discrete forms of mental illness and relate them to underlying organic forces. Instead, each conceived of mental illness in broad rather than specific terms and severed explanations of psychic disturbance from biological causes.

The vagueness and nonspecificity that marked biopsychosocial efforts were compatible with dominant postwar schools of psychiatric thought but could not be assimilated to the scientific culture of medicine that emerged in psychiatry during the 1970s. The assumption that mental illnesses differed in fundamental ways from physical illnesses became professionally unsustainable. As multiple professional groups, including clinical psychologists, social workers, counselors, clergymen, and others, came to offer variants of psychotherapy, the justification for psychiatry's authority over mental illnesses came to depend on using a medical model that distinguished the field from its competitors. Yet, analysis itself was fragmented into a variety of competing factions that possessed no theoretical core; none had a valid claim to a truly medical identity. American psychiatrists had thoroughly intertwined their legitimacy with the medical profession; when medicine wholeheartedly embraced a scientific model, the medical justification for analysis became unsustainable. But, first, the profession would have to deal with a chorus of critiques from a disparate range of detractors including anti-psychiatric intellectuals, members of the youth counterculture, biologically oriented psychiatrists, insurance companies, and a newly biomedical NIMH.

## Acknowledgments

The sections "The Neo-Freudians" and "Psychotropic Drugs in the 1950s and 1960s" are adapted from Horwitz, 2013. The section "World War II" is adapted from Horwitz, 2018.

## Notes

1. Malcolm (1977) provides a good overview of classical analysis during the psychosocial era.
2. E.g. Harrington, 2019, 46–71.
3. In contrast to the movement toward a more psychosocial emphasis in the United States, analytic thought in other regions retained an inner focus. In England Anna Freud, Melanie Klein, and Donald Winnicott developed influential theories of object relations that emphasized the enduring impact of mothering during the earliest periods of infancy. French analyst Jacques Lacan's structuralist perspective, which joined linguistic analysis to Freudian concepts, had a potent effect on many European and South American analysts.
4. The warm reception of dynamic psychiatry in the United States was paradoxical. Freud only visited the United States once and did not have a warm view of the country, its culture, or its people.
5. Meyer later repudiated many of Freud's core ideas, especially those concerning sexual influences.
6. Harrington, 2019, 36.
7. Grob, 1994, 42–46.
8. Whooley, 2019, 65.
9. Meyer, 1957, 118.
10. Lunbeck, 1994, 11.
11. Hale, 1995, 51.
12. Fenichel, 1945/1996, 6.
13. Fenichel, 1945/1996, 526.
14. Makari, 2012, 122–23; Hartmann, 1939.
15. Layton, 1985; Kohut, 1971, 1977; Kernberg, 1995.
16. In 1914 Freud wrote an essay, "On Narcissism," which his biographer Peter Gay (1988, 545) called a "crucial turning point" for him. Freud postulated that during the earliest stage of human development babies were their own primary love object. Thus all infants were narcissistic and only directed libidinous energy outward to other people in later phases of life. This narcissistic orientation was part of the regular course of human development that was present in everyone and so was natural, not pathological. Most people, however, overcame their self-absorption as they grew older so that an adult's narcissistic personality could signify a mental disorder.
17. Lunbeck, 2012, 218.
18. Lunbeck, 2014, 7.
19. Rangell, 1995, E11.
20. Lasch, 1979. The nebulous concept of "borderline" personality, where patients have more severe pathology than displayed among ordinary neuroses but less severe conditions than the psychoses, also became popular.
21. E.g. Hale, 1995; Horwitz, 2002; Lunbeck, 1994.
22. Grob, 1987, 417.
23. Horney, 1937.

24. Horney, 1937, 15.
25. Horney, 1937, 54.
26. Horney, 1937, 79.
27. Ross, 2012, 170.
28. Fromm, 1941/1969, 318.
29. Political philosopher Herbert Marcuse (1955) attacked the neo-Freudians from a radical direction, arguing that Freud's emphasis on sexuality and aggression offered a more promising way to critique society than the mild reforms that Fromm and Horney suggested.
30. McLaughlin, 1998.
31. Grob, 1991a. Some smaller scale efforts to screen out soldiers who were psychologically unfit for military service arose in the final stages of World War I.
32. Grinker & Spiegel, 1943, 115.
33. Beebe & Apple, 1958.
34. Swank & Marchand, 1946, 243–44.
35. Keegan, 1976, 329.
36. Shephard, 2000, 326.
37. Shephard, 2000, 59; Jones & Wessely, 2005, 21.
38. In fact, little evidence suggested that these wartime therapies actually allowed the bulk of traumatized soldiers to return to combat (Jones & Wessely, 2005, 79, 84).
39. Harrington, 2019, 84.
40. Grob, 1991a, 42.
41. *Statistical Manual*, 1942.
42. APA, 1952.
43. Menninger, 1963, 32, 9.
44. Wilson, 1993, 400.
45. Menninger, 1963, 325.
46. APA, 1952, 12.
47. APA, 1952, 31.
48. APA, 1952, 33–34.
49. APA, 1968, 40.
50. APA, 1968, 39.
51. APA, 1968, 40.
52. APA, 1968, 42.
53. Menninger, 1947.
54. Rosenberg, 2007.
55. Grob, 1991a, 100.
56. Hale, 2000, 82.
57. One exception was the responsibility of the federal government for psychically incapacitated veterans of World War I. Between the two world wars the United States spent nearly $1 billion on veterans' psychiatric illnesses (Finley, 2012, 92). Other exceptions included running St. Elizabeth's mental hospital in Washington, DC, providing psychiatric services in federal prisons, and screening immigrants for mental illnesses.

58. Group for the Advancement of Psychiatry, 1950.
59. Menninger, 1949, 2.
60. Roberts, Halleck, & Loeb, 1966 quoted in Herman, 1995, 254–55.
61. Rennie, 1955, 10; Grob, 1991a; Torrey, 2014.
62. Quoted in Staub, 2011, 36; Grob, 1987, 417.
63. Herman, 1995, 248.
64. Grob, 1991a, 66–7. Indeed, the biologically oriented psychiatrists at Washington University who would become influential in developing the third edition of the DSM were unable to get grant funding from the NIMH during the 1950s and 1960s (Decker, 2013, 55).
65. Hale, 2000, 246.
66. Manderscheid et al., 1986; see also Menard, 2012, 196; Taylor, 2013, 58; Herman 1995, 3, 262.
67. Grob, 1991a.
68. Selye, 1956.
69. Smith, 1986.
70. Holmes & Rahe, 1967.
71. Srole et al., 1962.
72. See especially Healy, 1997; Herzberg, 2009; Tone, 2009.
73. Tone, 2009, 75.
74. Tone, 2009, 28.
75. Greene & Herzberg, 2010; Herzberg, 2009, 81.
76. Smith, 1985.
77. Parry et al., 1973.
78. Smith, 1985, 187.
79. Herzberg, 2009, 128; Tone, 2009, 156, 137; Smith, 1985, 187, 120.
80. Smith, 1985, 27.
81. Shapiro & Baron, 1961; Cooperstock & Lennard, 1979; Raynes, 1979.
82. Friedan, 1963/2001, 11.
83. See especially Healy, 1997.
84. Rieff, 1966; Zaretsky, 2004, 120.
85. Others responded to this situation through joining political movements, most prominently, fascism and communism.
86. Rieff, 1959, 162; Zaretsky, 2004, 5.
87. Cowley, 1951, 64.
88. Hale, 1995, 63.
89. Rangell, 1996, E3.
90. Grob, 1991a, 258.
91. Freud, 1917/1957, 304.
92. Rickels, Klein, & Bassan, 1950.
93. Grob, 1991a; Lunbeck, 1994; Hale, 1995.
94. Kadushin, 1969, 103.
95. Burnham, 2012, 4.
96. Zaretsky, 2004, 82.

97. Staub, 2011, 13.
98. Quoted in Roudinesco, 2016, 426.
99. Quoted in Halliwell, 2013, 67.
100. See especially Gabbard & Gabbard, 1999.
101. Quoted in Burnham, 2012, 158.
102. Malcolm, 1977, 22.
103. Herman, 1995, 238.
104. C. Brock Chisholm, 1948, quoted in Torrey, 2014, 22.
105. Menninger, 1963, 33.
106. Wilson, 1993, 400.
107. Quoted in Whooley, 2019, 143.

# 6
# The Decline and Fall of Dynamic Psychiatry

> Much Madness is divinest Sense—
> To a discerning Eye—
> Much Sense—the starkest Madness—
> 'Tis the Majority
> In this, as all, prevail—
> Assent—and you are sane—
> Demur—you're straightway dangerous—
> And handled with a Chain
>
> —Emily Dickinson, Poem No. 435

After psychiatry's ascendancy during the two decades after World War II ended, the profession entered its most troubled period. From the emergence of the anti-psychiatry movement in the mid-1960s through the resurrection of a biomedical model in the *Diagnostic and Statistical Manual of Mental Disorders* (third edition, DSM-III) in 1980, the field endured a time of continual crisis. The general culture shed its earlier infatuation with analytic ideas and turned sharply against the discipline. The medical profession, biologically oriented psychiatrists, and third-party insurers, too, came to reject psychodynamic approaches. The National Institute of Mental Health as well shed its initial psychosocial emphasis in favor of a strong biological focus. Another government agency, the Food and Drug Administration (FDA), forced drug companies to stop advertising their products as remedies for general distress and mandated that they show efficacy in treating specific diseases.

The high pedestal that dynamic psychiatry rested on in the postwar period swiftly crumbled. New cultural, political, economic, and social forces arose to create a biomedical diagnostic template that replaced the delegitimated psychosocial approach. Changing conceptions of normality and abnormality, the causes of mental illness, and the relationship of mental and physical illnesses accompanied the transition from one paradigm of mental disorder to another, dramatically different one.

## The Abnormal Becomes Normal

For millennia, from ancient writings through the 1950s, commentators clearly distinguished the mad from others, usually based on the incomprehensibility of their behaviors. During the first few decades of the 20th century, dynamic psychiatrists and others blurred the boundaries between normality and mental illness. With rare exceptions, however, they did not challenge the gap between the psychoses and other behaviors.[1] This situation drastically changed during the 1960s when a diverse group of critics, loosely known as "anti-psychiatrists," emerged. They did not so much criticize psychiatry for its vast expansion of abnormality as they questioned the validity of calling *any* human action a "mental illness." Indeed, their critiques specifically focused on the legitimacy of viewing the psychoses as mental disorders. Their writings powerfully influenced both cultural and institutional views of mental illness, although their ultimate effect was far from their intentions.

### Anti-Psychiatry

From the end of World War II to the mid-1960s, psychodynamically oriented psychiatry enjoyed extraordinary prestige in American society. At that time, however, its legitimacy was called into question from a number of quarters. Many of the most influential attacks stemmed from critics who questioned the very reality of mental illness. Their work helped turn cultural attitudes of the psychiatric profession from highly favorable to sharply negative.

Renegade psychiatrist Thomas Szasz's *The Myth of Mental Illness* (1961) provided the first, and most influential, critique of psychiatry. Szasz (1920–2012), an émigré from Hungary, asserted that starkly different principles underlay explanations of human actions and organic material. When they were applied to human behavior, concepts of normality and abnormality were arbitrary value judgments: "What people now call mental illnesses are, for the most part, communications expressing unacceptable ideas, often framed in an unusual idiom."[2] This completely differed from their application to bodily diseases, which definitionally stemmed from some physical lesion that is an "abnormality of cells, tissues, organs or bodies."[3] Because mental illnesses did not display such lesions, they were "myths" that were actually ethical judgments of disliked behaviors.

Szasz asserted that the term "mental illness" had no valid use in medicine and so should be abolished as a way of describing human actions.[4] For him claims that mental illnesses were genuine diseases allowed psychiatry to illegitimately partake in the prestige and power of the medical profession and to justify its authority over socially deviant behaviors. In fact, tags of mental illness were arbitrary designations that bolstered professional interests at the expense of labeled patients. Behaviors that were called "mental illnesses" involved values that "are and must be the legitimate concern of everyone and fall under the special competence of no particular group."[5]

Ironically, Szasz's sharp partition of mental from physical conditions echoed the DSM-I's central principle: this manual's core division was between disorders that were due to an impairment of brain tissue function and those that arose from psychogenic origins. In the case of organic conditions, it was clear what constituted a disorder—some aspect of brain functioning was defective. But the DSM did not answer this question for conditions that were not associated with physical causes: what quality made the "psychogenic" psychoses, neuroses, and personality conditions *disorders* in the first place? Because these conditions definitionally had no organic structure or function, Szasz could plausibly claim that their placement in the diagnostic manual resulted from value considerations without any medical grounding.

For Szasz, whose politics stemmed from the libertarian right, psychiatry extended authoritarian state influence in order to apply coercive control over nonconformists. His major interest was in the way the psychiatric profession exercised illegitimate power over people they considered to have mental disorders. He was most concerned with protecting "the autonomy, integrity, responsibility, and freedom of the individual."[6] Szasz particularly attacked involuntary commitments to mental hospitals that, he claimed, imprisoned people who had not broken any laws. Forced hospitalization represented the most serious violation of civil liberties that existed in American society. In 1969, he teamed with L. Ron Hubbard, the founder of the church of Scientology, to form the Citizens Commission on Human Rights in order to expose "psychiatry's abusive and coercive practices."[7]

Another renegade psychiatrist, R. D. Laing (1927–1989), was a second influential anti-psychiatrist. In contrast to Szasz's institutional focus, which pitted psychiatrists against patients, Laing focused on how madness arose from familial conflicts between parents and their children. Unlike Szasz, Laing did not deny the reality of mental illness, but he insisted that it was the product of social relationships and communications. He strove to show how psychosis could be understood as an attempt to come to terms with difficult environmental—specifically, familial—circumstances. In particular, people with schizophrenia made rational responses to irrational domestic contexts; their conscious or unconscious interactions with parents led them to become mad. Allying with countercultural heroes such as poet Allen Ginsberg and drug guru Timothy Leary, he glorified the mad for possessing special insights: "Madness need not be all breakdown. It may also be break-through. It is potential liberation and renewal as well as enslavement and existential death."[8]

Sociologist Erving Goffman (1922–1982) was another perceptive critic of the psychiatric profession. He linked Szasz's concerns about the oppressive nature of mental institutions with Laing's emphasis on familial dynamics. Like the two psychiatrists, Goffman focused on the situations of people who were labeled as "psychotic," not on those with less severe diagnoses. His fieldwork at St. Elizabeth's mental hospital in Washington, DC, led him to view the psychiatric symptoms of mental patients as products of their circumstances, not of their mental illnesses. Goffman traced their plights to betrayals by family members who usually initiated hospitalizations after disruptive and unruly behaviors. Psychiatrists associated with mental hospitals then reinforced familial desires for commitment.[9]

For Goffman, labels of mental illness were initially products of contingencies such as social class, the degree of visibility of behavior, or even how far an individual lived from a mental hospital. Patients then entered institutions that interpreted their actions in the hospital through the assumption that they were mentally ill, which created a self-fulfilling prophecy. Goffman, however, believed that symptoms were more likely to be responses to the pathological nature of hospital life itself than aspects of some individual condition. The effect of Goffman's work, like Szasz's and Laing's, was to return attention to the plights of persons with serious mental illnesses, who had fallen off the radar of dynamic psychiatry.

Another American sociologist, Thomas Scheff (b. 1929), also developed a major critique of the psychiatric viewpoint on mental illness. In his 1966 book, *Being Mentally Ill*, Scheff defined what are usually called "psychiatric symptoms" as "residual rule-breaking."[10] Residual rule-breaking referred to norm-violating behaviors that lack explicit cultural labels. "Mental illness" is a label that observers use to explain those rule-violating behaviors that they are unable to interpret through any culturally recognizable category. Psychiatric symptoms violate residual rules; they are not intrapsychic disturbances of individuals. Because Scheff's empirical studies of residual rule-breaking analyzed commitment hearings to mental institutions, he reinforced the tendency of anti-psychiatrists to critique the most serious forms of mental illness.

An especially important figure who undermined notions of mental illness was the extraordinarily popular French philosopher Michel Foucault.[11] Foucault (1926–1984) argued that classifications of mental illness emerged in a complex field of power relations in Europe during the 18th century. Mental disorders had little to do with disease but involved issues related to the control of deviant behavior. The mental institutions that appeared at that time replaced the leprosariums that had previously contained nonconformists. Foucault challenged the notion that asylums were beneficent facilities and argued that they functioned to monitor and regulate people—including not just the mentally deviant but also witches, religious heretics, criminals, and the idle poor—who transgressed social norms and did not fit into productive social roles. Asylums confined those who threatened existing power structures and at the same time promoted the interests of the psychiatric profession. Those who were called "insane" were not so much victims of a disease as of those who supposedly tried to cure them.

Psychologist David Rosenhan (1929–2012) was another notable representative of the anti-psychiatric movement. In 1973, he published a study in the prestigious journal *Science*, "On Being Sane in Insane Places," which was the most dramatic and influential single study turning the culture against psychiatry. His research, now seen as a landmark in the critique of psychiatric diagnosis, directly challenged the ability of psychiatrists to distinguish normality from psychosis. Rosenhan had eight seemingly normal individuals present themselves at hospitals and report only auditory hallucinatory symptoms (they claimed to hear voices saying things like "thud," "dull," and "empty") but otherwise act and speak normally.[12] All of these pseudo-patients were admitted and classified as psychotic (almost all as schizophrenic), and they retained these labels for an average of nineteen days before being released, even though they immediately reverted to acting normally while in the hospital. Adding insult to injury, fellow patients did identify several pseudo-patients as likely normal.

To understand the views prominent at the time, consider a few of sentences in the introduction to Rosenhan's article:

> Normality and abnormality, sanity and insanity, and the diagnoses that flow from them may be less substantive than many believe them to be. . . . Based in part on theoretical and anthropological considerations, but also on philosophical, legal, and therapeutic ones, the view has grown that psychological categorization of mental illness is useless at best and downright harmful, misleading, and pejorative at worst. Psychiatric diagnoses, in this view, are in the minds of observers and are not valid summaries of characteristics displayed by the observed.[13]

Based on his results, Rosenhan concluded: "It is clear that we cannot distinguish the sane from the insane in psychiatric hospitals."[14]

In the 1960s and early 1970s, the antipsychiatry critics were not marginal eccentrics but major figures in an intellectually prominent cultural movement. Their critiques of the concept of mental illness and of mental hospitals had tremendous influence not only within academia but also in the broader culture. Movies, which had idealized psychiatry during the postwar period, turned sharply against the profession. The film *One Flew Over the Cuckoo's Nest*, which won the top five Oscars in 1975 (best picture, best director, best actor, best actress, best screenplay) was probably the best-known popular critique associated with the anti-psychiatry movement. Based on the best-selling novel of the same name by Ken Kesey, which had been published in 1962 and had sold over a million copies in that decade, the film became the seventh-highest grossing film ever at the time, bringing in almost $300 million worldwide.

The film portrayed the exploits of an affable nonconformist, Randle P. McMurphy, who innocently requested a transfer from a prison to a psychiatric ward because he mistakenly believed that he would be free of coercive control while in the hospital. McMurphy is not mentally ill but is a rebel who refuses to show sufficient deference to those in authority. By the end of the film, he is brutally lobotomized. The film's message was that inpatient psychiatric treatment was akin to imprisonment; psychiatrists and other staff at these facilities functioned, essentially, as prison guards.

Paradoxically, the highly coercive picture of mental hospitalization that the film painted was already thoroughly anachronistic in 1975. By that time, the major problems that persons with serious mental illnesses faced had to do with homelessness and the lack of mental health services; institutional commitments were very hard to implement. Nevertheless, allied to the argument that mental illness was not real but a label used to control nonconformists, the film was perceived as a powerful critique of psychiatric theory and practice.

Notably, all of the major anti-psychiatrists directed their fire against the labeling of the psychoses, which had been always been consensually accepted as legitimate forms of mental illness. With the exception of Laing, who focused on families, all were primarily concerned with people who were committed to mental institutions. They had little to say about the bread and butter neurotic conditions that prevailed in outpatient practices or the distressing impairments that dominated discussions of mental health and illness from the1940s through the 1960s. Their take-home message was that mental illness itself was a fiction perpetuated by a self-interested and oppressive psychiatric profession. Psychiatrists and other mental health professionals functioned to enforce conformity among people who might behave idiosyncratically but who were not ill.[15]

For a brief and unique period of time during the 1960s and early 1970s, the domain of abnormality contracted rather than expanded. All types of conditions that had fallen within the legitimate realm of psychiatry came into question. As the title of Rosenhan's article reflected, even people with psychoses were seen as rational actors responding to insane conditions. Mental hospitals, not their residents, were mad. The work of the anti-psychiatrists did not cause but it did help facilitate the mass exodus of patients from inpatient institutions. They were unconcerned with what sort of care would replace the services found within mental hospitals. Subsequent generations of people with serious mental disorders would have to fend for themselves, often joining the ranks of the homeless or incarcerated.

## A New Generation Emerges

A second profound shift in conceptions of and responses to mental illness resulted from the emergence of a youth counterculture in the mid-1960s. By this time, the intellectuals and bohemians who had been among the most enthusiastic promoters of analysis in the 1920s and 1930s had turned sharply against it. While psychoanalysis was initially seen as a cutting-edge therapy that challenged dominant social norms, its critical side nearly disappeared in later decades. Far from identifying psychoanalysis with sexual liberation, rebels in the 1960s—if they thought about analysis at all—saw it as reinforcing repressive sexual and social roles. For this generation, dynamic psychiatry had abandoned the radical qualities that had comprised much of its former allure. They viewed analysis as a technique used to enforce conformity to traditional social norms. Moreover, psychoanalysis had traditionally been identified with Jewish clients and practitioners. As Jews became increasingly assimilated within American society, dynamic approaches lost much of their appeal to them.[16]

Dynamic psychiatry was incompatible with the values of the group that came of age in the 1960s. Young people at the time were most interested in exploring alternative lifestyles oriented around sexuality, drugs, rock music, and radical politics. The 1960s generation viewed analysts as upholders of discredited and conventional familial, gender, and erotic roles. They identified psychiatry with promoting oppressive family structures, retrograde attitudes toward sexuality, and repressive lifestyles. Far from its early attraction to rebels, sexual libertines, bohemians, and other unconventional groups during its formative period, psychoanalysis seemed to be a bastion of traditional values.

In contrast, the anti-psychiatrists' message was attuned to the anti-authority ethos of the time, and it found a receptive audience among many college students and intellectuals. The anti-psychiatric focus of the 1960s generation took several forms.[17] One was resistance to all forms of established authority including military, governmental, and corporate—as well as psychiatric—institutions. Another resonant theme was that social definitions of "normal" often represented acceptance of the militarism, racism, and economic oppression that they felt characterized mainstream society. In contrast, madness could often be an appropriate response to insane social conditions. For many young people, society itself was responsible for making individuals crazy. Psychologist Abraham Maslow captured this aspect of the times when he asked:

> Adjusted to what? To a bad culture? To a dominating parent? What shall we think of a well-adjusted slave? . . . It seems quite clear that personality problems may sometimes be loud protests against the crushing of one's psychological bones, of one's true inner nature. What is sick then is *not* to protest while this crime is being committed.[18]

Far from their original appeal as heralds of a liberating creed, psychiatrists came to be viewed as agents that upheld outmoded social norms.[19] The counterculture not only rejected what psychiatry viewed as "normal" but also often celebrated what it defined as "abnormal."

During the 1970s, the cultural climate began to turn away from the goal of self-liberation that marked the previous decade to a more group-focused emphasis on how gender, sexuality, and race shaped identity. With some individual exceptions, the feminist, gay liberation, and Black power movements that embodied the new collective orientation were either apathetic toward or actively hostile to dynamic psychiatry. The attacks of second-wave feminists were especially harsh.[20] Prominent critics including Betty Friedan, Shulamith Firestone, Phyllis Chesler, and Germaine Greer demonized Freud and later analysts as patriarchal misogynists who blamed mothers for causing their children's mental illnesses. From being the object of uncritical hero worship during the 1930s, 1940s, and 1950s, Freud's reputation sharply plunged, and he was treated with derision. One leading feminist, Kate Millett, asserted that Freud was "beyond question the strongest individual counter-revolutionary force in the ideology of sexual politics."[21]

The definition and treatment of homosexuality proved to be especially embarrassing for psychiatry. Contrary to the general thrust of Freud's own view, which did not regard homosexuality as an illness, mainstream analysts had come to consider it as abnormal. Psychiatrists associated homosexuality with the narcissistic failure to resolve "the mother-child symbiosis that precedes the Oedipal period."[22] The DSM-II used it as the primary example of "sexual deviation," characterizing it as follows:

> This category is for individuals whose sexual interests are directed primarily toward objects other than people of the opposite sex, toward sexual acts not usually associated with coitus, or toward coitus performed under bizarre circumstances. Even though many find their practices distasteful, they remain unable to substitute normal sexual behavior for them.[23]

Although many psychiatrists felt that treating homosexuals as mentally ill was far more humane than punishing them as criminals, a number of them actively strove to convert gay people to heterosexuality, sometimes to the point of using electroshock and other extreme therapies. A newly energized

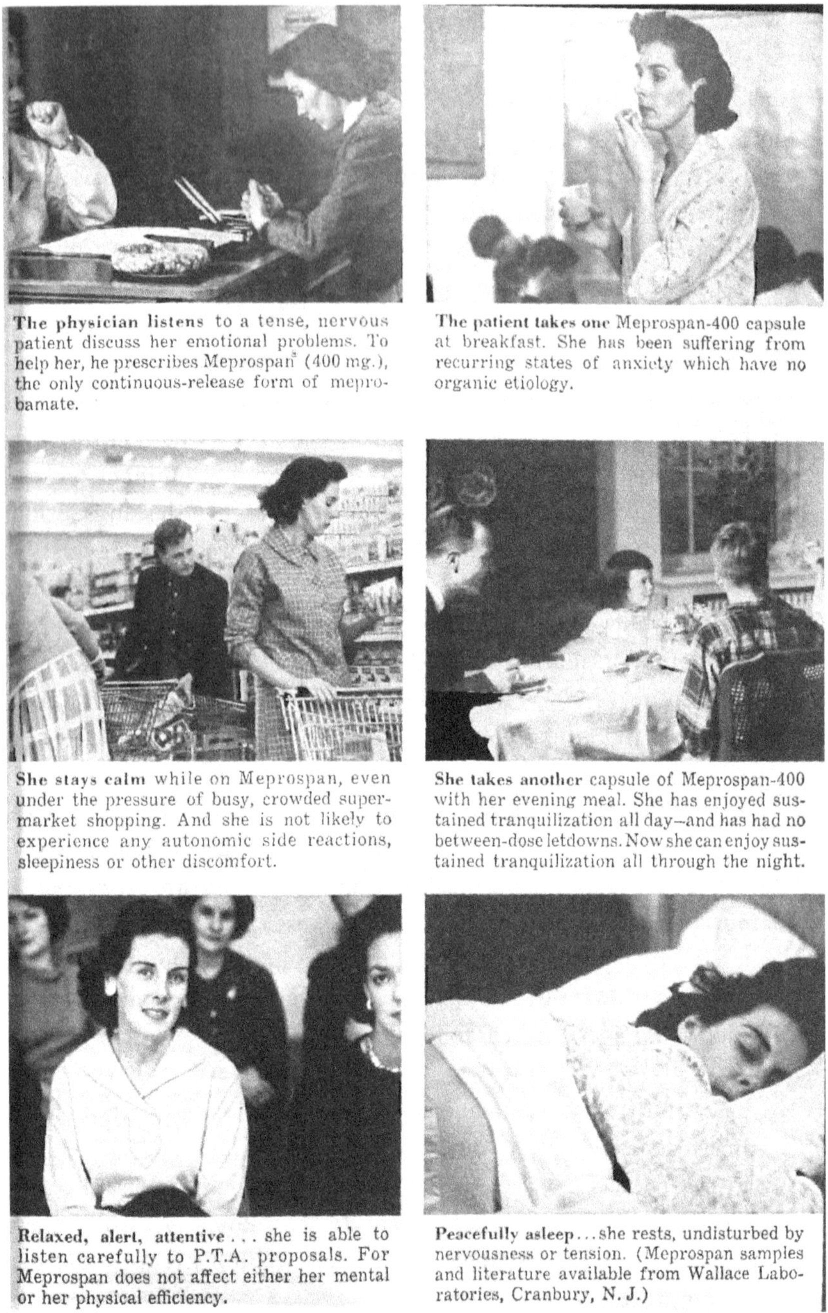


**Figure 6.1** Masked psychiatrist John Fryer, posing as Dr. Anonymous at the 1973 meeting of the American Psychiatric Association, was a leader of the protest that led to the removal of the homosexuality diagnosis from the DSM-II. Photograph taken by Kay Tobin Lahusen. New York Public Library Digital Gallery Image ID: 1606144.

gay liberation movement targeted the American Psychiatric Association (APA) for an attack that culminated in a very public anti-psychiatry demonstration at the group's annual meeting in 1970. Three years later, the organization voted to remove homosexuality from the diagnostic manual, although it retained a category of "sexual orientation disturbance" that was limited to people who were uncomfortable with their sexual identity and who desired psychiatric help.[24]

The highly visible controversy over homosexuality seemed to demonstrate several things about psychiatry and its definitions of mental illness. It shone a bright light on the extent to which psychiatric diagnoses were riddled with value-laden judgments over what kinds of sexual activities were normal or disordered. Further, it made psychiatry appear to be a stronghold of conventional behavior at a time when the youth culture was promoting the exploration of a wide variety of sexual identities. Perhaps most embarrassing, using a vote to determine what was or was not a mental illness seemed to show the highly contingent nature of psychiatric diagnosis itself. The humiliating storm over the nature of homosexuality drove the prestige of the psychiatric profession to an even lower point. The situation was so dire that the APA warned its members, "Our profession has been brought to the edge of extinction."[25]

The 1960s generation had little use for psychiatrists who they perceived upheld, rather than challenged, conventional social norms, especially sexual norms. For many people, the debacle over whether homosexuality was a mental illness exemplified how irrelevant psychiatry had become in deciding what appropriate or inappropriate behaviors were. Terms such as "abnormal," "deviant," or "sick" came to be applied to society, not to individuals. The very subject matter of psychiatry—mental illness—was called into question. If mental illnesses were myths, the existence of the psychiatric profession itself would be in peril.

Dynamic psychiatry, which had been such a central part of American culture since World War II ended, had lost its general appeal. In 1966, Leo Rangell, the president of the American Psychoanalytic Society, bemoaned a "change in the hospitality, ranging up to sharp hostility, in the scientific and intellectual community, in medicine and in the public press."[26] Its troubles were not confined to the anti-psychiatrists and the counterculture but also penetrated the psychiatric profession itself.

## Mental Illnesses Lose Medical Authority

At the same time as dynamic psychiatry lost its claim as a credible source of definitions of normality and abnormality, it also came to lack standing as a branch of medicine. The anti-psychiatric critique that rejected any resemblance between mental and physical illnesses was widely accepted. In contrast to bodily functioning, mental illnesses belonged to the world of meanings, thoughts, and actions. They were identified through deviations from social norms and values, which were arbitrary and culturally determined. There was, that is, nothing medical about mental illnesses or even any reason to call them "illnesses" at all.

Just as the authority of dynamic psychiatry was fading, the social context of psychiatric practice was also dramatically changing in ways that challenged the viability of its conceptions of mental illness. The dynamic system of classification was becoming less and less satisfactory within the new professional, regulatory, and political environments that psychiatry faced. As it passed through the 1970s, the field needed to find some way to call the conditions it treated "diseases" or "disorders" rather than more amorphous "reactions" to psychosocial circumstances. Immense external and internal pressures forced the psychiatric profession to overthrow the dynamic model and adopt a more medically relevant system.

### Competing Therapies

The growing challenge from non-psychiatric professional competitors was a central factor contributing to psychiatry's crisis of legitimacy during the 1970s. Dynamic psychiatrists had successfully promoted psychotherapy as a treatment for a wide variety of psychic disturbances that fell somewhere in between sanity and madness, but in doing so they also spawned their own competition. There was nothing explicitly *medical* about dynamic psychiatry. Medical training seemed irrelevant for understanding processes such as repression, childhood sexuality, and the symbolic interpretation of symptoms that were at the heart of analytic treatment. Its therapy primarily consisted of talking to patients, a technique that non-medical professionals were equally able to learn and practice. If psychiatrists were unable to claim that the conditions they treated were genuine diseases, they had no more medical legitimacy than the many non-medical clinicians who had entered the profitable and growing psychotherapeutic marketplace.

Psychiatry had to defend itself from rival practitioners who were making inroads into its most lucrative clientele: college-educated, middle- and upper-income patients. By 1980, the National Medical Care Utilization and Expenditure Survey reported that there were 28,000 psychiatrists, 50,000 psychologists, and 300,000 social workers, with the latter two groups having increased their ranks by 700% since 1950.[27] Psychologists were providing as much outpatient treatment as psychiatrists, each profession supplying about one-third of the total, while social workers and primary care physicians provided another third. Psychiatrists' use of talk therapy might help many patients with mild mental illness, general anxiety, or mood problems. But, critics charged, clinical psychologists, social workers, marriage and family counselors, and others could employ this type of therapy more cheaply but with the same degree of success.[28]

Lacking medical justification, there were few reasons for patients to patronize dynamic psychiatrists rather than consult alternative practitioners. "Apart from their training in medicine," Harvard psychiatrist Thomas Hackett worried, "psychiatrists have nothing unique to offer that cannot be provided by psychologists, the clergy, or lay psychotherapists."[29] Psychologists, in particular, contested psychiatric authority over the practice of psychotherapy.[30] Over the course of the 20th century they had developed new cognitive and behavioral models of treating the various neuroses. Their techniques were time-limited, easy to apply, and, at least in the short run, highly effective. The emergence of cognitive/behavioral therapies provided a major alternative to the dynamic concepts that dominated psychiatric models and practice.[31]

Making matters worse, a new wave of therapies that stood apart from any professional group had emerged as part of the 1960s counterculture. The decade spawned a potpourri of various treatments, many driven by opposition to science and reason, including encounter groups, meditation, gestalt, primal scream, transcendentalism, and numerous sexual therapies. These practices loosely promoted goals of self-discovery, being in touch with one's feelings, and living in the present moment. Gurus such as Alan Watts, Fritz Perls, Baba Ram Dass, and Werner Erhart became prominent alternatives to Freud and his heirs.[32]

In the 1970s, popular self-help manuals also became a booming business. The enormous sales of pop psychology books such as Thomas Harris' *I'm OK, You're OK* and Eric Berne's *Games People Play* emphasized the need for people to change themselves and their unhealthy behavior patterns.[33] Individuals were capable of altering their own dysfunctional behaviors without professional assistance from psychiatrists or other professionals. The flourishing of self-help therapies added to the crisis of dynamic psychiatry.

## Third-Party Payment

An entirely different line of attack on psychiatric practice stemmed from insurance companies that bemoaned the field's lack of financial accountability and demonstrated effectiveness. For the first half of the 20th century, most clients of dynamic psychiatry paid for their therapy as an out-of-pocket expense so that clinicians were not generally answerable to third parties. This situation began to change in the 1960s, when many medical insurance plans started to include psychotherapy as a partially reimbursable expense. Between 1965 and 1980, the number of psychiatric patients with private insurance coverage nearly doubled, growing from 38% to 68%.[34] During the 1970s, the federal Medicaid program joined private insurance coverage as a prominent source of payment for psychotherapy.[35]

The economic basis of the therapeutic relationship was no longer solely between therapists and their clients but involved private and public third-party payers. This situation contributed to pressures to change the dynamic model: the lack of specific diagnoses and the symbolic mechanisms of dynamic psychiatry did not fit an insurance logic that would only pay for the treatment of discrete disorders. Third-party payers required not only that clinicians deal with genuine diseases but also that they provide some sort of accountability for the outcomes of their treatments. While psychiatrists struggled in their efforts to pass insurance and governmental tests of cost-benefit analysis, researchers conducted empirical studies that cast doubt on the long-term efficacy of psychotherapies.[36] These techniques were consuming larger amounts of total health care spending but lacked persuasive scientific evidence that they worked effectively.

Insurance companies and the government, both of which paid large amounts of money for mental health services, took note of the problems of psychotherapy, psychoanalysis in particular, that effectiveness research revealed. They viewed psychotherapies as a financial bottomless pit; patients could spend years in therapy, requiring potentially uncontrollable resources. The private and public institutions that were paying for therapy were becoming increasingly skeptical about psychiatry's legitimacy. Blue Cross Vice-President Robert J. Laur's summary was representative:

> Compared to other types of [medical] services there is less clarity and uniformity of terminology concerning mental diagnoses, treatment modalities, and types of facilities providing care. . . . One dimension of this problem arises from the latent or private nature of many services; only the patient and the therapist have direct knowledge of what services were provided and why.[37]

The continued willingness of both insurance companies and the federal government to reimburse mental health treatment became contingent on demonstrations that psychiatric treatments were effective, financially accountable, and, most of all, directed at real disorders. In 1980, the Senate Finance Committee proposed limiting government support of mental treatment to therapies that the FDA judged to be "safe and effective on the basis of controlled clinical studies which are conducted and evaluated under generally accepted principles of scientific research."[38]

Psychiatrists faced a major professional dilemma. Third-party payers wanted answers to basic questions: Did patients in psychotherapy have a medical illness? Was psychotherapy cost-effective compared to alternative treatment methods? Were its lengthy treatments necessary or even helpful?[39] If psychiatrists could not demonstrate that their talk therapies produced superior results to those of psychologists, social workers, and counselors, who could undercut them in price, they would have to define psychiatry's exclusive contributions and jurisdiction in other ways, especially through focusing on drug-related therapies that other professions were unable to apply.

## Drug Companies Need Diseases

Drug companies as well as psychiatrists came under intense pressure to show medical legitimacy. During the 1950s and 1960s, the popularity of the minor tranquilizers stemmed from their use with general life stresses and protean distress conditions, with little consideration of whether or not they treated explicit disease states. A review at the time concluded that "only about 30% of use is in identified mental disorders and the remainder covers the rest of medicine."[40] Accordingly, pharmaceutical companies presented the minor tranquilizers to physicians and psychiatrists as drugs that treated a variety of nonspecific complaints, including anxiety, tension, depression, and psychic stress that resulted from common social problems.[41]

After the spectacular success of the tranquilizing drugs in the 1950s and 1960s, however, the industry faced serious problems during the 1970s. At this time, government regulators began to enforce more stringently the legislative requirement dating from 1962 that drug companies could only market their products for particular biomedical conditions.[42] The FDA began to question the broad marketing techniques of drug companies and insisted that they provide proof that their products were efficacious treatments for specific mental disorders. In 1971 FDA Commissioner Charles Edwards sent a letter to the sponsors of all new psychotropic drug applications that stated: "These drugs are approved for the treatment of clinically significant anxiety, depression, and/or other mental conditions, but, of late, advertisements have also promoted their use in the treatment of symptoms arising from the stresses of everyday living, which cannot properly be defined as pathological."[43] Increasing regulatory pressure forced drug companies to stop touting their products as remedies for general problems in living and ordinary distress. Instead, they had to show that they targeted discrete mental disorders.

A sharp reaction against the tranquilizing drugs developed in the general culture as well as among federal regulators. Feminist groups, in particular, became vocal opponents of their perceived use in controlling women's lives as well as their addicting potential. Leaders of this movement such as Betty Friedan singled out the tranquilizers as functioning to resign housewives to their subordinate status. Activist groups began to mount legal challenges against the benzodiazepines. Their lawsuits, which claimed that these drugs were highly addictive, drew considerable media attention. At the same time, feminist and patient advocacy groups launched public relations campaigns featuring messages about the dangers of benzodiazepine use. News stories trumpeted reports about addiction among both ordinary women and celebrities including First Lady Betty Ford. Several well-publicized Congressional hearings reinforced the impression of these drugs' addictive and suicidal potential.[44]

As a result of these pressures, in 1975 the FDA classified Librium and Valium as Schedule IV drugs, which limited the number of prescription refills, required strict reporting, mandated insertion of a statement that the drugs could be abused, provided strong criminal sanctions for illegal sales, and ordered the inclusion of an information packet that stated: "Anxiety or tension associated with the stress of everyday life usually does not require treatment with an anxiolytic (antianxiety) drug." A sharp plunge in sales resulted: prescriptions for tranquilizers plummeted from 103.2 million in

1975 to 71.4 million in 1980.[45] The marketers' dream drugs had become nightmares. Drug companies had to search for new ways to market their products as remedies for specific diseases, not for the relief of general psychosocial stresses.

## A Crisis in State Support

The deinstitutionalization of mental patients that began in the mid-1950s and accelerated during the 1960s deepened psychiatry's crisis. From a peak of 559,000 individuals in 1955, the number of institutionalized patients declined modestly to around 475,000 a decade later, a decrease of 15%.[46] At this point, the states greatly accelerated the discharge of severely and persistently mentally ill patients from public mental asylums; the number of patients plunged to 138,000 in 1980, a decline of almost 60%.[47] While 77% of mental patients were institutionalized in 1955, just 8% were inpatients by 1975.[48] Moreover, this shrinking group was spending far less time in hospitals. Historian Gerald Grob observes: "Before 1965 many patients spent years, if not decades, in asylums, after 1970 length-of-stays began to be measured in days or weeks."[49]

A variety of forces contributed to what eventually became the virtual abandonment of state mental hospitals. One factor was the introduction of chlorpromazine, an antipsychotic medication, in 1954. This drug calmed agitated patients and controlled the most serious symptoms of psychosis and so facilitated their release to community settings. While chlorpromazine helped enable deinstitutionalization, the federal government's introduction of new public policies, especially Medicare and Medicaid in 1965, had an even greater impact on the pace of this process.[50] These programs did not pay for institutional treatment and so incentivized states to discharge patients from the mental hospitals they funded into community programs that could receive federal payments. In addition, civil rights advocates, many influenced by Szasz's arguments, successfully argued for expanding legal protections against involuntary commitments to mental institutions. Consequently, entry into these settings became considerably more difficult. The cumulative effects of new psychotropic drugs, federal policies, and legal efforts were remarkably effective in emptying public inpatient facilities.

At the same time that hundreds of thousands of former patients were entering the community, the Community Mental Health Centers (CMHCs) that were at the center of the National Institute of Mental Health's (NIMH) efforts during the 1960s pursued a broad mandate focusing on a combination of outpatient care for mildly disturbed individuals combined with ambitious psychosocial programs. The NIMH had developed CMHCs as alternatives to state-run mental hospitals, but, in fact, they rarely provided services to the former patients with serious mental illnesses who were exiting from mental institutions. This group only comprised from about 4% to 6% of all CMHC patients. Instead, the new community facilities were helping people who might have some neurosis or problem of living but rarely treated those with serious mental illnesses. The largest categories of CMHC patients had "social maladjustment or no mental disorder" (22%) and "neuroses and personality disorders" (21%).[51] The community treatment programs the NIMH promoted were not reaching those who most needed them.[52]

Psychiatry was unable to fill the gap in treatment that the emptying of mental hospitals created. Dynamic psychiatrists, following Freud's example, typically refrained from treating psychotic patients. They had largely established themselves in outpatient practices that catered to urban, cosmopolitan intellectuals amenable to lengthy and expensive analytic treatments. Consequently, they were ill suited to deal with the conditions of schizophrenia, degenerative brain disorders, alcoholism, or drug addiction common among formerly institutionalized patients.[53] Most important, their office-based therapies had no relevance for the housing, food, jobs, and need for social interaction that former patients required in the community. The necessity for these services was especially acute because many discharged patients had no family members who were willing to take care of them, provide them with places to live, and insure they took their medications.

Many deinstitutionalized patients ended up on the streets or in the criminal justice system. They joined a new demographic group of young adults with severe and chronic mental illnesses who had never been institutionalized but who drifted in and out of medical emergency rooms, short-term psychiatric wards, and correctional facilities.[54] This new cohort of younger mentally ill also included many persons with substance abuse and alcohol problems, which made them especially unattractive and frustrating clients for mental health professionals to treat. The dynamic model that psychiatry had become identified with was spectacularly unequipped to respond to the major social problems that mental illness created in the era of deinstitutionalization.

If the realm of psychiatry truly was "a mission to change the world" as one president of the APA bemoaned, the members of Congress who had to approve the NIMH budget were not going to finance this mission.[55] Their political and economic threats compounded the pressures psychiatrists faced from professional competitors, private and public insurance providers, and the FDA's questioning of the lack of particular biomedical targets for popular drug treatments. Exacerbating the crisis that dynamic psychiatrists faced, several developments within psychiatry and medicine were making the field's amorphous conditions increasingly untenable.

## Medical Opposition to Dynamic Psychiatry

At the same time as psychiatry faced growing pressure from professional competitors, third-party payers, drug companies, and the fallout from deinstitutionalization, severe internal fissures also arose. The dominance of dynamic psychiatry in the United States during the half century between 1920 and 1970 was tied to its fierce identification with the medical profession.[56] Freud himself was often antagonistic to medicine and saw no reason for analysis to be limited to medically trained physicians. In the United States, however, dynamic psychiatrists justified their system of knowledge and protected themselves against professional competitors such as psychologists and social workers through their status as physicians. The close identification with medicine that helped raise the field's prestige during the first half of the 20th century worked to undermine its position in the 1960s and 1970s.

Changing medical norms hoisted dynamic psychiatrists by their own petard. By the 1960s, mainstream medicine had come to rely on a scientific model that used precisely defined disease entities, large statistical studies, and double-blind placebo trials. It emphasized objective knowledge over clinical intuition, the study of groups over individual cases, and demonstrations of efficacy according to scientific methods over claims of insight. Biomedically oriented doctors did not view dynamic psychiatrists as real physicians. They dismissed the case studies that provided the core empirical base of dynamic psychiatry as anecdotal and unscientific. Indeed, the symbolic, private, and interior essence of the dynamic therapeutic approach was in many ways the opposite of the objective, public, and reproducible emphasis of the classical scientific method.

The inability to define and measure the entities they studied became a particular thorn in the side of dynamic psychiatry. Because the DSM-I and DSM-II did not provide specific definitions for each of their diagnoses, psychiatrists had to rely on their clinical judgment to assess how well each patient fit a particular category. This led to great disparities in the application of diagnostic labels. For example, the well-known U.S.-U.K. Diagnostic Project, the results of which were published in 1972, examined the ways that psychiatrists in these two countries diagnosed mental disorders. The study demonstrated an alarming lack of agreement between American and British psychiatrists as well as among psychiatrists within each group. For example, after viewing videotapes of two English and American patients, American psychiatrists made diagnoses of schizophrenia in 85% and 69% of cases compared to only 7% and 2% of British psychiatrists. In contrast, British clinicians were far more likely than their American counterparts to diagnose manic depression.[57]

The U.S.-U.K. study was the most prominent example of research that showed a remarkable lack of diagnostic agreement in cases in which psychiatrists received the same information.[58] These studies challenged the reliability not only of distinguishing closely related diagnostic categories, such as one affective disorder from another, but also of differentiating between larger categories, such as affective versus anxiety disorders or psychosis versus neurosis. The threat of such gross unreliability and, by implication, invalidity was not only an acute embarrassment to clinical expertise but also a challenge to the scientific status of psychiatry. It was becoming more and more clear that psychiatry was in desperate need of a reliable classification of diagnoses that would replace the amorphous categories of the DSM-II.

## Biological Psychiatry

Attacks on the dominant psychodynamic paradigm from inside the profession exacerbated the crisis in psychiatry. Disenchanted psychiatrists who sought to bring the field into the mainstream of scientific inquiry led a powerful new movement that was based on the traditional biomedical model. They embraced the rapidly developing fields of biological psychiatry and psychopharmacology, neither of which had much in common with dynamic theory or practice. Biological psychiatrists looked to brains, not to minds, as ways of uncovering the roots of mental disorders. Psychopharmacologists strove to target drugs toward specific and distinguishable conditions not for general states such as "neuroses" or "stress."[59]

Biological psychiatrists, who disparaged the analysts' focus on the unconscious, lack of empirical rigor, and neglect of the brain, were becoming increasingly prominent within the profession. Their dynamic counterparts gave short shrift to traditional systematic research, instead relying on a combination of established theory and insights gained from their clinical encounters. In contrast, the emergent research-oriented group scorned the analysts' casual view of measurement and insisted that psychiatry could only obtain legitimacy through becoming a thoroughly scientific discipline. Toward this end, they advocated for expanding scientific research on mental disorders, studying the efficacy of various treatments, clearly demarcating different mental disorders, and increasing diagnostic reliability among clinicians. Perhaps most importantly, the critics contended, when specific etiologies of mental disorders could not be empirically ascertained, they should not be attributed to any cause at all (and especially not to psychoanalytic concepts such as "libido" or "oedipal conflicts"). Instead, diagnostic systems should focus on the observable symptoms of each disorder.[60]

A new somatic paradigm accompanied the upsurge of biomedically oriented research. Psychiatrist Joseph Schildkraut's (1934–2006) catecholamine hypothesis of affective disorders, which proposed that depression might be associated with an absolute or relative deficiency of particular neurotransmitters, was the best-known theoretical work. This article, which was published in the *American Journal of Psychiatry* in 1965, "captured the imagination of the field and helped us to understand the biology of psychiatric disorders."[61] Schildkraut's work, which became the most frequently cited article in the history of this journal, served as a model for the exploration of the neurobiological grounding of mental disorders.[62] "There are now substantial indications that serious mental illnesses derive from chemical, rather than psychological, imbalances," prominent biological psychiatrist Seymour Kety proclaimed eleven years later.[63]

The changing view of psychopharmacology provided the second pillar for the new biological psychiatry. In the 1950s and 1960s the capacious array of problems that the minor tranquilizers treated was in accord with the dynamic perspective. Its nonspecific view of diagnosis did not handicap clinicians, who prescribed these drugs promiscuously across the whole range of neurotic and other problematic conditions. Over time, however, the growing popularity of drug treatments both facilitated the fall of analysis and propelled psychiatry's movement toward a far more biological model.

Although analysts often employed drugs in their practices, the use of medication was difficult to reconcile with their core theoretical tenets.[64] One reason was that the somatic basis of drug treatments did not accord with the psychic principles of dynamic thought and practice. Another was that the growing reliance on drugs conflicted with the dynamic belief that anxiety was therapeutically valuable because it motivated patients to explore their repressed thoughts and experiences. Yet, the tranquilizers reduced the amount of experienced anxiety so that chemical suppression could discourage the search for the underlying causes of neuroses.

By the late 1970s, psychiatrists who did not employ drugs were exceptions; even two-thirds of psychoanalysts were prescribing medications.[65] As the psychiatric profession came to organize itself around drug treatments, the influence of Freudian views of neuroses became marginalized. Analytic treatments also became a luxury: why should people undergo long, expensive, and painstaking periods of analysis that had uncertain results when a pill could provide more immediate and effectual relief for symptoms? Psychoanalysts could not match the ease and efficiency of medication-based therapies. Ultimately, psychoanalysis seemed to be a thoroughly impractical mode of treatment.

Research on psychotropic drugs was also entering a new stage. In the 1950s psychopharmacologists fortuitously discovered a host of targeted drugs: chlorpromazine for schizophrenia, lithium for bipolar conditions, and monoamine oxidase inhibitors and imipramine for depression. All of these medications stemmed from accidental findings: none were grounded in any neurological theory. By the following decade, drug research was centered on deliberate attempts to develop drugs for specific mental disorders.

The trend toward specificity in drug development, testing, and prescribing was far more compatible with Kraepelinian than dynamic ways of thinking about mental illnesses. Scientifically oriented psychiatrists called for new, stringent standards for demonstrating the effectiveness of treatments. They advocated for applying to psychotherapy the same principles the FDA used to test the efficacy of drugs: quantitative and comparative studies based on matched samples of patients who were uniformly diagnosed, randomly assigned, and treated with standardized procedures. Impartial observers who not involved in the treatment should judge the outcomes of each procedure.[66] The incompatibility of dynamic psychiatry with scientific norms was increasingly problematic not just for biologically oriented psychiatrists but also for medical schools that formed the institutional base of the profession.

## Medical Schools Reject Dynamic Psychiatry

Freud did not believe that analysts must be physicians and urged psychoanalytic institutes to admit non-physicians into their programs: "As long as I live I shall balk at having psychoanalysis swallowed by medicine."[67] Unlike Freud, however, American analysts insisted that practitioners must have medical degrees; their training institutes did not admit anyone who did not possess this credential. Since 1923 training programs affiliated with the American Psychoanalytic Association required all candidates to be physicians. Trainees had to first obtain a medical degree and then enter a specialized analytic institute to undergo an analysis themselves. This onerous process took an average of between six to eight years after graduation from a medical school.[68]

Psychoanalytic institutes were divorced from medical schools and employed decidedly non-medical educational techniques. Analytic training had far less to do with teaching a specific body of verified scientific knowledge than with immersing trainees in nonreplicable, intersubjective processes. At its core was a long-term, in-depth analysis conducted by an initiated practitioner, a procedure that more closely resembled trial periods for priests than for physicians.[69] Psychoanalyst Michael Balint likened training analyses to a "primitive initiation ceremony."[70] Another analyst, Abram Kardiner, noted: "For those who are not analysts, this description [of psychoanalysis] can only be compared to a religious dogma. And a source that blocks its own growth tends to resemble a cult."[71] Freud understood the appearance that this requirement gave, stating apologetically in 1933:

> You can believe me when I tell you that we do not enjoy giving an impression of being members of a secret society and of practicing a mystical science. Yet we have been obliged to recognize and express as our conviction that no one has a right to join in a discussion of psychoanalysis who has not had particular experiences which can only be attained by being analyzed oneself.[72]

Nevertheless, their decidedly non-medical model did not prevent psychoanalysts from enjoying great prestige and status through the middle decades of the 20th century.[73]

American analysts faced the fundamental quandary of insisting that psychoanalysts must possess medical degrees that were necessary to insure their legitimacy but that were essentially irrelevant for their practices. Dynamic theory and treatment were infused with holistic, psychosocial conceptions of mental illness that had little in common with specific, biologically grounded, organic conditions. Indeed, the sort of rote learning typically found in medical education was exactly the wrong sort of preparation for analysts who needed to interpret the idiosyncratic problems of their clients.[74] From the 1920s onward, the basic irony of American psychoanalysis was to require training that was thoroughly distinct from its basic tenets.

Beginning in the early 1960s and intensifying in the following decade, concerns over the unscientific nature of dynamic psychiatry arose in medical schools.[75] Within medicine, dynamic psychiatry suffered from its inability to precisely define the phenomena it studied, its indistinct methods, and its rejection of somatic treatments in favor of talk therapies. The lack of a body of biomedical conditions was an unsustainable situation for the supposedly medical discipline of psychiatry. Medical schools came to reject analytic approaches, which became marginal to academic research and teaching.

While dynamically oriented psychiatrists had dominated departments of psychiatry in the 1950s, by the 1970s dynamic advocates were largely exiled from these departments.[76] Academic psychiatrists ridiculed psychoanalysis and held its practitioners in contempt.[77] As their authority faded in the academy, the number of psychiatrists identifying with psychoanalysis fell rapidly. The significant decline in the number of medical students entering psychiatry was especially worrisome: while about 10% of medical students entered psychiatry during the 1960s, by the end of the 1970s this proportion had fallen to less than 3%.[78]

Once their institutional base in medical schools crumbled and the number of new psychiatrists drastically shrunk, analytic institutes had to rethink their admittance policies that excluded all aspirants who lacked medical credentials. They increasingly became open to all types of clinicians, not just physicians. By the 1980s, new analysts were primarily non-medically trained clinicians, especially psychologists and social workers. The fall of psychoanalysis from its heady postwar period within medicine was complete.[79]

Freud and the theories he developed and inspired were the towering influences on the study of mental illness for much of the mid-20th century. Subsequently, the refusal of psychoanalysts to shed their insularity, to accept scientific norms, and to adapt to the biological turn in psychiatry resulted in their professional marginalization. By 1990, psychoanalysis was virtually extinct within psychiatry and just one of numerous subcultures within the mental health professions more generally.

## The Discrediting of the Psychosocial Model

For much of the postwar period psychiatrists successfully promoted a wide variety of social and psychic causes of mental illness. Most notably, the NIMH upheld and stimulated the view that a potpourri of psychological and environmental circumstances led to mental disturbances. The agency provided generous funding for social scientists and psychiatrists to pursue psychosocial research. During the 1970s, the dynamic view of what factors caused mental illness came under siege from two different directions.

On one side, anti-psychiatrists derided dynamic views, insisting that there was no point in looking for the causes of a condition that was a social construction. Social factors played a huge role in singling out who was labeled as "mentally ill," but they could not produce a condition that did not exist. On the other side, biological psychiatrists attacked analysts for failing to accept a somatic basis for mental illnesses. The psychosocial approach accepted virtually all causes of mental illness, except for biological ones. The etiologically oriented DSM-I and DSM-II segregated organic illnesses from all others, leaving them in the domain of asylum psychiatrists who themselves were isolated from the rest of the profession. At the same time, biological psychiatrists were developing new theories of mental illnesses as brain-based states. The practice of psychiatry, too, was turning more and more toward the use of drugs and away from talk therapies. As biological frameworks for understanding human behavior became more culturally acceptable and professionally necessary, dynamic psychiatry's incompatibility with neurobiological causes became a serious liability.

By the 1970s the NIMH had become disenchanted with psychosocial approaches to mental illness. At the same time as this agency was implementing policies oriented toward community treatment, it was facing serious political problems in dealing with Congress and the executive branch of government. Its environmental focus during the 1950s and 1960s boomeranged when a more conservative Republican administration assumed power in the late 1960s. From 1965 through 1972, funding for the NIMH decreased by 5% each year.[80] After a decade of prolonged struggle between the supporters and opponents of the NIMH's sweeping agenda, Ronald Reagan's ascendency to the presidency in 1980 led to the demise of the agency's focus on the psychosocial study of mental illness. The NIMH repudiated its now toxic focus on social problems and turned its attention and funds toward establishing a biomedical, diagnostically oriented, and far more politically acceptable research agenda.[81] The NIMH became, in the words of its then-director, Herbert Pardes: "a more scientific institution focused on basic biology and behavioral science, major clinical disorders, diagnosis, treatment, and epidemiology."[82]

The psychosocial emphasis of the profession had become a millstone. Psychiatry was in a period of deep professional crisis. In 1976 the president of the APA, Alan Stone, concluded that social psychiatry and social activism were "carrying psychiatrists on a mission to change the world, [but] had brought the profession to the edge of extinction."[83] The field would soon shed its externally oriented focus in favor of a radically internal, brain-based approach.

## Conclusion

During the 1960s, many issues converged to place the dynamic paradigm in jeopardy. Some of these factors stemmed from professional pressures: psychiatry's marginal status within the medical profession, the growing influence of biologically oriented psychiatrists, the need of clinicians to obtain reimbursement from the increasingly doubtful private and public funders of therapy, and growing threats from nonphysicians such as clinical psychologists, counselors, and social workers. Others involved broader changes in the social environment: the anti-psychiatric attacks that discredited psychiatry in the general culture, the emergence of a youth counterculture that identified the profession with the social control of deviant behavior, the need of drug companies to target their products as treatments for specific disorders, and the problematic politics that surrounded the embattled NIMH. During the 1970s, growing concern over the cost-ineffectiveness of the field came to the fore. The convergence of these factors forced psychiatrists to change their conceptions of what mental disorders were and how best to treat them. Only a revolutionary transformation in the extant diagnostic system could accommodate a biomedical paradigm that had become the only legitimate way to pursue medical research and practice.

By the mid-1970s, the legacy of analysis within psychiatry was on life support. Whatever influence dynamic thought still possessed was found in isolated analytic institutes or academic humanities departments.[84] Yet, in a remarkably short period of time, psychiatry was able to resurrect itself under a new veneer that restored its influence within medicine, the federal government, and the culture at large. This reinvention centered on a new DSM that fundamentally redefined what mental disorders were and how they should be diagnosed and treated. Subsequently, the field would focus on brains, genes, and psychopharmacology.

## Acknowledgments

Portions of this chapter are adapted from Horwitz, 2002, and Mayes & Horwitz, 2005.

## Notes

1. Swiss psychiatrist Carl Jung, who believed that analysts could successfully treat schizophrenia and other psychoses, was the most prominent exception.
2. Szasz, 1991, 19.
3. Szasz, 1997, 12.
4. The biological revolution in psychiatry, medicine, and culture had no impact on Szasz's thought. A half century later, he was still making identical arguments (Szasz, 2011).
5. Szasz, 1970, 41.
6. Szasz, 1960, 62.
7. Quoted in Staub, 2011, 113.
8. Laing, 1970, 110.
9. Goffman, 1961.
10. Scheff, 1966.
11. Foucault, 1965.
12. In 1887, journalist Nelly Bly anticipated Rosenhan's study. Bly convinced psychiatrists that she was mentally ill and required hospitalization. She remained in the hospital for ten days and upon release wrote a well-known expose of her experiences and a popular book, *Ten Days in a Mad-House.*

13. Rosenhan, 1973, 250.
14. Rosenhan's article drew many scathing critiques including one from Robert Spitzer (1975).
15. The basic gestalt of the anti-psychiatrists persists at present in what is generally called the "recovery" model of service delivery. This client-led movement emphasizes the positive aspects of experiences that are usually called "mental disorders" and argues that consumers, rather than mental health professionals, should control treatment activities. See Davidson, 2013; Rose, 2019.
16. Shorter, 1997, 187.
17. Staub, 2011, 120.
18. Maslow, 1962, 8. Also quoted in Herman, 1995, 273.
19. Two exceptions were philosophers Herbert Marcuse (1955) and Norman O. Brown (1959) who argued that Freud's early work could be the basis for liberated lifestyles based on the release of libidinous sexual energy.
20. Tomes, 1994. The proportion of women, who had comprised nearly a third of analysts in the 1930s, had sharply declined to less than 10% in the 1950s (Zaretsky, 2004, 297).
21. Quoted in Ross, 2012, 186. A number of feminists who were more sympathetic to dynamic ideas including Juliet Mitchell, Nancy Chodorow, and Elisabeth Young-Bruehl became analysts during the 1970s and 1980s (Whitebook, 2017, 3).
22. Herzog, 2017, 75.
23. APA, 1968, 44.
24. Bayer, 1981.
25. Quoted in Lieberman, 2015, 115–16.
26. Quoted in Zaretski, 2004, 313.
27. U.S. Department of Health and Human Services, Agency for Health Care Policy and Research, 1980.
28. Greenberg, 1980.
29. Hackett, 1977, 434.
30. Buchanan, 2003.
31. E.g. Barber, 2008.
32. Gitlin, 1993, 424–27.
33. Staub, 2011, 154–55.
34. Hale, 1995, 341.
35. Mechanic, 1998.
36. Frank et al., 1994; Starr, 1982.
37. Wilson, 1993, 403.
38. Marshall, 1980, 35.
39. Hale, 1995.
40. Blackwell, 1973, 1638.
41. Herzberg, 2009.
42. Smith, 1985.
43. Quoted in Smith, 1985, 189.
44. Gabe & Bury, 1991; Gabe, 1990; Tone, 2009.
45. Tone, 2009, 203; Smith, 1985, 32.

46. Grob, 1995.
47. Hale, 1995.
48. Goldman et al., 1983.
49. Grob, 1994, 287.
50. Gronfein, 1985.
51. Torrey, 2014, 77.
52. Grob & Goldman, 2006.
53. Redlich & Kellert, 1978.
54. Grob, 1994.
55. Stone, quoted in Wilson, 1993, 402.
56. The next three paragraphs are adapted from Horwitz, 2002, 58–59.
57. Cooper et al., 1972.
58. E.g. Temerlin, 1968.
59. Guze, 1989; Healy, 1997.
60. Feighner et al., 1972.
61. http://www.nytimes.com/2006/07/08/us/08schildkraut.html. Interestingly, Schildkraut (1965) proposed that epinephrine, not serotonin, was the neurochemical related to depression.
62. Harrington, 2019, 198.
63. Quoted in Harrington, 2019, 165.
64. Scull, 2015, 384.
65. Grob, 1991a, 299.
66. Hale, 1995.
67. Quoted in Gay, 1988, 491. See especially Freud 1926/1990.
68. Malcolm, 1984, 151. Malcolm's pseudonymous subject, Aaron Green, spent fifteen years in his two training analyses.
69. Malcolm, 1981.
70. Quoted in Zaretsky, 2004, 293.
71. Kardiner, 1977, 121.
72. Freud, 1933/1965, 68.
73. The vast cultural influence of psychoanalysis belies the small number of analysts. During the heyday of psychoanalysis in the mid-1950s, less than a thousand analysts practiced in the United States (Hale, 1995, 289).
74. Rieff, 1966, 101. For Karl Menninger, probably the most prominent American psychiatrist during the post–World War II period, psychiatry had more in common with Judeo-Christian religious values than traditional medical concerns with disease: "Essentially, both are trying to do the same thing: to make people more comfortable and to save them from evil. . . . They both use verbal techniques and employ their own personalities to accomplish their results. They both recognize the value of confessional catharsis . . . For both psychiatrist and priest, love is the greatest thing in the world" (Quoted in Friedman, 1990, 123).
75. See especially Hale, 1995, Ch. 17.
76. Burnham, 2012, 6.
77. Nesse, 2019, 187.

78. Taylor, 2013, 114.
79. See especially Malcolm, 1984.
80. Wilson, 1993, 403.
81. Grob, 1991a.
82. Quoted in Harrington, 2019, 175.
83. Wilson, 1993, 402.
84. Ross, 2012, 188.

# 7
# Diagnostic Psychiatry

> There is a closer relation between the Medical Sciences and the conditions of Society and the general thought of the time, than would at first be suspected.
>
> —Oliver Wendell Holmes, *Currents and Counter-Currents in Medical Science* (1861, 7)

Psychiatry faced a major predicament as it entered the 1970s: it lacked the disease conditions that would provide the field with medical legitimacy. This difficulty was not limited to its want of credibility within medical schools. It also called its financial foundation into question as insurers questioned whether they were paying to treat real disorders. Moreover, the field's major source of support for training and research—the National Institute of Mental Health (NIMH)—was under increasing political pressure to turn away from its expansive psychosocial agenda and prove that it dealt with serious forms of mental disorder. The pharmaceutical industry, another leading backer of psychiatry's endeavors, confronted a federal mandate to demonstrate that its products targeted genuine diseases. Psychiatry, in short, needed to show that the problems it treated were comparable to the physical ailments found in other medical specialties.

It was becoming increasing clear that dynamic psychiatry, whose very essence was to divorce mental from organic conditions, could not provide the necessary diseases. This approach neglected the biological factors that were at the core of medicine, instead examining the particular life experiences leading mental illnesses to arise and persist. Analysts paid scant attention to overt symptoms but instead strove to uncover the deeper meanings that surface manifestations disguised. The diagnoses in the extant DSM-II were infused with untested and inferred psychosocial assumptions about their causes but provided meager information about how to measure each state. Analytic treatments, which featured intense interpersonal relationships between clinicians and their patients, did not lend themselves to double-blind, controlled studies.

The publication of DSM-III by the American Psychiatric Association (APA) in 1980 marked a thoroughgoing change in thinking about mental illness. In one stroke, psychiatry discarded one intellectual paradigm that had little concern with diagnosis and adopted an entirely new system of classification that imported a model from medicine where diagnosis is "the keystone of medical practice and clinical research."[1] The promulgators of the new DSM overthrew the broad, vague, and often etiologically oriented concepts that were embodied in the DSM-I and DSM-II, reclaiming the diagnostic tradition that dominated 19th-century psychiatry before Freud developed his transformative conceptions. The new manual assumed that different observable clusters of symptoms indicated distinct underlying diseases. Its fundamental principle was to define mental disorders through the use of symptom-based entities without reference to their causes.

I call this model "diagnostic psychiatry" because all of the major functions of the mental health professions—framing the causes behind, prognoses of, and treatments for mental illnesses—follow from their classification into discrete entities. The DSM-III created a powerful standardized system of diagnoses that reigned virtually unchallenged for the following three decades.[2] It allowed psychiatry to reorganize itself from a discipline where diagnosis played a marginal role to one where it was the basis of the specialty. Contrary to common beliefs, the DSM-III itself did not expand the range of mental abnormality or impose a biological causal system on mental disorder. Instead, its major impact was to convert the person-centered frame of dynamic psychiatry into the specific diseases that continue to dominate discourse regarding mental illness. The DSM-III revolutionized conceptions of mental disorder through transforming conditions that had been thought to be distinctively psychosocial into ones that were disease-like states.

## Creating a Medical Specialty

For most of its history perhaps the most central debate in psychiatry was the extent to which the field studies minds or brains.[3] The dynamic model left no doubt that its subject matter was psychic, not organic. Despite their insistence that psychoanalysts must possess medical degrees, dynamic psychiatrists had little interest in brain-related processes, which they were happy to leave to the dwindling group who practiced in asylums. They regarded personal experiences as inherently intra- and intersubjective and so without counterpart in physical medicine.

At the heart of dynamic thinking was the idea that mental illnesses were inseparable from the life histories and often idiosyncratic understandings of particular individuals. In contrast, the conception of disease specificity is at the center of medical thinking. Historian Charles Rosenberg explains:

> This modern history of diagnosis is inextricably related to disease specificity, to the notion that diseases can and should be thought of as entities existing outside the unique manifestations of illness in particular men and women: during the past century especially, diagnosis, prognosis, and treatment have been linked ever more tightly to specific, agreed-upon disease categories.[4]

The view that brain-based diseases can be abstracted from their individual and social contexts and studied as objects that have distinct causes, courses, and responses was not only different from but was diametrically opposite to the basic tenets of dynamic psychiatry.

The DSM provided psychiatry the credentials it required to become a legitimate medical field. "Psychiatry is a branch of medicine," Gerald Klerman, one of the highest-ranking psychiatrists in the federal government at the time, unambiguously proclaimed.[5] The major entities found in the manual—for example, major depressive disorder (MDD), the various anxiety disorders, the personality disorders, eating disorders, post-traumatic stress disorder (PTSD), and the like—are now consensually recognized as mental illnesses. It contains the criteria that virtually all medical and other clinical students learn, that mental health professionals use to diagnose patients, and that serve as the basis of psychiatric research. "[The] vast majority of US psychiatric trainees, clinicians, and researchers, now, could hardly imagine any other way to approach psychiatric diagnosis," psychiatrist Kenneth Kendler observes.[6] Moreover, the DSM provides the official guidelines defining mental illness for social institutions including insurance companies, government funding agencies, courts, schools and universities, drug companies, and the military—in short, every group concerned with the subject. They also form the basis for the financing of mental health care, hospital and clinic records, and research funding.

Diagnostic psychiatry's great accomplishment was to convince not just psychiatrists but also social institutions and the public at large that the entire range of mental illnesses were genuine diseases. The DSM conditions provide the template for the self-definitions that people make of their distress and the frame for cover stories in national magazines, themes of television shows and best-selling books, and science reporting in the news media. This widely accepted framework for mental disorder is a creation of the diagnostic revolution that arose in 1980 with the publication of the DSM-III.

## A Model for the DSM-III

The social and cultural context surrounding psychiatry in the 1970s led the times to be ripe for a scientific revolution in the classification of mental disorders. The first step came in 1974, when the APA appointed a leading research psychiatrist, Robert Spitzer, to head a task force charged with revising the DSM-II. Spitzer (1932–2015) had played an important role in brokering the contentious issue of removing the diagnosis of homosexuality from that manual. This effort led him to develop an intense interest in what qualities constituted legitimate mental disorders. The most pressing problem that Spitzer needed to resolve was psychiatry's erosion of credibility that attacks on the meaningfulness of diagnosis had created. Although he had psychoanalytic training, Spitzer saw the dynamic view's unverified theory and resistance to empirical testing as the major obstacles to psychiatry's attaining scientific status.[7] The central element in Spitzer's vision of psychiatry, pursued in his research efforts in the 1960s and 1970s and culminating in the DSM-III in 1980, was to develop a reliable system of classification in which different diagnosticians would generally arrive at the same diagnosis based on the same clinical information.[8]

Spitzer had an alternative to the dynamic paradigm. He was a close collaborator with a group of research psychiatrists at Washington University in St. Louis, led by Samuel Guze (1923–2000) and Eli Robins (1921–1994). During the era when psychodynamic perspectives dominated the psychiatric profession, the Washington University Department of Psychiatry was an outpost of traditional medically minded thinking. Guze later summarized this group's perspective:

**Figure 7.1** Research psychiatrist Robert Spitzer led the DSM-III revolution in diagnostic classification. Eve Vagg. New York State Psychiatric Institute.

> The nature and development of mental functions is the centre of psychiatric interest, just as the nature and development of bodily functions generally are the centre of medical interest. Psychiatry is a branch of medicine, which in turn is a form of applied biology. It follows, therefore, that biological science, broadly defined, is the foundation of medical science and hence of medical practice. The other disciplines can and must make their contributions, but they cannot displace biology from its critical role.[9]

Robins and Guze set out five criteria for valid psychiatric classifications: (a) clinical description, (b) laboratory studies, (c) exclusion studies to separate people with different conditions, (d) follow-up studies, and (e) family studies.[10] As each of these validators became increasingly accurate, etiologically precise, and homogeneous, diagnostic categories would follow. The Washington University group also expected that a well-defined classificatory system would result in improved treatments for mental disorders. Lithium might be the treatment of choice for bipolar disorders, phenothiazine for schizophrenia, or imipramine for depression. Reciprocally, finding that a particular treatment worked for some condition helped to specify what diagnosis was indicated.[11]

Spitzer and the Washington University psychiatrists insisted that the only solution to the diagnostic confusion regarding mental disorders was to give up the use of any presumed underlying pathology, unconscious dynamics, or life history as the starting point for definitional criteria.[12] They traced their roots to Emil Kraepelin, not to Sigmund Freud. Kraepelin's approach to psychiatric classification assumed that mental disorders are best understood as analogues to physical diseases and that the classification of mental disorders demands careful observation of visible symptoms and their course over time instead of inferences based on unproven causal theories.[13] Rather than focusing on any underlying causes for mental disorders, Kraepelin stressed the need to classify them according to their unique symptoms, course of development, duration, and eventual outcome. Spitzer summarized: "With its intellectual roots in St Louis instead of Vienna, and with its intellectual inspiration drawn from Kraepelin, not Freud, the task force was viewed from the outset as unsympathetic to the interests of those whose theory and practice derived from the psychoanalytic tradition."[14]

To achieve their goals, the DSM-III Task Force needed a reliable system of diagnosis that could differentiate the etiology, prognosis, and treatment responses of various conditions. The Washington University psychiatrists had already developed operational criteria for fourteen disorders, known as the "Feighner criteria" after the resident who was the first author of the 1972 article that described them.[15] The members of this group were researchers, not clinicians; they designed their diagnostic criteria for use in academic investigations.

The Feighner criteria rested on several assumptions. Most fundamentally, they assumed that psychiatry was a branch of medicine based on scientific principles, on diagnosis and classification, and on statistical methods. In this model, the major prerequisite of a scientific discipline was to order phenomena into distinct diseases. Such entities must be directly observed, not inferred from nonobservable mechanisms such as the unconscious. This emphasis on visible phenomena presumably ensures that the resulting classification system is factual and objective. Careful observation of overt symptoms was the most important aspect of classification because the other goals of a good diagnostic system—etiology, prognosis, and treatment—followed from accurate description. The Feighner criteria thus elevated to a central position exactly the symptom-based elements that dynamic psychiatry had dismissed as misleading surface manifestations of underlying conditions.

The aim of this group was to create discrete and homogeneous categories of pathology that were natural entities equivalent to physical disorders.[16] For them, an ideal system of classification was mutually exclusive; no entity overlaps with another entity within the same system. Different disorders consisted of co-occurring symptoms whose presence is not coincidental. Each disease entity is presumed to be separable from other disease entities, not different manifestations of similar underlying dynamics. Every aspect of the Feighner criteria—grounding classification in overt symptoms, rejecting unproven etiological assumptions, and expecting different disorders to respond to distinct treatments—diverged from the vague and overlapping diagnoses of dynamic psychiatry.

The conceptualization of anxiety provides a good example of the discrepancies between the dynamic and diagnostic approaches. For dynamic psychiatrists, anxiety was the core phenomenon underlying all forms of neuroses. It was a very broad and universal aspect of human experience that stemmed from conflicts between unconscious desires and external frustrations. Furthermore, it lay behind *every* form of neurosis so was not associated with any particular condition.[17] In contrast to the dynamic view, the Feighner criteria for a diagnosis of an anxiety disorder required a specified number of overt symptoms. At least four symptoms from among dyspepsia, palpitations, chest pain or discomfort, choking or smothering sensation, dizziness, and paresthesias (tactile disorders) must be present during the majority of anxiety attacks. A diagnosis requires the presence of at least six of these attacks, separated by at least a week from each other. Symptoms cannot result from physical exertion, a life-threatening situation, or a medical condition. The Feighner criteria thus transformed anxiety from a very general concept underlying many sorts of overt signs into a discrete, quantifiable disorder based on manifest symptoms.

The Washington University group viewed the Feighner criteria as a first step toward the development of a more established biomedical system of diagnosis. They repeatedly made statements such as: "These criteria are not intended as final for any illness."[18] They expected that their measures would change substantially as other groups employed and modified them. Their initial presentation distilled the experiences and intuitions of the Washington University and like-minded researchers. One resident stated, "We all sat around a table and simply made these criteria up from the old Kraepelin stuff."[19] While the resident was presumably exaggerating, the research base behind these admittedly preliminary diagnoses was scant. Consider that the criteria for depression, which would become the basis for psychiatry's core diagnosis from 1980 through the present, almost entirely stemmed from a single study of one hundred hospitalized patients called "manic-depressive" and fifty medically sick controls.[20]

The Washington University model and the Feighner Criteria that embodied it did not rest on a well-established and empirically tested body of findings but were more of an aspirational standard. Despite the fact that these researchers realized that their diagnostic criteria were highly exploratory and presented them as a first step that awaited future validation, by 1989 the article in which the Feighner criteria first appeared was the single most-cited article in the history of psychiatry.[21] In many respects, the limited range of psychiatric diagnoses in the Feighner criteria provided valuable checks over the boundless phenomena of dynamic psychiatry. In marked contrast to the subsequent DSM-III, they encompassed only fourteen conditions. A major psychiatric text based on these criteria contained only eleven diagnoses.[22] The Washington University group did not envision that their symptom-based model would provide the framework for the entire array of the nearly three hundred diagnoses that arose in the DSM-III.

## Creating the DSM-III

A full third of the members of the DSM-III Task Force were psychiatrists trained at Washington University and most of the others were close collaborators with this group.[23] Their two major goals were, first, to create a reliable classificatory system based on the manifest symptoms that each condition displayed and, second, to purge the new manual of psychodynamic etiological inferences.[24]

Spitzer and his collaborators were particularly concerned with improving the reliability of psychiatric diagnosis, which had become a central reason for the profession's crisis of legitimacy. Reliability refers to the degree of consistency with which appropriately trained members of a field will apply the same definition to observable phenomena. The nature of the DSM-II diagnoses insured that the field had no standardized way to classify the phenomena it described, explained, and treated. Academic training varied considerably from one university and psychiatric institute to another, and the practices of each psychiatrist varied according to his or her educational background and personal beliefs. Psychiatrists used clinical intuitions to judge how well each patient fit a particular diagnosis. Unsurprisingly, studies such as the U.S.-U.K. project discussed in the previous chapter had shown diagnostic reliability to be appallingly bad.[25]

The low reliability of psychiatric diagnoses was not just a professional issue but had been at the center of highly publicized efforts to discredit the field itself. The conclusion of the Rosenhan study that psychiatry did not know how to define mental illness was widely accepted: "[P]sychological categorization of mental illness is useless at best and downright harmful, misleading, and pejorative at worst. Psychiatric diagnoses, in this view, are in the minds of observers and are not valid summaries of characteristics displayed by the observed."[26] The poor reliability of psychiatric diagnosis posed an acute challenge not only to psychiatrists' clinical expertise but also to the scientific status of psychiatry itself.[27]

The focus on reliability of diagnoses was the perfect intellectual tool to provide a scientific foundation for psychiatry. It would presumably yield the consensual meanings, common communicative tools, and diagnostic agreement that would establish the field's medical credentials.[28] Showing that different observers could agree on whether a particular cluster of symptoms indicated a particular mental disorder allowed the proponents of diagnostic psychiatry to contrast their reliable system with the vague, unmeasurable mechanisms of dynamic psychiatry. The emphasis on manifest symptoms allowed psychiatry to appear more scientific, regardless of whether these symptoms actually fit a categorical model. The emphasis on improved reliability thus became the key rationale for the claim that diagnosis could be scientific without having to demonstrate why any of the classified entities ought to be considered instances of a valid mental disorder.[29]

In addition to emphasizing reliability, Spitzer and the DSM-III Task Force focused on shedding the causal suppositions of the DSM-I and DSM-II. The dynamic approach of these manuals segregated organic psychoses from all other conditions, which they assumed arose through a combination of psychological and social factors. For example, the definition of depression in the DSM-II stipulated that this "disorder is manifested by an excessive reaction of depression due to an internal conflict or to an identifiable event such as the loss of a love object or cherished possession."[24] Spitzer and his colleagues disdained these causal assumptions:

> It was the unanimous opinion of the Committee that etiology should be a classificatory principle only when it is clearly known, and that conventional speculations about etiology should be explained if they must appear. . . . A diagnosis should be made if the criteria for that diagnosis are

met. . . . It is hoped that this will stimulate appreciation, among psychiatrists, of the distinction between the known and the assumed.[30]

For them, the etiologically driven DSM-II brought too many unproven conditions within the realm of psychiatric diagnosis.

Contrary to the claims of its proponents, the new diagnostic framework of the DSM-III did not arise from a new knowledge base. The major empirical test of the DSM-III criteria stemmed from the NIMH-sponsored field trials that Spitzer and his colleagues conducted between 1977 and 1979. They involved about five hundred psychiatrists who used preliminary drafts of the DSM-III to diagnose more than twelve thousand patients. About three hundred psychiatrists were paired, and their evaluations compared for consistency. Evaluators were volunteers, and all offers to participate were accepted; they did not comprise anything like a random sample. The results seemed to be mildly encouraging although very little was published about them and the available data were often conflicting.[31]

Perhaps most important, the field trials did not compare the new symptom-based system to any other possible classification system or to previous DSMs but took for granted that the proposed criteria for the DSM-III provided the optimal method for classifying mental disorders. The triumph of the categorical diagnostic system was not based on evidence that it was more adequate than the dynamic or any other alternative. Moreover, the field trials were conducted *after* the Task Force had already developed the diagnostic criteria for various disorders; the criteria did not emerge from empirical tests.[32] The trials did not lead to any revisions of the criteria but ratified the document that Spitzer and his task force had already developed.[33] The new system imposed a standardized, symptom-based system on phenomena that had previously been considered as indefinite and highly overlapping without demonstrating that it actually improved the process of psychiatric diagnosis.

In fact, the successful implementation of diagnostic psychiatry resulted from the efforts of the most effective politician in the history of the psychiatric profession, Robert Spitzer. In a process whose outcome no one could have predicted, Spitzer transformed a field whose identity centered on the psychosocial aspects of mental illness to one that thoroughly embraced the assumption that mental disorders were a subset of physical disorders. How did this happen?

## Robert Spitzer: The Master Politician

When the APA appointed Robert Spitzer to head the task force that was charged to revise the DSM-II in 1974, it went virtually unnoticed because no one at the time paid much attention to the DSM.[34] Spitzer was best known for his creation, in collaboration with Eli Robins, of the Research Development Criteria, which elaborated the Feighner criteria to develop twenty-five diagnostic categories for research purposes.[35] Nothing in Spitzer's background, however, would have led anyone to believe that he would become "the most influential psychiatrist in the last quarter of the [20th] century."[36]

The goals of the DSM-III Task Force and of the Washington University group were similar in some ways and different in others. On the one hand, they shared a common commitment to destroy the untested and probably untestable dynamic assumptions that underlay the current system of classification. Both groups also staunchly supported a thoroughly empirical, a-theoretical system of measurement. Of paramount importance for each, any new diagnostic system must be a reliable one so that different users would be able to make the same diagnosis for the same patient. On the other hand, the Washington University psychiatrists were primarily researchers, who formed a small, albeit prestigious, part of the profession. Their goal was to create entities solely and explicitly for research purposes. Spitzer's mandate as head of the Task Force was far more ambitious: to create a diagnostic system that would serve the purposes not only of the psychiatric research community but also of the much larger and rapidly expanding ranks of psychiatrists and other professionals with clinical practices.

The divergent concerns of psychiatric researchers and clinicians, as well as the competing interests of psychiatrists and other mental health professionals, dictated that the drafting of the DSM-III would involve a vehement political struggle. As Klerman observed, "DSM-III is a political document in many ways. It appeared in response to some of the ideological and theoretical tensions within the profession of psychiatry. It also has been caught up in the rivalries and tensions among the various mental health professions—psychiatrists, social work, psychology."[37]

One major controversy that arose during the Task Force's deliberations involved Spitzer's attempt to define "mental disorder."[38] Neither of the first two DSMs had tried to explain what general concept united all of the conditions they described. Spitzer, however, believed that such a definition was necessary to counter anti-psychiatric claims that psychiatry was unable to define its subject matter and that the concept of mental disorder was inherently a value-laden and culturally relative one.

With Kraepelin's and the Washington University researchers' assumptions as his guiding framework, Spitzer initially proposed including the definition that "mental disorders are a subset of medical disorders" in the manual.[39] Yet, this statement was a red flag to many analysts as well as to psychologists, social workers, psychiatric nurses, counselors, and other professional groups that used psychiatry's diagnostic system. Spitzer did not anticipate the ferocity of opposition from non-psychiatric professional organizations, especially the American Psychological Association. These groups viewed the claim for the medical nature of mental disorders as a power play by psychiatry to preempt the mental health field and lay exclusive rights to diagnosing and treating mental disorders. Theodore Millon, a distinguished psychologist who was one of the few non-psychiatrists on the Task Force, wrote that "to attribute marital conflict or delinquency . . . to a biological defect, to biochemical, nutritional, neurological, or other organic conditions . . . is to sell our psychological birthright for short term gain."[40] The conflict reached the point that the psychologists' professional organization threated to initiate legal action against the APA if the DSM-III included the claim that mental illnesses were medical conditions.

Although non-psychiatric mental health professionals were not members of the APA and, therefore, were not officially able to ratify or reject the new manual, they nevertheless were major players in the mental health landscape. If these groups opposed the manual, its credibility and usefulness would be diminished. Although Spitzer initially took a confrontational approach to the psychologists' objections to the proposed definition, he realized that his opposition was nonproductive, and he acceded to their demands. Later, he explained that the initial proposal "would only alienate our psychology colleagues who had made so many important contributions to the field of psychopathology."[41]

Spitzer reframed the concept of mental disorder through the phrase "in the DSM-III each of the mental disorders is conceptualized as a clinically significant behavioral or psychological syndrome."[42] This resolution satisfied the various professional groups without seriously compromising the Task Force's attempt to construct a well-defined system of disorders. The chair of the Psychological Liaison Committee responded to Spitzer's decision, writing: "We were delighted to hear that you have decided not to include in DSM-III any reference to the classification being a subset of medical disorders. This decision is most constructive, and will do much to gain wide-spread acceptance of the revised nomenclature, and to further good relations among the various disciplines."[43] Spitzer's tactical accession to non-psychiatric groups allowed the revision process to continue with minimal interprofessional conflict but with no damage to its central goals.

A second controversy involved the status of the personality disorders in the new manual. These conditions were a staple of dynamic thinking and were highly consequential for clinicians (see Chapter 5). Yet, they were of little interest to researchers, many of whom scorned the very notion of "disordered personality." Prominent psychiatrist Aaron Beck considered the concept of personality disorder itself to be "so artificial and removed from observables, that it is probably of little utility and, even worse, it is probably a misleading fiction."[44] Their holistic nature rendered them difficult to square with the DSM-III's precise, symptom-based system. Aside from antisocial personality disorder, none had appeared in the Feighner criteria. In addition, it was nearly impossible to separate what was a "disordered" from a "normal" personality. Some conditions that enamored clinicians, in particular borderline personality, could hardly be defined at all. Leading diagnostician Donald Goodwin asserted: "The borderline syndrome is a mess . . . the borderline syndrome stands for everything that is wrong with psychiatry and the category should be eliminated."[45] Yet, it would have been politically disastrous for the Task Force to eliminate conditions that were such a critical aspect of clinical practice.

Spitzer's solution to the conflict between researchers who disdained the personality disorders and clinicians who found them essential was to place eleven distinct personality disorders on a separate axis, Axis II, from other conditions, which were listed on Axis I. Although this result made researchers uneasy, it satisfied the far larger constituency of practitioners.[46]

A third imbroglio involved the Task Force's proposal to eliminate the term "neurosis" in the new manual. This concept was a particular target for Spitzer and his allies who intended to expunge it from the manual because, in their opinion, it had no empirical basis. It was, instead, a sloppy and unmeasurable term that had outlived its usefulness.[47] Yet, although it was exceedingly broad and ambiguous, neurosis was the most fundamental notion of dynamic psychiatry. Eliminating the concept that was at the very heart of its thinking would virtually be a death blow to dynamic psychiatry. Not surprisingly, the proposal generated "a tidal wave of protest from the analysts."[48] Faced with enormous political opposition, the DSM-III was in serious danger of not being approved by the APA Board of Trustees unless neurosis was included in some capacity.

Again, Spitzer deftly responded to a crisis. He developed an ingenious political compromise between hardline members of the Task Force who adamantly opposed any mention of neurosis and analysts who were equally insistent about the need to include it in the manual. In what Spitzer called a "neurotic peace treaty," categories such as "dysthymic disorder (or depressive neurosis)" were included in the DSM-III after prolonged and bitter negotiations.[49] Neurosis would be used to describe disorders rather than to explain them. As one of Spitzer's allies, Donald Klein, observed: "The neurosis controversy was a minor capitulation to psychoanalytic nostalgia."[50] Once the Task Force reinserted "neurosis" into the final draft of the DSM-III, the APA's Board of Trustees formally approved it for publication in 1980.

In retrospect, Spitzer admitted that the politics associated with the development of the new manual was a source of unavoidable embarrassment for many psychiatrists:

> The entire process of achieving a settlement seemed more appropriate to the encounter of political rivals than to the orderly pursuit of scientific knowledge. On each side of the controversy, it was held that important scientific truths were at stake, and yet the situation had demanded, of those who found themselves in opposition, the adoption of strategic postures and the employment of the technique of politics. . . . Scientific politics is not a mere replica of more ordinary politics, but it is politics nevertheless. . . . That this dispute took on a political form and that it was at times passionately fought should therefore come as no surprise.[51]

Spitzer's genius combined a strategic vision of creating a more scientific diagnostic system with a wily sense of tactics about how to realize his ambitious goal. It enabled a small group of researchers to grasp control of the psychiatric profession from a much larger corps of dynamically oriented clinicians. At the time, however, the notion that the DSM-III would become *the* textbook for all mental health professionals, as opposed to the "little manual" (less than 140 pages) its two predecessors had been since 1952, was unanticipated.[52]

## The DSM-III

The DSM-III was a radical departure from the earlier DSM-I and DSM-II. These manuals provided short descriptions of disorders that emphasized the psychic mechanisms that presumably underlay pathology. The drafters of the DSM-II explicitly rejected the notion that mental disorders were fixed disease entities.[53] In contrast, the DSM-III emphasized categories of illness rather than blurry boundaries between normal and abnormal behavior, dichotomies rather than overlapping conditions, overt symptoms rather than underlying etiological mechanisms, and static manifestations of symptoms rather than their dynamic unfolding.[54] In each respect, it thoroughly transformed thinking in psychiatry.

Kraepelin had only focused on two categories of mental disorder: dementia praecox and manic depression; the Feighner criteria defined just fourteen conditions. The DSM-III's revolutionary change, which persists to this day, was to apply a medical framework to every condition that it inherited from dynamic psychiatry. The manual developed explicit definitions of several hundred diagnostic entities that remain the standard for what counts as a mental disorder.[55] Its goal, as Millon explained, was to "embrace as many conditions as are commonly seen by practicing clinicians."[56]

The DSM-III contained 265 discrete diagnoses, organized into sixteen general categories (e.g., schizophrenic disorders, affective disorders, anxiety disorders, etc.). It placed the eleven personality disorders in a separate section, which it called "Axis II." Like the conditions found on Axis I, the personality disorders were defined by their overt symptoms. Three other axes for physical disorders (III), psychosocial stressors (IV), and level of patient functioning (V) were also present but were never extensively used. More important than the number of different entities and the number of axes, however, was the radically different way the DSM-III defined each diagnosis compared to the DSM-I and DSM-II.

Harking back to 19th-century conceptions of mental disorders, the DSM-III assumed that psychological symptoms were not dependent on subjective interpretations but instead afflicted people in comparable ways as medical diseases.[57] All diagnoses in the new manual adhered to the principle that they could be identified through their symptoms without reference to lived experiences. "Psychiatry now recognizes," proclaimed Nancy Andreasen, one of the key members of the DSM-III Task Force, "that the serious mental illnesses are *diseases* in the same sense that cancer or high blood pressure are diseases."[58]

## Anxiety and Depression in the DSM-III

The anxiety disorders provide one example of how highly specific and detailed lists of symptoms replaced the earlier cursory and vague definitions. In the DSM-II, "Anxiety is the chief characteristic of the neuroses."[59] Anxiety lay behind *all* of the ten subcategories of neurotic disorders that were briefly defined in the "neuroses" category. In contrast, the DSM-III carved anxiety conditions into nine different conditions. These included three types of phobic disorders: simple phobia, social phobia, and agoraphobia, which itself was split into conditions with or without panic attacks. Other categories were panic disorder, generalized anxiety disorder, obsessive-compulsive disorder, PTSD, and conditions not otherwise specified. The rejection of any unifying etiology behind the various anxiety diagnoses thoroughly distinguished the DSM-III from the first two DSM manuals.

The carving up of the anxiety neuroses into distinct conditions had no precedent in psychiatric history. Even Freud, whose early work described the symptomatology of most of these anxiety conditions, had a unifying theory that brought them together as different manifestations of common underlying processes. The DSM-III, however, created the impression that clear-cut disorders existed that had little relationship to each other, to other categories in the manual, to personality dispositions, or to more generalized distress and problems in living.

The criteria for depression, which the DSM-III called "major depressive disorder," provide a second example that thoroughly contrasts with the definitions of the anxiety disorders. For most of Western history, depression had been equated with melancholia, a very serious disorder marked by thoroughgoing lethargy, despondency, hopelessness, worthlessness, and gross misinterpretations of reality often accompanied by delusions, hallucinations, and suicidal thoughts.[60] Unlike anxiety, melancholic disorders were not central to psychodynamic theory or practice and were more closely associated with hospitalized patients than with outpatients. Correspondingly, the DSM-II considered the major affective disorders to be psychoses and segregated them from depressive neuroses, which were epiphenomenon of anxiety and largely submerged into the broader category of the psychoneuroses.

Before the DSM-III, while researchers concurred that a separable, psychotic form of depression existed, they could not agree about the nature of non-psychotic types of depression. Researchers who studied depression used various classifications that ranged from a single to as many as nine or more separate categories.[61] Disputes abounded over whether depression should be classified according to its symptoms, etiology, or response to treatment. Others conceived of neurotic depression as more closely resembling a personality or temperament type than a disease condition.[62] Like the other major diagnoses in psychiatry at the time, opinions regarding definitions of depression at the end of the 1970s featured an extraordinarily broad range of unresolved conflicts on how to best measure this condition.

The MDD diagnosis that emerged in the DSM-III, which closely resembled the Feighner criteria, resolved all of these disputes in a stroke. It required the presence for at least a two-week period of persistent dysphoric mood or loss of interest or pleasure in usual activities and four or more additional symptoms out of the following eight: (a) weight gain or loss or change in appetite; (b) insomnia or hypersomnia (excessive sleep); (c) psychomotor agitation or retardation (slowing down); (d) decreased sexual drive; (e) fatigue or loss of energy; (f) feelings of worthlessness or excessive or inappropriate guilt; (g) diminished ability to think or concentrate or indecisiveness; and (h) recurrent thoughts of death or suicidal ideation or suicide attempts.[63] Yet, an MDD diagnosis did not require the presence of any of the final three more serious symptoms.

The capacious treatment of depression as a single, extraordinarily heterogeneous condition sharply contrasted with the division of the anxiety disorders into many distinct entities.[64] The MDD criteria could encompass melancholics who suffered lengthy periods where they could not get out of bed, felt worthless, and were suicidal as well as people who felt depressed for two weeks after a minor setback but then quickly recovered. Philosopher Dominic Murphy observes:

> It seems unlikely that the same underlying causes explain an irritable adolescent who sleeps late, diets frantically, and lies around the house all day threatening to commit suicide on the one hand, and a sad middle-aged man who can not settle down to any of his normal hobbies, hardly sleeps, eats more and more, can not make love to his wife, and feels worthless.[65]

One group of researchers calculated that the MDD definition yields 227 diagnostic combinations or up to 1,497 different diagnoses if criteria are split into their individual symptoms![66]

Another divergence between the anxiety and the depression criteria was that the DSM-III allocated the most general symptoms of distress—for example, change of appetite, insomnia, fatigue—to MDD. In contrast, the diagnostic criteria for each of the various anxiety disorders were more specific, centering on expressions such as intense fears of specific objects or situations, obsessions and compulsions, and post-traumatic stress. Third, the duration criteria for the anxiety conditions were considerably longer than those for MDD. Most anxiety diagnoses required "persistent" symptoms, usually of at least six months' duration, which ruled out diagnoses of short-lived responses to stressful conditions.[67] In contrast, symptoms that endured for a mere two weeks met the MDD qualifications.[68] Transient responses to stress, therefore, could meet diagnostic criteria for depression but not for any of the anxiety conditions.

Finally, the disparate treatment of the contextual basis of anxiety and depression facilitated diagnoses of depression over those of anxiety. A very high proportion of patients enter mental health and primary medical care settings with psychosocial problems of stress that are often the proximate reasons for their symptoms. Yet, the diagnostic criteria for the anxiety diagnoses were hedged with qualifiers that distinguished them from contextually appropriate symptoms. For example, only "irrational" or "unreasonable" fears met the criteria for phobias, thus excluding proportionate and reasonable fears. Or, panic disorders had to occur "unpredictably" and could not be responses to life-threatening situations.[69] Diagnoses of anxiety, therefore, ruled out appropriate responses to dangerous situations. In contrast, many patients reacting to stressful psychosocial contexts could meet the MDD criteria, which lacked any contextual qualifiers aside from bereavement. Because it was so easy to meet its a-contextual criteria, after 1980 MDD would dethrone anxiety and become the brightest light in the firmament of the new diagnostically oriented psychiatry.

## The Exceptional Case of PTSD

The criteria for the anxiety, depressive, and other DSM-III conditions emerged from internal discussions involving the members of the various task forces, other psychiatrists and, occasionally, other mental health professionals. The creation of the new diagnosis of PTSD was the sole major exception to the intraprofessional concerns that generated the other DSM-III diagnoses. Indeed, the Washington University group opposed the creation of the PTSD diagnosis. In particular, Lee Robins and others complained that its criteria involved an external cause and so conflicted with the a-causal principle that guided the other diagnoses in the new manual.[70] In addition, they believed that a separate PTSD diagnosis was unnecessary because almost all trauma victims already met criteria for other diagnoses such as depression or another anxiety condition.

Nevertheless, the PTSD diagnosis successfully entered the DSM-III because of vigorous advocacy from veterans' groups and their allies. The Task Force members were keenly aware of the intense sympathy in the broader culture for veterans who were struggling with emotional problems. Sociologist Wilbur Scott concluded that PTSD is in DSM-III "because a core of psychiatrists and veterans worked consciously and deliberately for years to put it there. They ultimately succeeded because they were better organized, more politically active, and enjoyed more lucky breaks than their opposition."[71]

The PTSD diagnosis that emerged in the DSM-III required the "existence of a recognizable stressor that would evoke significant symptoms of distress in almost everyone." The symptom criteria for the diagnosis required, first, that the trauma be reexperienced through intrusive recollections, recurrent dreams, or feelings of reoccurrence. Second, sufferers had to show numbed responsiveness through diminished interest in activities, detachment from others, or constricted affect. Finally, they had to display at least two symptoms from among hyperalertness, sleep disturbance, survivor guilt, trouble concentrating, avoidance of reminders of the traumatic event, and intensification of symptoms following such reminders.[72] Ironically, in light of the manual's thoroughgoing rejection of dynamic assumptions, these criteria closely resembled Freud's conception of trauma that emerged from World War I. Although PTSD was an anomaly within the DSM-III classification system, it would soon become a hugely popular diagnosis in both the mental health professions and the broader culture.

## The Acceptance of the DSM-III

The DSM-III triggered a critical change of political power within American psychiatry.[73]Its basic feature was to employ a model that equated visible and measurable symptoms with the presence of diseases. This grounding in overt symptoms allowed psychiatry to employ a standardized system of measurement that was congenial to research-oriented psychiatrists who needed to measure mental illness in reliable and reproducible ways. Yet, the new manual seemed to be a poor fit for the bulk of practitioners who were concerned with the particularities of the patients they treated but had little interest in whether their diagnoses conformed to those of other clinicians.[74]

Although at the time it appeared that clinicians had lost a power struggle with researchers, they quickly came to see the benefits of the new manual for their own interests. The new diagnoses allowed all psychiatrists to justify themselves through a disease model that was uniquely suited to provide medical legitimacy. In addition, clinicians of all theoretical persuasions could use the new manual because, PTSD notwithstanding, it defined illnesses solely through symptoms without regard to causes and so was theory-neutral. Perhaps most importantly, the new DSM framework, which encompassed nearly three hundred diagnoses, covered the expansive range of problems that therapists saw in their practices. This feature allowed practitioners to obtain reimbursement from third-party insurers and so secured their loyalty to the new system. Clinicians realized that the DSM-III's standardized diagnoses solved the crisis they faced in getting paid for their services.[75]

The DSM-III almost entirely emerged as the result of intraprofessional concerns. Although participants were conscious of pressures from other groups, insurance companies in particular, outside interests played a negligible role in creating the classification.[76] Shortly after the DSM-III was published in 1980, however, a variety of external forces saw how well the new biomedical diagnostic categories fit their own agendas.

## Extra-Professional Interests

With the notable exception of the PTSD diagnosis, the development of the DSM-III was remarkably free of participation of nonprofessional interests such as the pharmaceutical industry, insurance companies, lay advocacy organizations, or other third parties. Yet, these groups soon realized how valuable the DSM-III diagnoses could be for their own ends. Drug companies, the NIMH and its allied researchers, and advocacy groups had particularly powerful impacts on the rapid growth of the manual's influence and its institutionalization within the broader culture.

Although the pharmaceutical industry did not directly influence the construction of the DSM-III, it quickly recognized how valuable the manual could be in helping it overcome its own crisis of legitimacy. The DSM-III provided the specific, symptom-based disease entities that allowed drug companies to meet the FDA mandate that psychotropic drugs must only target legitimate disorders (see Chapter 6).[77] It also effectively recast the psychic effects of problematic social conditions as genuine diseases, thus neutralizing the critiques of the tranquilizing drugs from feminist and other opposition groups.

Another major consequence of the DSM-III categorizations was to make depression a more promising target than anxiety for the new class of medications—the selective serotonin reuptake inhibitors (SSRIs)—that came on the market in the late 1980s. Although they were called "antidepressants," the SSRIs have little relationship to MDD or any other particular diagnostic category. These drugs act very generally to increase levels of serotonin in the brain that raise low mood states, lower levels of inhibition, and decrease anxiety. They are used nonspecifically to treat not only major depression but also the various anxiety disorders, eating disorders, and alcohol and drug problems as well as a diffuse array of other conditions.

Peter Kramer's best-selling *Listening to Prozac: A Psychiatrist Explores Antidepressant Drugs and the Remaking of the Self* (1993) propelled the SSRIs—Prozac in particular—into the cultural spotlight. It cemented the link of these drugs with the treatment of depression. Kramer associated antidepressant treatment with miraculous transformations of selves, focusing on how the new class of drugs empowered and enhanced its users. Yet, while Kramer emphasized the general impact of the SSRIs on changing personalities, he connected the condition that the drugs transformed with "depression."[78]

The FDA's loosening of restrictions on direct-to-consumer drug advertisements in the late 1990s both enhanced the popularity of the SSRIs and reinforced their link to depressive illness. Advertisements for Prozac focused on its use in treating major depression, relentlessly pushing the view that "depression is a disease" linked to deficiencies of serotonin in the brain. At the same time, they used the imagery depicted in Kramer's book showing women becoming "better than well" while cheerfully fulfilling both work and family roles. Ads typically connected the most general symptoms of MDD—sadness, fatigue, sleeplessness, and the like—with common situations involving interpersonal problems, workplace difficulties, or overwhelming demands from social roles. The visual imagery was remarkably similar to the earlier advertisements for Miltown, Valium, and Librium that depicted the negative psychic consequences of social demands; stressful interactions with children, spouses, and bosses; and frustrating situations such as dealing with traffic jams.[79] The new ads retained the association of social stress and psychic disturbance while reversing the cause and effect relationship: they portrayed social tensions as *consequences* of the disease of depression, rather than its source. The idea that mental disorders resulted from chemical imbalances became widely accepted.[80]

Antidepressants were prescribed for mood, anxiety, and many other disorders alike, gaining unchallenged control of the market that the tranquilizers once held. General physicians were particularly likely to prescribe the SSRIs, using them for a cornucopia of complaints including hypertension, nerves, fatigue, back pain, and sleep difficulties.[81] When antidepressants are used to treat such an array of symptoms, they all come to be seen as signs of "depression." Paradoxically, the DSM's highly specific categories proved to be the salvation of drug companies whose products act in decidedly generalized ways.

In addition to the pharmaceutical industry, diagnostic psychiatry also redeemed the NIMH. By the 1970s, this once influential federal agency was becoming peripheral to mental health policy. Its former mission to prevent a broad range of psychosocial problems had proven to be politically lethal. Its sponsorship and promotion of the DSM-III meant that the NIMH now confronted disease conditions, a new direction that was amenable to the conservative as well as the liberal legislators who had to approve its budget.[82] The NIMH became an especially important purveyor of the new diagnostic categories to the general public through both educational campaigns and surveys of the presence of the major DSM mental illnesses in the population (see Chapter 9).

The NIMH's new biomedical agenda was also advanced through its association with a powerful lay advocacy group, the National Alliance for the Mentally Ill.[83] The central tenet of this organization, which was mostly comprised of parents with children who had serious mental illnesses, was that mental disorders were brain-based diseases with biological causes. It fiercely opposed dynamic conceptions that linked these conditions to family processes. They were effective allies of the NIMH in urging Congress to enhance federal funding for biomedically oriented research that was grounded upon the DSM categories.

The efforts of Spitzer and his various task forces successfully resolved the crisis of legitimacy among psychiatry and associated interests. The publication of the DSM-III in 1980 supplanted a classification that had little resemblance to a disease-based model with one that was thoroughly infused with medical assumptions.

## Did the DSM-III Expand the Range of Pathology?

The DSM-III's central change was to make mental illnesses seem comparable to physical ones. The manual had much less impact in transforming definitions of normality and abnormality. Postwar psychiatry featured a dynamic model that had already vastly expanded the range of mental pathology. Gerald Grob observed that this period "reflected an extraordinary broadening of psychiatric boundaries and a rejection of traditional distinctions between mental health and mental abnormality."[84] Between 1945 through the 1960s, socially oriented psychiatrists considered that "all social, psychological, and biological activity affecting the mental health of the populace is of interest."[85] The NIMH, too, framed its mission as promoting extremely broad notions of mental illness within the general population.

Because the range of conditions previous diagnostic manuals viewed as signs of mental illness was so expansive, in one sense the DSM-III did not lead to what sociologists often call a growing "medicalization" of psychiatry.[86] Medicalization refers to the process through which problems that in the past would have been seen as crime, deviance, immorality, bad habits, or other disreputable conditions come to be considered as diseases that health professionals ought to treat. This extension of medical authority makes psychiatrists the culturally legitimated agents that deal with mental disorders.[87] The proponents of the medicalization thesis point to the exponential growth in the number of specific diagnoses when the DSM-III replaced the DSM-II (from 106 and 182 in the first two editions to 265 in the third one) as evidence for the expansion of a medicalized perspective regarding the nature of normality and abnormality.

Diagnostic psychiatry did medicalize mental illnesses in the sense of making them seem like "real" diseases that were independent of the particularities of individual patients. However, to the extent that medicalization involves a movement from some nonmedical conception into the legitimate realm of a medical specialty, the DSM-III did not expand the range of medicalized pathology.[88] Instead, as Spitzer noted: "The larger number of diagnostic categories in the DSM-III as compared with DSM-II reflects the clinical and research need for greater specificity in describing behavioral syndromes."[89] The more numerous DSM-III diagnoses compared to previous manuals reflected the recategorization as discrete disease entities the psychic consequences of a wide range of problems—poor marriages, troubled children, failed ambitions, general nervousness, and diffuse anxiety—that dynamic psychiatry had already medicalized.[90]

In addition, dynamic assumptions had pervaded the treatment of deviant behavior in schools, juvenile courts, and child guidance clinics since the 1920s.[91] Community studies during the 1950s and 1960s also used continuous scales that considered reports of even a single symptom as a "mild" form of disorder. Dynamic psychiatry, not its diagnostic successor, initiated the tremendous growth of medicalization over the course of the 20th century. The unique contribution of the DSM-III revolution to medicalization processes was to turn vaguely formulated notions such as "neuroses" into specific, symptom-based disorders. In general, however, the expansion of abnormality was *not* a major aspect of the DSM-III revolution.

## A Theory-Neutral Manual

The dynamic model was explicitly psychosocial, grounding its psychogenic disorders in a diffuse array of both internal and external forces. While its conditions could emerge from unconscious conflicts, they could also result from external stressors and losses. It assumed that anyone who was faced with enough environmental pressure could succumb to mental illness. Similarly, American culture during the 1950s and 1960s was amenable to both psychological and social portrayals of neuroses. For example, psychotropic drugs were touted for their ability to ease the mental pain of both interior and situational factors.[92]

When the dynamic model turned inward, however, it found psychology, not biology. The DSM-II (and DSM-I) strictly segregated the class of organic brain syndromes from nonorganic psychotic and neurotic disorders and concentrated on the latter groups. Whenever a brain-related malfunction seemed apparent, the condition should be classified as organic. Therefore, biology was unrelated to the functional psychoses, neuroses, or personality disorders.

The pressures for psychiatry to embrace a biological perspective, which had periodically marked the field but had diminished in the post–World War II period, intensified during the 1970s. Psychiatrists faced intense demands to show that they were genuine physicians and not simply glorified counselors or social workers. The Washington University group, following Kraepelin, assumed that all the major mental illnesses had a biological basis. Nevertheless, while the research psychiatrists who midwifed the DSM-III were far more biological than psychosocial in their outlooks, the manual's definitions themselves did not dictate any particular causal attributions. Conditions that met the symptom criteria for a mental disorder could emerge because of defective brains, intrapsychic experiences, or stressful environments.

Although psychiatry turned sharply away from a psychosocial toward a biological framework in the aftermath of the DSM-III, the manual itself had little to do with this transformation. The DSM-III facilitated this process through placing biological causes on an equal footing with the previously dominant psychosocial model, but it did not cause the sharp turn to organicity that emerged after 1980.

## Conclusion

The success of the DSM-III (and subsequent DSMs) is often attributed to the power of scientific knowledge. Its advocates equate its classifications with the growth of objectivity, truth, and reason within psychiatry. According to Gerald Klerman, the movement from the DSM-I and DSM-II to the DSM-III was a "victory for science."[93] Likewise, Melvin Sabshin, the executive officer of the APA, called it a great triumph of "science over ideology."[94] For the proponents of the DSM-III, "the old psychiatry derives from theory, the new psychiatry from fact."[95] Another prominent diagnostic psychiatrist asserted that "scientific evidence" rather than the charismatic authority of "great professors" stood behind the classificatory systems of the DSM-III and its successors.[96] This reverence of the DSM-III and denigration of its predecessors persists. In 2015 a former president of the APA wrote that the DSM-III created "a system rooted in observation and data rather than tradition and dogma."[97]

This chapter presented a contrasting explanation for the triumph of diagnostic psychiatry. The wholesale transformation in classification that arose in the DSM-III provides a near-textbook case of Thomas Kuhn's model of scientific revolution.[98] In Kuhn's view, a conversion from one school of scientific thought to another school is untaken in order to resolve a state of crisis in the previously dominant paradigm. The new model gains acceptance because it is better able to justify and legitimate the social practices of the relevant discipline.

Although the adequacy of Kuhn's model for explaining scientific change in general has been widely debated, it does fit the discrediting of dynamic psychiatry and ascendancy of diagnostic psychiatry in the last decades of the 20th century. The thoroughgoing crisis in psychiatry that began in the 1960s, accelerated during the 1970s, and was resolved in the 1980s exemplifies Kuhn's model of scientific change.[99] The DSM-III was not a product of novel scientific knowledge. Indeed, original research did not play a major role in producing the new paradigm. The development and application of diagnostic psychiatry to the domain of mental illness was less the triumph of science over ideology than the use of the ideology of science and medicine to justify changes that resolved a deep institutional crisis. The DSM-III and its successors solved the professional difficulties that psychiatry faced in the 1970s and provided the accepted view of mental illness that has persisted through the present. As Spitzer later noted, the manual's success stemmed from the fact that it "looks very scientific. If you open it up, it looks like they must know something."[100]

The symptom-based model of mental illness that emerged in the DSM-III exemplifies how scientific revolutions can emerge not just from the discovery of new facts but also from changing worldviews. The shift from dynamic to diagnostic psychiatry provided the profession with a standardized system of measurement that overcame the field's loss of legitimacy in the 1970s. It helped silence the critics of the previous system, who claimed that mental illnesses could not be defined in any objective sense and allowed research-oriented psychiatrists, a small but highly influential group in the profession, to measure mental illness in reliable and reproducible ways. Because the manual defined illnesses solely through symptoms without regard to causes, it was theory-neutral and could be used by clinicians of all persuasions. For these practitioners, who comprised the vast majority of the psychiatric profession, the new diagnostic system legitimized claims to be treating real diseases and, most important, allowed them to obtain reimbursement from third-party insurers. The manual also delivered the pharmaceutical industry the specific diseases that it needed in order to bring their products to market. As well, it provided the major federal agency concerned with mental illness, the NIMH and the researchers it sponsored, with seemingly objective objects of explanation that allowed them to fend off political critics and establish their scientific credentials.

Within a very short period of time, psychiatry shed a paradigm grounded in individual particularities and adopted one that emphasized objective diseases. The general public, not to mention psychiatric researchers and clinicians, now view mental illnesses as comparable to physical ailments. In one of the intellectual marvels of the 20th century, the DSM-III thoroughly relegitimated a discredited discipline. Although the new diagnostic paradigm resolved the field's crisis of legitimacy that threatened to destroy it before the DSM-III revolution, new developments were soon to arise in the scientific community that raised serious questions over the adequacy of the DSM's symptom-based diagnostic paradigm.

## Acknowledgments

Portions of this chapter are adapted from Horwitz, 2002, Horwitz, 2011, and Horwitz, 2013.

## Notes

1. Goodwin & Guze, 1996.
2. Because the basic framework that arose in the DSM-III has remained unchanged in subsequent revisions of the manual through the DSM-5 in 2013, I sometimes use the term "DSM-III" to refer to all of these versions.
3. Kendler, 2015, 145.
4. Rosenberg, 2007, 13.
5. Klerman, 1978, 104.
6. Kendler, 2015, 187.
7. Spitzer, 1978; Bayer & Spitzer, 1985.
8. Spitzer & Fleiss, 1974; Kirk & Kutchins, 1992.
9. Guze, 1989, 319.
10. Robins & Guze, 1970.

11. Klein & Davis, 1969.
12. Wilson, 1993.
13. Kraepelin occasionally broke this principle and offered causal speculations for some conditions (Decker, 2007). Spitzer himself denied being a "neo-Kraepelinian" on the grounds that he assumed neither that distinctive pathology underlay different syndromes nor that mental disorders were ultimately due to physical brain diseases, both tenets of Kraepelin's approach.
14. Bayer & Spitzer, 1985, 188.
15. Feighner et al., 1972. The diagnoses were the primary affective disorders of depression and mania, secondary affective disorders associated with some other psychiatric or medical diagnosis, schizophrenia, anxiety neurosis, phobic neurosis, hysteria, antisocial personality disorder, alcoholism, drug dependence, mental retardation, homosexuality, transsexuality, organic brain syndrome, and anorexia nervosa, as well as a residual category of undiagnosed disorder.
16. Robins & Helzer, 1986.
17. Fenichel, 1995/1946.
18. Feighner et al., 1972, 57.
19. Shorter, 2013b, 9.
20. Cassidy et al., 1957.
21. Feighner, 1989.
22. Goodwin & Guze, 1996.
23. Blashfield, 1982.
24. Bayer & Spitzer, 1985.
25. E.g. Spitzer, 1978; Spitzer & Fleiss, 1974; see especially Kirk & Kutchins, 1992.
26. Rosenhan, 1973, 250.
27. In fact, as Spitzer (1975) and many others showed, Rosenhan's study was seriously flawed and did not demonstrate anything important about actual diagnostic practices. Nevertheless, perhaps more than any other single study, it shaped public consciousness about the nature of psychiatric knowledge.
28. Kirk & Kutchins, 1992.
29. The DSM Task Force did not address the issue that was at the heart of Kraepelin's view: changing symptoms over time could indicate the unfolding of various stages of a single illness process rather than unreliable measurement (Foulds, 1976, 13).
30. Minutes of the Task Force quoted in Wilson, 1993, 405.
31. Decker, 2013, 251.
32. Spitzer & Williams, 1988; Kirk & Kutchins, 1992, 121–31; Decker, 2013, 251–75.
33. Kirk & Kutchins, 1992; Decker, 2013, 257, 287.
34. Decker, 2013, 96.
35. Spitzer, Endicott, & Robins, 1978, 82.
36. Decker, 2013, 313.
37. Klerman, 1987, 3.
38. Decker, 2013, 187–95.
39. Spitzer, Sheehy, & Endicott, 1977, 4.

40. Millon, 1986, 45.
41. Quoted in Kirk & Kutchins, 1992, 192.
42. Millon, 1986, 45.
43. Decker, 2013, 194.
44. Quoted in Decker, 2013, 196.
45. Quoted in Decker, 2013, 198–99.
46. Spitzer, 2001. The discomfort of researchers was well founded: the personality disorders remain highly overlapping, difficult to define, and inseparable from normal, if undesirable, character types. These were exactly the problems of the earlier dynamic manuals that the DSM-III was meant to overcome.
47. Millon, 1983.
48. Shorter, 2013b, 11.
49. APA, 1980, 220; Bayer & Spitzer, 1985.
50. Wilson, 1993, 407.
51. Bayer & Spitzer, 1985, 195.
52. Madow, 1976.
53. APA, 1968, 122.
54. Although the manual states: "In DSM-III there is no assumption that each mental disorder is a discrete entity with sharp boundaries (discontinuity) between it and other mental disorders, as well as between it and No Mental Disorder" (APA, 1980, 6), it defines each of its diagnoses through criteria that yield sharply bounded entities between each other and between them and normality.
55. E.g. Wilson, 1993; Kirk & Kutchins, 1992; Horwitz, 2002. While the DSM-III is often labeled as "neo-Kraepelinian," in some ways, its symptom-based approach *negates* Kraepelin's model. For Kraepelin, the careful observation of symptoms was only the starting point of diagnosis. Symptoms were not important in themselves but only became of interest when they could predict the future course and outcome of the condition. "A diagnosis," Kraepelin (1910, 1–2) maintained, "should mean much more than a mere grouping of the momentary symptoms. It must contain a more or less definite view of the origins and of the presumed further course of the diagnosed case." Yet, the various DSM criteria have little to say about Kraepelin's central focus: the prognosis of each separable diagnosis.
56. Millon, 1986, 39.
57. Berrios, 2015, 36.
58. Andreasen, 1984, 8.
59. APA, 1968, 39.
60. APA, 1952, 25. Horwitz & Wakefield, 2007; Shorter, 2013a.
61. Kendell, 1976.
62. Eysenck, 1970.
63. APA, 1980, 213–14.
64. Horwitz, 2011.
65. Murphy, 2006, 329.
66. Ostergaard, Jensen, & Bech, 2011.

67. APA, 1980, 227.
68. APA, 1980, 213.
69. APA, 1980, 227–30.
70. Helzer, Robins, & Davis, 1976.
71. Scott, 1990, 308.
72. APA, 1980, 238.
73. One indicator of this transformation was that the DSM-III mentioned the Group for the Advancement of Psychiatry, which had played such a prominent role in postwar psychiatry, just once and in a disparaging manner (APA, 1980, 469).
74. Two analysts provided perhaps the most colorful portrayal of this situation: "the elimination of the past by the DSM-III Task Force can be compared to the director of a national museum destroying his Rembrandts, Goyas, Utrillos, van Goghs, etc. because he believes his collection of Comic-Strip Type Warhol's (or what have you) has greater relevance" (quoted in Decker, 2013, 183).
75. Most of the researchers who developed the DSM-III criteria were funded by their academic positions and research grants and so were not dependent on the insurance payments that were of paramount concern to clinicians.
76. Spitzer wrote a memorandum to the Task Force indicating that clusters of symptoms must be called "disorders" or "syndromes" so that psychiatrists could be reimbursed for their treatment (Wilson, 1993, 405).
77. Healy, 1997.
78. Kramer, 1993.
79. Tone, 2009; Herzberg, 2009.
80. Pescosolido et al., 2010.
81. Mojtabai & Olfson, 2008a.
82. Kirk, 1999.
83. McLean, 1990.
84. Grob, 1987, 417.
85. Roberts, Halleck, & Loeb, 1969, quoted in Herman, 1995, 254–55.
86. E.g. Hale, 1995; Shorter, 1997; Wilson, 1993.
87. Conrad & Schneider, 1992; Conrad, 2005.
88. Wakefield, 2001.
89. Quoted in Kirk & Kutchins, 1992, 192.
90. E.g. Hale, 1995; Herman, 1995; Lunbeck, 1994.
91. Danziger, 1990.
92. E.g. Tone, 2008; Herzberg, 2010.
93. Klerman et al., 1984, 539.
94. Sabshin, 1990, 1272.
95. Maxmen, 1985, 31.
96. Kendler, 1990.
97. Lieberman, 2015, 129.
98. Kuhn, 1970. As an anonymous reviewer of this manuscript suggests, the DSM-III revolution does not completely fit Kuhn's model because he also emphasized how

paradigms ultimately change because too much data accumulates that is inconsistent with the older model. The case of the DSM-III did not so much involve new data as a wholesale new reconceptualization of the phenomenon of mental disorder.

99. Horwitz, 2002, 56–57.
100. Quoted in Greenberg, 2019, 32.

# 8
# Biology Re-Emerges

> All mental processes are brain processes, and therefore all disorders of mental functioning are biological diseases. The brain is the organ of the mind. Where else could mental illness be if not in the brain?
>
> —Eric Kandel, in Weir, "The Roots of Mental Illness." 2012, 30

The DSM revolution's major accomplishment was to provide psychiatry with the disorders it required to become a legitimate medical specialty. After 1980 psychiatrists could take the equivalence of mental and physical disorders for granted. The DSM-III, however, did not dictate any particular cause of mental disorder. It classified each diagnosis through its symptoms, not by what factors led symptoms to emerge. Indeed, the manual's theoretical neutrality was a key reason for why the diverse factions within psychiatry and other mental health professions accepted it. The next transformation in views of mental illness involved yoking the DSM's symptom-based diagnoses to the view that mental disorders were brain diseases produced by malfunctioning neurochemical systems and problematic genes.

In the aftermath of the DSM-III's publication in 1980, psychiatry reembraced the basic assumption that marked the field during the 19th century: all mental processes derive from biological properties of the brain. "You, your joys and your sorrows, your memories and your ambitions, your sense of personal identity and free will, are in fact no more than the behavior of a vast assembly of nerve cells and their associated molecules," Francis Crick, the co-discoverer of DNA, proclaimed in 1994.[1] Another Nobel prize-winner, psychiatrist Eric Kandel (b. 1929), succinctly summarized, "Your brain makes you who you are."[2] Prominent psychiatrists contrast this state with the ignorance of their dynamic predecessors: they speak of a "sea change" in psychiatry "as it matured from a psychoanalytic cult of shrinks into a scientific medicine of the brain."[3]

Since 1980, psychiatry has replaced the biopsychosocial model with a bio-bio-bio model that emphasizes brains, genes, and medications.[4] A single-minded focus on psychopharmacology has supplanted the pluralist combination of psychotherapy, psychosocial interventions, and drug treatments that characterized the field during the postwar period. Over a short time period, the biological study of mental illness evolved from a marginal and discredited enterprise to become the dominant model in not just psychiatry but also popular culture.

## The Rise, Decline, and Resurgence of Biological Explanations

Beliefs that mental illnesses are brain diseases reflect not just current fashion but also a long historical tradition. The contemporary importance placed on the biological foundations of mental disorders revives earlier biological thinking in psychiatry. As previous chapters indicated, before dynamic psychiatry developed at the turn of the 19th century, psychiatrists insisted that various types of madness were organic diseases, arose from specific brain mechanisms, and had distinct symptoms with fixed clinical courses and outcomes. They searched, albeit usually unsuccessfully, for the brain profiles that produced different mental disorders.[5]

During the last half of the century, Charles Darwin's new evolutionary theory exerted a towering influence over the scientific zeitgeist, including psychiatry.[6] Darwin (1809–1882) viewed mental phenomena as organically grounded traits that evolved from earlier forms of life. The principle of natural selection, which showed how various psychic qualities had different survival values in diverse environments, guided this process. Those characteristics that most helpfully enabled individuals to adjust to their circumstances were more likely to survive and spread to future generations than those that were less adaptive. Darwin's theory thus emphasized the intrinsic interplay between natural and environmental forces.

In the late 19th and early 20th centuries, however, Darwin's theories became associated with the influential movement of social Darwinism.[7] Darwin's own approach had little similarity with the racist undertaking that adopted his name. Although he shared many common Victorian notions of Western superiority, Darwin emphasized the similarities more than the divergences among human groups. The social Darwinists, however, viewed cultural differences as rooted in biological differences among individuals, social classes, and races. Applying the concept of natural selection to human societies, they unabashedly proclaimed the preeminence of White, Western, especially northern European, cultures and the inferiority of non-White, non-Western groups.[8]

The work of Darwin's half-cousin, Francis Galton (1822–1911), exemplified the spirit of the social Darwinists. Galton founded the field of eugenics, a word he coined in 1883 from the Greek term for "well-born."[9] Eugenics was a movement that strove to improve the quality of the human race through encouraging reproduction among people who possessed superior genes and suppressing it among those with inferior ones. Biological differences became ideological tools to justify, first, Western colonialism and, later, Nazi atrocities.

Galton also initiated the debate over the extent to which nature or nurture was related to human qualities. Although Darwin had emphasized the interrelationships between natural and environmental forces, Galton strove to separate nature from nurture. "Nature," Galton explained, "is all that a man brings with himself into the world; nurture is every influence without that affects him after his birth."[10] He believed that heredity determined most forms of human behavior so that unchangeable innate qualities led some people to be more talented than others, stating that: "When nature and nurture compete for supremacy . . . the former proves the stronger."[11] Galton's major legacy for research about the biology of mental illness was his insistence that inherited and acquired traits were distinct; the greater importance of nature meant the lesser significance of nurture as well as vice versa.

Over the course of the 20th century, biological theories in general became linked to reactionary political thought and were stigmatized and discredited. The association of eugenic thought with the Nazi movement and the resultant Holocaust destroyed the cultural viability of biological thinking about human behavior from the 1930s through the 1960s. As a result, organic explanations largely dropped out of the intellectual landscape during the middle part of the century.[12]

The rise of the social sciences after their founding in the late 19th century also served to suppress biological explanations of human action. Emile Durkheim's proclamation that sociology rested "wholly on the basic principle that social facts must be studied as things, that is, as realities external to the individual" was emblematic.[13] Beginning in the 1920s, sociologists, anthropologists, and psychologists promoted extreme environmentalist approaches that minimized biological influences on social behavior. Following Durkheim's example, sociologists strove to develop a field that was independent of biology (and psychology). Likewise, John Watson and B. F. Skinner, the leading figures of behavioral psychology, were absolute environmentalists who denied the relevance of innate factors and internal processes, stressing how people displayed whatever behaviors their environments reinforced. Anthropologists, too, militantly opposed the recognition of inborn human traits. "We are forced to conclude," Margaret Mead wrote, "that human nature is almost unbelievably malleable, responding accurately and contrastingly to contrasting cultural conditions."[14]

During the 1970s a major counterreaction against the sociocultural emphasis emerged as the organic outlook that dominated 19th century thought slowly returned to prominence. E. O. Wilson's tome, *Sociobiology* (1975), was a landmark in the return of this view. This influential volume synthesized scholarship on how genes determined most animal behaviors and concluded with a controversial chapter suggesting that human values also might have universal, natural underpinnings. In later works, Wilson expanded on this perspective, attempting to demonstrate how "all tangible phenomena, from the birth of stars to the workings of social institutions, are based on material processes that are ultimately reducible, however long and tortuous the chains, to the laws of physics."[15] In this view, all thought and experience must originate from some organic basis. Similar-minded works from widely read authors including Richard Dawkins, Daniel Dennett, Steven Pinker, and many others, indicated that biologically based views of human behavior had returned to respectability.[16] The field of psychology, too, sharply reacted against behaviorism, turning toward the study of cognitive and brain processes.

Thought about mental illness generally reflected these dominant cultural and scientific trends. Between the 1920s and the 1960s, few researchers or clinicians outside of mental institutions showed any interest in the brain or other biological underpinnings for mental disorders. Twin and adoption studies provided an exception. During the period when analytic and psychosocial models dominated psychiatry, the biological tradition persisted through an undercurrent of inquiries that used twins and adoptees as ways to distinguish internal natural qualities from external learned aspects of behavior.

Elaborating on Galton's pathbreaking research, one way this school inferred the relative power of nature and nurture was through the comparative degree of concordance in traits between MZ twins, which share all of their genes, and DZ twins, which share an average of half of their genes. Both types of twins presumably grew up in the same environment.[17] To the extent that genetic factors influence some mental illness, MZ twin pairs should have twice the concordance rate as DZ twins. Conversely, if environmental factors predominate, both types of twins should exhibit comparable levels of the condition in question. In general, studies of schizophrenia and bipolar disorder indicated that MZ twins showed considerably higher concordance compared to DZ twins.[18] In contrast, twin research typically indicated that, although there were heritable components to nonpsychotic conditions, environmental effects equaled or exceeded genetic influences.[19]

Adoption studies provided a second method for disentangling the influence of inherited from acquired traits. This research capitalizes on the fact that the parents who transmit genes to an adopted-away child are distinct from those who raise the child. The extent to which adopted children most closely resemble their birth or their adoptive parents indicates the relative power of genes and environments. Overall, research examining adoptees found that about half the influence on many traits is genetic and half environmental.[20] For the vast majority of people who were not twins or adoptees, individuals with schizophrenia and bipolar disorder were also more likely than the general population to have siblings or parents with the same conditions.[21]

On the whole, however, the hegemony of psychosocial views in the postwar period insured that biologically oriented students of mental illness stood outside of mainstream research and practice. Those who embraced organic approaches had low status in the profession and were often relegated to positions in state hospitals.[22] During the period when dynamic approaches dominated thinking about mental illness, broader cultural currents insured that biological views of human functioning would have little credence.

## The Rise of Neuroscience

Biological research on mental disorders began to enter a new era in the 1950s and 1960s as natural scientists made tremendous advances in understanding genes and brains. Studies of twins and adoptees indicated that some mental disorders were likely to have innate components, but these methods were intrinsically incapable of specifying what genetic mechanisms accounted for these findings. Crick and Watson's identification of DNA in 1953 had the promise of showing how the structure and function of particular genes led to the inherited aspects of mental disorders. Around the same time, the discovery of the medications chlorpromazine, lithium, and imipramine for the treatment of schizophrenia, bipolar disorder, and depression, respectively, suggested not just that mental illnesses were grounded in neurons and neurochemicals but also that each condition was linked to distinct brain sites. By the mid-1970s and early 1980s, neuroimaging techniques such as computerized axial tomography scans and magnetic resonance imaging (MRI) allowed neuroscientists previously unthinkable opportunities to visualize and measure the operation of living brains.[23]

In recent years the area of molecular genetics has exploded through the development and use of new methods for genotyping and sequencing. Genetics is widely recognized as among the most exciting academic disciplines; the deciphering of the human genome in 2003 is hailed as one of the major achievements in the history of science.[24] Likewise, innovative techniques that accompany this accomplishment provide neuroscientists with unprecedented opportunities to move beyond the study of isolated genes and analyze huge genomic data sets and possible polygenic impacts.

Like genetics, the field of neuroscience grew rapidly: the number of academic departments that are dedicated to the field expanded from a smattering in 1980 to over two hundred in 2011. At the same time, membership in the area's professional society increased from five hundred in 1969 to over forty thousand in 2018.[25] Publications regarding the brain have grown from thirteen thousand in 1970 to over sixty thousand each year since 2010.[26] President George H. W. Bush declared the 1990s would be "the

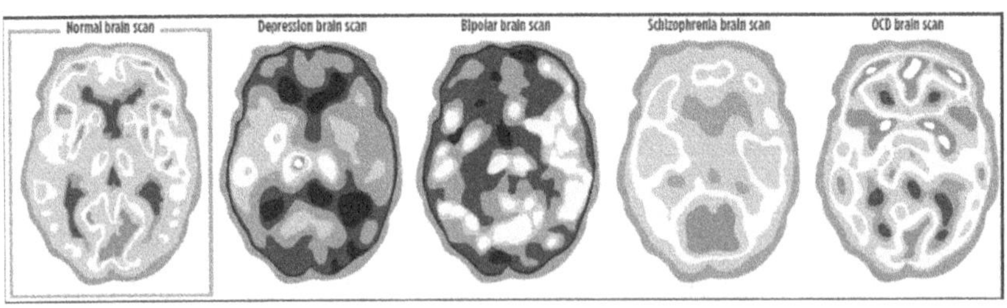


**Figure 8.1** PET scans compare the brain activity of people diagnosed with various mental disorders with the brain activity of those without such diagnoses. U. S. Government Office of the Director of National Intelligence. Mental Fitness Course.

decade of the brain" and by 2005 the National Institutes of Health were providing about $4 billion to support the field.[27] In 2013, President Obama rolled out the BRAIN initiative (Brain Research through Advancing Innovative Neurotechnologies) and allocated $100 million of funding for it. The goal of this enterprise is "to revolutionize our understanding of the human mind and uncover new ways to treat, prevent, and cure brain disorders like Alzheimer's, schizophrenia, autism, epilepsy and traumatic brain injury."[28]

The dawn of the neuroscientific era was marked by tremendous optimism in psychiatry. In 1984, Nancy Andreasen (b. 1938) wrote a popular book, *The Broken Brain: The Biological Revolution in Psychiatry*, where she confidently asserted regarding mental disorders:

> *These diseases are caused principally by biological factors, and most of these factors reside in the brain* . . . a large amount of evidence has been amassed suggesting that mental illness is caused by biochemical abnormalities, neuroendocrine abnormalities, structural brain abnormalities, and genetic abnormalities.[29]

Since the publication of the DSM-III, the view that mental illnesses are brain diseases has dominated psychiatric curricula, research, and scholarship. The title of Samuel Guze's 1988 article, "Biological Psychiatry: Is There Any Other Kind?" summarized the prevailing zeitgeist. Neurobiological approaches bring the most recognition, prestige, and research funding for those who study mental illness. Within a few decades psychiatry had radically recreated itself from a psychosocially oriented discipline to a profession focused on the study of brains and genetics.[30]

The current prominence of the biological view is such that even psychoanalysts justify their efforts through slogans that claim analysis changes brains: "We now have solid scientific evidence to suggest that the so-called 'talking cure,' originally devised by Freud, literally alters the way in which the neurons in the brain are connected to one another," analyst Susan Vaughan asserts.[31] In the 2015 epilogue to her canonical book, *Trauma and Recovery*, psychiatrist Judith Herman goes even further, asserting that neuroimaging techniques reveal the specific brain regions where unconscious traumatic memories reside: "Advances in neurobiology have documented the effects of trauma on the brain that cause 'repressed memories.' "[32]

Biogenetic explanations have spread from scientific work to dominate popular images of mental illness. Media coverage is far more likely to focus on new findings about the biological roots of mental disorder than findings from other perspectives.[33] Since 1980 popular magazines have increasingly used biomedical portrayals and endorsed medications as treatments of choice for mental illnesses.[34] Laypeople now use terms such as "chemical imbalance" to explain the reasons for their mental illnesses, "Katrina brain" to describe their responses to natural disasters, and "dopamine hit" to refer to pleasurable experiences.[35] In widely publicized testimony regarding the confirmation of Brett Kavanaugh to the Supreme Court in 2018, Christine Blasey Ford asserted that thirty-year-old memories of his sexual assault were "indelible in the hippocampus."[36] Genes have also become consumer commodities: millions of individuals use inexpensive services such as 23andme, Ancestry.com, and MyHeritage that sequence their genomes and assess their chances of developing various diseases, including mental disorders. The information they receive can become an integral part of the way they understand their selves and their experiences.[37]

By the end of the 20th century, changes in both the general and scientific cultures had totally transformed the intellectual landscape as biological views had regained preeminence in explaining human behavior. "At present the road to salvation is presumably through biological psychiatry, neuroscience, and genetics," Gerald Grob observed about psychiatry. [38] Neuroscientific researchers assumed that applying technological leaps in imaging and sequencing to the study of mental disorders would result in unprecedented new understandings. They took for granted that the DSM classifications accurately reflected underlying brain and genetic states and so did not develop any new taxonomy. Yet, the DSM revolution and the symptom-based diagnoses it spawned have turned out to be millstones for the neurobiological study of mental illness.

## Neurobiological Findings

Understanding the biological underpinnings of mental disorder requires not only knowledge about genes, neurotransmitters, and neurochemicals but also a clear idea of what condition embedded in the brain is being genetically transmitted. In 1942 psychiatrist Felix Brown succinctly summarized this necessity: "The part played by heredity in the development of the psychoneuroses is one of the fundamental unsolved problems in psychiatry, but the chief difficulty is to define the condition the heredity of which one is attempting to trace."[39] Accurate specifications of phenotypes (the presence or absence of some disease) are essential for the discovery of genotypes (the underlying genes that are connected to various disorders). This is especially important because, in contrast to most physical diseases, almost all mental illnesses lack objective markers such as x-rays, blood tests, or heart monitors, which can confirm or deny the presence of some pathology.[40] Classifications of disorders thus play an outsized role in studies of mental disorder compared to other medical conditions because they are typically the only tool available for confirming the existence of a pathological state. The DSM was the sole resource that neuroscientists possessed when they defined what condition they studied.

After 1980, interest in the genetic etiology of the DSM's specific entities flourished. Initially, the manual's specific and well-bounded categories seemed to provide clear definitions that had the promise of showing "the condition the heredity of which one is attempting to trace." Genetic researchers in the 1980s and 1990s expected to find the gene or genes that predicted the emergence of the various DSM disorders. For example, the discrete DSM-III categories of anxiety and depression, which earlier manuals had intertwined, implied that the presumed genetic mechanisms beneath each condition were distinct from the mechanisms that gave rise to the other. Yet, if the symptoms that marked the various DSM diagnoses did not correspond to different genotypes, neurobiological researchers would be led on a fruitless search for underlying causal processes that do not exist in nature. To date, results do not provide much confidence that the DSM categories have discrete underlying biological causes. Indeed, they indicate the opposite: the manual's distinct diagnoses are extraordinarily heterogeneous internally, highly overlapping with other entities, more continuous than categorical, and share common rather than specific vulnerabilities.

The development of genome-wide association studies (GWAS) in the first decade of the 21st century provided researchers with the opportunity to study hundreds of thousands or even millions of alleles. Given the estimates from MZ/DZ comparisons in twin studies, which had no capacity to specify particular genes, investigators initially expected that the GWAS would reveal the gene or genes that produced such large assessments of genetic influences. Instead, this research produced startling results: no single gene explains more than a small fraction of the genetic variation for any mental disorder.[41]

Some disorders, such as schizophrenia, display many genetic variants that each has a weak effect. GWAS research indicates that well over a hundred different genes increase the risk of susceptibility to schizophrenia. Others claim the existence of over a thousand such genes.[42] Cumulatively, genetic components account for only around 7% of the variance in individual differences in symptomatic presentations of schizophrenia, far less than estimates from twin studies had suggested.[43] Moreover, part of this small amount stems from rare mutations. Genomic studies also show that hundreds of genes might be associated with autism spectrum disorders.[44] Other conditions, such as major depressive disorder (MDD) and anorexia, have no replicated genetic associations at all.[45] To the extent that mental disorders are inheritable, they arise from complex interactions among many genes with weak cumulative influences. "We do not have and are not likely to ever discover 'genes for' psychiatric illness," Kenneth Kendler concludes.[46] Even more bleakly, psychiatrist Randolph Nesse observes: "This is the most important—and most discouraging—discovery in the history of psychiatry."[47]

A second central finding of biogenetic research concerns the internal heterogeneity of the presumably homogeneous DSM conditions. For example, studies consistently indicate that MDD, the core condition of diagnostic psychiatry, is not a single entity. What the DSM criteria call "major depressive disorder" consists of an as yet unknown number of diverse conditions. "There is every reason to believe," a recent review concludes, "that among the range of individuals currently subsumed under the diagnosis of Major Depression are those with distinct disease states mediated by very different underlying pathophysiological mechanisms."[48] A second review summarizes: "no variable has been clearly shown to define a more heritable or genetically homogeneous subtype of depression." This appraisal notes "no clinician or researcher believes that MDD is a single 'illness.'" It concludes: "Despite more than two decades of sustained investment in psychiatric neuroscience, the fundamental pathology underlying depression remains elusive, and the diagnostic criteria for depression remain descriptive in nature."[49] Even the supposedly homogeneous condition that Kraepelin identified as "dementia praecox" turns out to be far more heterogeneous than the German pioneer believed. While, as Kraepelin insisted, many people with schizophrenia show a deteriorating course as they age, large proportions stabilize, improve, or even recover as they grow older.[50]

Perhaps the most important conclusion emerging from biogenetic studies is that, contrary to the DSM assumption of disorder specificity, genes for virtually all psychiatric disorders are nonspecific. Each disorder does not correspond to a distinct gene or group of genes but instead shares a large amount of genetic vulnerability with other conditions.[51] Conversely, any given genetic variant that is tied to one diagnosis is also associated with multiple other diagnoses.[52]

Consider the findings from family studies about the co-occurrence among different relatives of the most common mental illnesses, depression and the various anxiety conditions. One major study concluded that depression "in the proband was associated with anxiety disorders, but only slightly with depression or alcoholism, in the relatives; anxiety in the proband was associated with major depression and alcoholism in relatives, but only slightly with anxiety disorders in the relatives."[53] Indeed, families with a depressed member have elevated rates of bipolar disorders, drug addiction, alcohol abuse, and eating disorders, among others.[54]

Anxiety, too, which the DSM-III carved into nine distinct conditions, seems more related to common, rather than distinct, processes.[55] Neuroscientist Joseph LeDoux indicates that the major forms of anxiety disorders reflect the "activation of one and the same underlying anxiety response."[56] Many studies of a single anxiety condition also find that rates of other disorders—especially depression—are higher among family members than rates of other anxiety disorders.[57] "[G]eneralized anxiety disorder and major depressive disorder are the same thing genetically," behavioral geneticist Robert Plomin concludes.[58] In addition, as has been proposed since Hippocratic times, results from biogenetic studies suggest that high levels of anxiety can reflect a personality type.[59] Such temperaments, as the French psychiatrist Jean-Etienne Esquirol emphasized in the early 19th century, are not themselves pathological but render people more vulnerable to develop many kinds of psychopathology.[60]

The most general conclusion from genetic research is that broad vulnerabilities, not specific mechanisms, underlie mental disorders. For example, a review of 537 studies with 21,427 participants compared functional MRI (fMRI) images of schizophrenia, bipolar disorder, MDD, anxiety disorders, and obsessive-compulsive disorder with nondisordered controls. It found many differences between the diagnosed and the control groups but none that were specific to any particular diagnosis. This analysis concluded: "The abnormalities in brain networks and network-regions we can observe with fMRI reflect disorder-general conditions that facilitate the emergence and persistence of symptoms but are insufficient for explaining symptomatic variability across disorders."[61] An even larger study of 265,218 patients and 784,643 controls published in *Science* in 2018 showed widespread genetic overlap between attention-deficit/hyperactivity disorder (ADHD), bipolar disorder, MDD, and schizophrenia.[62] Its senior author concludes: "The tradition of drawing these sharp lines when patients are diagnosed probably doesn't follow the reality, where mechanisms in the brain might cause overlapping symptoms."[63]

Epidemiological research reinforces these findings. It shows that over two-thirds of depressed people suffer from some anxiety disorder and over three-quarters also have some other mental illness.[64] People with an underlying susceptibility to develop symptoms of depression and anxiety share a common vulnerability that makes them highly reactive to life stressors.[65] For example, psychologist Robert Krueger examined the symptoms reported by respondents in a large national study without regard to particular categorical disorders. He found that all the symptoms of anxiety and affective disorders were aspects of a single broad condition, which he suggests might best be called "neurotic."[66] Likewise, the major American twin study found one common genetic risk factor predisposing to major depression, generalized anxiety disorder, and phobia and a second common factor for alcohol dependence, drug abuse/dependence, antisocial personality disorder, and conduct disorder.[67] The director of this study, Kenneth Kendler, concluded: "Genetic influences on [anxiety and depression] were largely nonspecific. That is, while genes may 'set' the vulnerability of an individual to symptoms of psychiatric distress, they do not seem to code specifically for symptoms of depression or anxiety."[68]

There is little evidence that depression or any of the particular anxiety disorders in the DSM have distinct patterns of family and genetic transmission.[69] Instead, findings suggest that a broad underlying etiological factor that might be called "neuroticism" seems to underlie a variety of anxious, depressive, and other distressing conditions. This trait indicates a tendency to experience a variety of negative emotions that encompass not just anxiety and depression but also fear, guilt, self-blame, and embarrassment, among others. Paradoxically, these results better correspond to the older concept of psychoneurosis (its psychosocial etiology notwithstanding) in the DSM-I and DSM-II than with the sharp categories of more recent DSMs. As earlier manuals portrayed them, mood and anxiety disorders appear to be interconnected aspects of a common distress-related syndrome. The findings of biogenetic studies indicate that mental disorders that used to be grouped under the "neurotic" category have fuzzy rather than discrete boundaries and stem from generalized vulnerabilities.[70]

These conditions contrast with a second set of equally wide-ranging "externalized" conditions such as ADHD, drug dependence, antisocial personality disorder, and conduct disorder that are marked by a lack of inhibition, the pursuit of self-interest, and social conflicts.[71] For example, studies consistently show that the 5-HT genetic variant is associated with general dimensions of impulsivity and aggression but is not specific to any particular mental disorder.[72]The various personality disorders show an even greater overlap with each other. People with one personality disorder are the exception; the rule is that they qualify for multiple diagnoses.[73]

Perhaps the most surprising finding from genetic research is that even those conditions that seemingly best fit the biomedical model—the psychoses—do not appear to be discrete entities. The foundation of Kraepelin's path-breaking work was that dementia praecox and manic depression had distinct symptoms, courses, and outcomes.[74] Genome-wide association studies, in contrast, demonstrate that schizophrenia and bipolar disorder share about two-thirds of their risk alleles.[75] Most patients have indications that typify both schizophrenic and bipolar states; moreover, their dominant symptoms shift across these conditions as they move through the life course. In addition, the same genes that elevate the risk of schizophrenia are associated with heightened risk of autism spectrum disorder.[76]

Other research indicates that families with a schizophrenic patient also have increased risk of having a member with an affective disorder; conversely, those with an affective patient show an increased risk of schizophrenia.[77] "[The] boundaries between schizophrenia and other psychiatric disorders are indistinct," one major review concludes.[78] Moreover, schizophrenia does not breed true across generations. Parental schizophrenia is as likely to be linked to conditions such as bipolar disorder, depression, or ADHD among children as is schizophrenia itself.[79] When people are followed long enough, they move in and out of diagnostic categories: "Today's patient with schizophrenia was yesterday's boy with conduct disorder or girl with social phobia (and tomorrow's elderly person with severe depression)," psychologists Avshalom Caspi and Terrie Moffitt conclude.[80]

Many neuroscientists now reject DSM (and Kraepelinian) views that separate schizophrenic and bipolar disorders and instead believe that they have common neurobiological origins. A recent evaluation summarizes: "The debate over whether the psychoses are best modeled as dichotomous or unitary concepts is fueled by mounting scientific evidence of their overlap in epidemiology, genetics, neuroanatomy and neuropsychiatry."[81] Former National Institute of Mental Health (NIMH) Director Steven Hyman suggests that very broad notions such as "psychosis" best characterize the results of neurobiological studies.[82] At the extreme, some researchers argue that a single dimension of general psychopathology unites all disorders.[83]

An inescapable conclusion from familial, genetic, and brain-based studies is that the DSM system artifactually divides a few general underlying vulnerabilities into multiple diagnoses.[84] In fact, neurobiological research uncovers overlapping characteristics of various psychiatric disorders and general as opposed to specific vulnerabilities that produce them. Genetic findings show no evidence for the inheritance of specific disorders, over and above the inheritance of more general liabilities to broad tendencies for neuroticism, externalization, or psychoses. Expansive susceptibilities that lie beneath many diverse conditions seem to characterize the genetic basis for mental disorders, a situation that is ill-suited for the current DSM diagnostic system. These results more closely resemble the conceptions of mental disorder in the first two DSM manuals than in those that followed. "We have been trying to map the landscape of mental disorders by drawing lines around clusters of symptoms as if they were islands, but mental disorders are more like ecosystems: areas of arctic tundra, boreal forest, and swamp blend into one another, defying crisp boundaries," Nesse nicely summarizes.[85] The central goal that Samuel Guze and Eli Robins, following Kraepelin, proposed for psychiatric classification—to develop homogeneous entities with distinct descriptions, explanations, and outcomes—remains almost entirely unrealized.[86]

## Drug Treatments

Drug treatments for both major and less severe mental disorders have been core components of psychiatric practice since the 1950s. The seeming specificity and effectiveness of the first generation of psychotropic drugs was one major stimulus for the rise of biological psychiatry. After 1980 drug companies became critical institutional sources of support for academic departments of psychiatry and replaced the NIMH as the major sponsor of research on drug efficacy. Because the initial psychotropic medications were no longer profitable after their patents expired, the pharmaceutical industry and allied psychiatrists searched for novel drugs to replace them. In the late 1980s older medications were abandoned for a plethora of new antipsychotics and antidepressants. Remarkably, by 2008 antipsychotic drugs had become the top-selling category of all prescription medications.[87]

Although different products have supplanted them in the marketplace, the initial drugs—chlorpromazine for schizophrenia, lithium for bipolar conditions, and imipramine and the monoamine oxidase inhibitors (MAOIs) for depression—still form the biological template for the most commonly prescribed products such as Abilify, Cymbalta, Zyprexa, Risperdal, and Seroquel. Despite the hyperbole that often surrounds the promotion of new medications, understandings of how they work have not surpassed those of the breakthrough treatments in the 1950s. "Our current drugs for the treatment of psychosis," a 2018 appraisal concludes, "are based on the serendipitous discovery of chlorpromazine and the isolation of D2 receptor antagonism as an underlying mechanism."[88] Likewise, sixty years after the discovery of antidepressant medications, neuroscientists still are not clear about how they work.[89] "All currently approved medications for depression," noted neuroscientist Eric Nestler writes, "are based on chance discoveries that were made more than six decades ago. . . . Despite several decades of research, the changes that the drugs induce in the brain that underlie their therapeutic actions still remain uncertain."[90]

Growing skepticism has arisen about the effectiveness of current medications; newer classes of drugs do not seem to be more efficacious than their precursors.[91] Despite the fact that more than sixty different types of antipsychotic drugs have entered the marketplace since chlorpromazine and Haldol were discovered in the 1950s, none show greater efficacy than the initial antipsychotics.[92] Although lithium, which has been widely employed since the 1960s, is now less likely to be prescribed than recent, far more costly medications, it remains an unsurpassed mood-stabilizing drug for treating bipolar conditions.[93] In addition, while current antipsychotic medications have different profiles of side effects than the initial drugs, it is not clear that they have more benign consequences overall than the original phenothiazines or lithium.[94]

The effectiveness of the later generation of the antidepressant serotonin selective reuptake inhibitors (SSRIs) and serotonin norepinephrine reuptake inhibitors (SSNIs) also does not surpass that of the early antidepressants. "Many antidepressant drugs have been developed since the 1950s," Steven

Hyman notes, "but none has improved on the efficacy of imipramine or the first MAOIs, leaving many patients with modest benefits or none at all."[95] Some observers even claim that rates of remission among patients who receive first-generation antidepressants such as the tricyclics exceed those taking newer classes of these drugs.[96] Because current antidepressants are designed for long-term use, they entail a high risk of severe side effects including weight gain, drowsiness, loss of sexual desire, and movement problems.

Many studies also indicate that much of the improvement that occurs with antidepressant treatments is attributable to placebo effects.[97] For example, one review that involved over nine thousand patients indicated that 42% of subjects who received an antidepressant improved compared to 30% who received a placebo.[98] Psychopharmacologist David Healy concludes that, in a typical group of ten patients who are given antidepressants or placebos, only one responds to the drug while nine either are unaffected or respond to the placebo.[99] Even these findings seem to be inflated because of biases where studies with positive results are far more likely to be submitted and published than those with negative results.[100] Analyses that take into account unpublished as well as published research show little difference in outcomes between those who receive an antidepressant or a placebo.[101]

Other evidence suggests that, while most consumers of SSRIs do not benefit from them, a smaller subgroup of persons with especially serious depressions do improve.[102] Recent well-publicized studies claim that the benefits of newer antidepressants are "minimal or non-existent, on average, in patients with mild or moderate symptoms" but are only "substantial" for those with severe depression.[103] The tremendous heterogeneity of the DSM diagnostic categories contributes to the failure of drug efficacy to exceed placebo response: it is difficult to ascertain if some subgroups benefit from a medication while others with the same diagnosis do not.

Another striking aspect of the ways in which the major classes of medication work is that they provide damning evidence *against* the DSM categorizations. One of the major tenets of diagnostic psychiatry was the belief that drug treatments worked in specific, rather than general, ways. This assumption stemmed from findings in the 1950s that chlorpromazine worked for schizophrenia, lithium for bipolar conditions, the tricyclics and MAOIs for depression, and the anxiolytics for anxiety but that these medications did not help other conditions. While earlier drugs were initially viewed as highly specific to particular disorders, their actions now seem far broader. Comparable to the nonspecific biogenetic etiology of psychiatric conditions, it appears that the major psychotropic drugs work in decidedly general ways.

In contrast to the notions of specificity that guided pharmacological approaches during the 1950s and 1960s and that strongly influenced the development of the DSM-III, the most common drugs are used across multiple diagnostic groups.[104] The comprehensive and nonspecific actions of the "antipsychotic" and "antidepressant" medications do not map onto particular conditions. For example, Healy cites "compelling evidence that the drugs called antipsychotics can be used to treat mood disorders, delirium, and anxiety disorders and that many antidepressants may be much more effective for anxiety states than for mood disorders."[105] Another expert summarizes: "On average, a marketed psychiatric drug is efficacious in approximately half of the patients who take it. One reason for this low response rate is the artificial grouping of heterogeneous syndromes with different pathophysiological mechanisms into one disorder."[106] Suspicions have grown that the DSM categories themselves are responsible for the stagnation in progress in drug development; no new, efficacious therapeutic agents have entered the market for the past thirty years.[107]

One of the most curious recent trends is the unexpected revival of psychedelic substances as treatments for mental disorders. Drugs associated with the 1960s counterculture and the recreational club scene since the 1980s are now heralded as groundbreaking responses to various mental illnesses. Ketamine, colloquially called "Special K," is a hallucinogenic drug that leads users to become detached from reality and that is also used as a "date rape" drug that incapacitates its victims. "[T]he discovery of ketamine," Eric Kandel reports, "has been hailed as the most important advance in depression research in the last half century."[108] Another psychiatrist makes a similar claim about a cannabis derivative: "CBD is the most promising drug that has come out for neuropsychiatric diseases in the last 50 years."[109] MDMA (ecstasy) is a prominent new treatment for post-traumatic stress disorder as is psilocybin (the active compound in hallucinogenic mushrooms) for depression, anxiety, and obsessive-compulsive disorder. Some professionals view Ayahuasca, a psychedelic drug used for centuries by indigenous tribes in South America, as an effective treatment for anxiety and addictions.[110] LSD, the signature drug of the 1960s counterculture, is undergoing a revival as a treatment for anorexia, addiction, and depression, among other conditions.[111] Pharmaceutical companies cannot be enamored with the revival of nonproprietary drugs that are used a handful of times rather than taken in daily doses across lifetimes.[112]

Despite the overall failure to achieve any breakthroughs in conventional drug treatments since the 1950s and the recent renaissance of psychedelic substances, many neuroscientists are convinced that knowledge about the brain will eventually increase to the point where they can target medications to fit the genomes of particular individuals. For example, identifying patients with different polymorphisms in their dopamine and serotonin systems might precede the prescription of particular antidepressant drugs.[113] In a book chosen for Oprah's Book Club, genetic epidemiologist Tim Spector proclaims: "In the future, when faced with a messy divorce, rather than resort to Prozac or gin and tonics, we could take specific highly tailored chemicals to epigenetically steady our sensitive genes until the crisis has passed."[114]

Eric Kandel goes further, asserting that individually tailored, gene-based medicine can be used to develop treatments well before any indications of a mental illness become apparent: "It is likely that personalized medicine, with its focus on clinical DNA testing—the search for small genetic differences in individuals—will reveal who is at risk of developing a particular disease and thus enable us to modify the course of that disease through diet, surgery, exercise, or drugs many years before signs and symptoms appear."[115] In his view, screening children, and perhaps even babies, for genetic traits related to, for example, depression or schizophrenia can identify those who are at risk, develop customized treatments for them, and in many cases prevent the disorder from ever arising at all.

At present such views reflect wishful thinking far more than realistic appraisals of the current state of knowledge. Existing medications rely on the same models that were accidentally uncovered in the 1950s. No new types of drugs have come on the market in recent years, nor are any in the pipeline.[116] The inertia surrounding the development of more effective and benign new psychotropic drugs has led most pharmaceutical companies to abandon efforts to develop innovative medications.[117] "There are very few . . . new ideas and almost nothing that gives any hope for a transformation in the treatment of mental illness," former NIMH Director Thomas Insel bleakly concludes.[118] The expectation that psychiatry will find some "penicillin for mental illness"[119] is as distant a goal now as it was sixty years ago.

## Why Has Neuroscientific Research Had Such Disappointing Results?

Three major problems remain unresolved in the turn to explain mental disorders through purely brain-based factors. The first stems from Galton's unfortunate legacy that views nature and nurture as competing forces. Another arises from the failure of neuroscientific research to seriously confront the ways that mental illnesses are in some ways distinct from organic diseases. Finally, their reliance on DSM criteria leads neuroscientific studies to conflate symptom-based conditions that are contextually explicable from those that arise from some dysfunction.

### Genes and Environments Are Connected

Galton's development of twin studies as a way to separate internal from external influences established an inauspicious precedent of viewing biological and social factors as in opposition to one another. These studies used heritability—the degree to which some trait is due to inherited genetic differences as opposed to environmental effects—to show the importance of internal vis à vis external causes of mental disorders. The legacy of isolating genetic from environmental forces has been a formidable barrier to understanding how mental disorders arise from the intrinsic interrelationships between biological vulnerabilities and problematic psychosocial conditions.

The failure of neurobiological research to discover powerful genetic effects partly stems from its Galtonian assumption that genetic impacts are independent from the environments in which they operate. In fact, the same genes that are expressed in some environments remain dormant in others. This inevitably reduces the overall association between genetic variation and the appearance of some mental disorder. The research of Krueger and colleagues, who use twins to study the heritability of neurotic personality traits, demonstrates the consequences of this neglect.[120] It found the typical result that across their entire sample of twins half of the variation in neuroticism was due to genetic forces and half to environmental ones. Yet, the influence of environments was far stronger in families with a high degree of conflict; conversely, genetic influences dominated among families that displayed less conflict. That is, the relative influence of genes depended on the social context in question, namely, the degree of discord within the family environment. Similarly, Pescosolido and colleagues show that the GASBA2 genotype, which puts men at higher risk for alcohol dependence, has much larger impacts on those who experienced material deprivation as children.[121] Conversely, this genotype does not lead to alcohol dependence among those with high amounts of familial social support. Standard models of heritability, which assume that genetic and environmental factors are distinct, are unable to detect these interactions.

A second blind spot in much genetic work is that typical interpretations of heritability inflate, often wildly so, the importance of inherited as opposed to environmental influences. This is because the heritability statistic only validly estimates the relative power of innate compared to acquired factors when rates of the condition in question are comparable across different settings. This might be the case for schizophrenia, a disorder that appears to have fairly similar prevalence across different societies as well as heritability estimates of as high as 0.80 .[122] Schizophrenia, however, is the exception, not the rule.

On average, studies of most mental disorders produce heritability rates of about 50% with the environment accounting for the remaining 50% chance of developing some condition. Yet, their prevalence shows vast differences across societies. Take depression, for example. One set of studies shows that its rates vary from a low of 3% of women in a rural area of Spain to a high of over 30% of women in an urban township in Zimbabwe.[123] Another summary of surveys in ten countries finds that depression rates range from 1.5% in Taiwan and 2.9% in Korea to 16% in Paris and 19% in Beirut.[124] Other mental disorders show equally steep divergences across cultures. For instance, social phobias diverge from 1.7% in Puerto Rico to 16% in Basel, Switzerland.[125] Rates of alcohol and drug dependence show even larger variations across national contexts. One cross-national study of alcoholism finds that the highest lifetime prevalence rate of 23% among a Native American population exceeds the lowest rate of .45% in Shanghai by more than 46 times![126] Amounts of disorders vary widely not only across groups but also over time. For example, 24-year-old Americans born between 1966 and 1975 were 800% more likely to have a substance dependence disorder than those born between 1936 and 1945.[127] The magnitude of such societal and temporal differences dwarf any inherited tendencies.

Because they ignore the vast variance in rates of mental disorders across space and time and only consider genetic influences within a single setting, researchers typically overinflate genetic and underestimate environmental influences. "The results are clear and consistent," one popular book concludes, "overall, heredity accounts for roughly 50% of the variation in the samples of people that have been tested, environmental influences for the other 50%."[128] A prominent twin researcher confidently asserted: "about two-thirds of the reliable variance in measured personality traits is due to genetic influence."[129] Another leading student of adoption claims: "Genetics is the most important factor shaping who we are. It explains more of the psychological differences between us than everything else put together."[130] In fact, except for conditions that have similar rates in different social contexts, studies of heritability provide seriously deceptive estimates of the extent to which innate or acquired factors account for mental illnesses. In most cases, social and cultural differences in rates of disorders are many orders of magnitude greater than genetic influences.

There are some promising signs that researchers are becoming increasingly aware that heritability estimates are misleading ways to look at the relative influences of genetic and environmental factors on mental disorders.[131] A growing trend, which the final chapter examines in more detail, rejects the Galtonian model that separates inherited from acquired traits and regards them as competing forces. Instead, researchers increasingly view genes and environments as inherently interactive. "Even in the genomic era," Alan Guttmacher and Francis Collins, the leaders of the National Human Genome Institute, emphasize, "it is not genes alone but the interplay of genetic and environmental factors that determines phenotype (i.e., health or disease)."[132]

## Are Mental and Physical Illnesses Comparable?

A second reason for the disappointing results of neurobiological research on mental disorders stems from the assumption—transported from the DSM—that mental illnesses are analogous to physical ones. Psychiatry's legitimacy is now fundamentally connected to its identity as a biomedical discipline. This requires it to consider mental illnesses as a subset of organically based diseases. Researchers routinely assume that mental disorders stem from molecular, cellular, and anatomical brain disturbances. "Mental illnesses are diseases that affect the brain which is an organ of the body just as the heart or stomach is," Andreasen summarizes.[133] While there is no doubt that the brain is a bodily organ, the question is whether it functions in the same ways as hearts or stomachs do. The brain could inherently be a different *type* of organ from other bodily systems so that consciousness and mental symptoms by their nature cannot be reduced to the actions of neurochemical systems.[134] Several factors argue for why brains are distinctive from other organs.

One reason regards the different role that social contexts play in defining mental compared to physical disorders. Context is typically not a major concern in judging whether some bodily system is working appropriately because it functions constantly and does not turn on and off. For the most part physicians do not need to understand the context in which an organic disturbance occurs; they only need to examine the performance of the system in question. A broken leg, for example, is a dysfunction regardless of the circumstances that led to it. Therefore, pathologists can accurately use the results of x-rays to determine if a break is present without knowing anything about how the fracture occurred.

In contrast, the assessment of mental functioning always involves contextual considerations.[135] As previous chapters showed, from Hippocratic times through the DSM-II, the "without cause" as opposed to "with cause" distinction separated normal from disordered psychic conditions. For example, the same symptoms that indicate a depressive disorder are natural when they emerge, in Robert Burton's words, after "every occasion of sorrow, need, sickness, trouble and grief."[136] Likewise, brain scans of soldiers going out on patrol in an Afghan neighborhood in which there has been recent violence and in which hypervigilance is essential would likely reveal brain activity in anxiety-generating centers at the same or higher levels as among people with anxiety disorders.[137] Yet, this anxiety would be an entirely normal response to extraordinarily threatening circumstances. The brain scans themselves, however, would not reveal whether the condition is a natural response to an extreme environment or an anxiety disorder. The same configuration of neurochemicals or electrical activity that is normal in the face of a direct threat can indicate a disorder when no danger exists. Because context is an inherent aspect of whether psychological processes are responding naturally, judgments about whether any symptom indicates a disorder must involve considerations of not only biological but also contextual criteria. This situation is less likely to arise in diagnosing physical illnesses.

Judgments about the presence of some mental disorder must be tied not only to contexts but also to cultural beliefs in ways that are rarely relevant for diagnoses of bodily conditions. For example, someone in our society who believes that devils, witches, or ghosts have taken control of his or her mind is likely to have a serious mental disorder. Similar beliefs in prior eras or in many current non-Western societies are more likely to reflect culturally legitimate expressions than mental disorders. For example, in 16th-century England: "As long as men continued to believe that Satan could appear to rational people, encounters with him could not be dismissed simply as symptoms of madness."[138] Diagnoses of mental illnesses, in contrast to those of organic conditions, must consider the cultural matrix of presenting symptoms.

Another fundamental difference between mental and physical disorders stems from the relationship between the underlying disease and its overt appearances. Symptoms provide direct indicators of physical diseases. The manifestations of cancer, heart disease, diabetes, and the like are unmediated results of these conditions. Their symptoms are comparable regardless of the cultural or social context in which they appear. This is not the case for most mental disorders, where manifest symptoms are often not straightforward products of brain processes.

Even when some organic process does underlie a mental disorder, a variety of external factors can shape what phenotypical expressions some genotype displays. Biological vulnerabilities for many disorders are molded to fit culturally normative ways of presenting symptoms. Consider depression and aggression. Certain cultures, such as the Amish, have such powerful restrictions on aggression that they almost never display overt aggressive behaviors.[139] A member of the Amish culture with a putative gene for anger might never engage in aggressive actions. Instead, the same genetic factors among the Amish can lead to an elevated probability of developing an affective disorder. Even within the same group, the cultural structuring of psychiatric disorders might lead women to express a genetic tendency to aggression through depression or self-harm while men with the same genotype might become violent rather than depressed.[140]

Anorexia nervosa provides another illustration of the difficulties in associating specific genes with specific forms of mental disorder. Symptoms of anorexia were rare before the 1970s. Although they are now spreading throughout the world, they are typically found in Western postindustrial societies among White, young females of relatively high social class backgrounds. Given the cultural and historical specificity of the disorder, it is unlikely that genes for anorexia underlie the symptoms that anorexics display. In other times, places, and cultures, people with the putative genes that produce anorexia in modern America would not have developed the same phenotypical expressions of these genes. The answer to the question of "what is a gene for anorexia a gene for?" is unlikely to be "anorexia."[141] Instead, a far more general trait such as compulsion or depression that has variable manifestations in different cultural contexts is likely to underlie the genetic component of anorexic symptoms. These considerations contrast with the characteristic operation of the biological processes that underlie physical diseases.

Social contexts can have a profound influence over not only the form that expresses a possible genetic tendency to disorder but also whether a disorder arises at all. People with genetic tendencies to various common mental disorders have varying chances of expressing these tendencies in different environments.[142] For example, the cultural restrictions against alcohol consumption among Mormons make the expression of genes that make people more prone to alcoholism far less likely to appear in this group than, for example, in members of the Navaho culture who do not face strict norms regulating drinking.[143] A Mormon who inherits a gene for alcoholism will therefore have a very different probability of developing this condition than a Navaho who inherits the same gene. Likewise, gambling disorders will not arise in groups where there are no opportunities to gamble. Only certain environments might activate genetic tendencies that would otherwise be dormant.

Therefore, many mental disorders—even those with strong genetic components—have expressions that are historically and culturally contingent. One-to-one correspondences between genetic and polygenetic factors with manifest symptoms are exceptions, rather than rules. This problem greatly complicates the genetic study of mental disorders because of the difficulty in reliably knowing what phenotypic measures are connected with what genes. Even when they exist, genetic predispositions are often so plastic that cultural rather than genetic influences shape what specific symptoms emerge. The variable relationships of manifest mental symptoms and biological foundations are highly problematic, so that the problem of how "to define the condition the heredity of which one is attempting to trace" is very difficult to overcome. In contrast to the study of physical diseases, biological studies of mental disorders face a difficult challenge of associating particular symptoms with particular genes.[144]

A third basic dilemma about the connection between mental and physical disorders concerns the nature of the relationship between brains and minds. Neuroscientific research has unquestionably shown that certain mental states are associated with certain brain states. Even before the development of biological psychiatry in the 19th century, many scientists assumed that mental conditions originated from underlying brain events. Like their predecessors, neuroscientists generally take for granted that brain processes have causal primacy over psychic phenomena. The actual relationship is likely to be far more complicated. While it is undoubtedly true that brain patterns accompany mental patterns, it is far less apparent whether some brain profile precedes the mental state or whether the mental state gives rise to the brain pattern.[145]

Explanations of mental disorders must be dynamic, multilevel, and multidimensional, encompassing not just neurobiological bases but also cultural expectations, social context, and individual interpretations, among other factors. The different roles that context and culture play in mental compared to physical conditions, the fact that physical but not mental symptoms are direct indicators of underlying diseases, and the intrinsically complex and multilayered relationship between brains and minds render statements like "mental illnesses are diseases of the brain" simplistic or meaningless. Such one-dimensional proclamations deflect attention from the extraordinarily complex relationships between mental and physical processes.

## Are Brain Processes Normal or Abnormal?

The fact that any phenomenon is considered to be biologically or brain based has no implications in itself for whether it is normal or abnormal, desirable or undesirable, good or bad.[146] Adequate conceptions of normality are foundational for accurate knowledge about what is pathological. To understand when a psychological mechanism is not operating appropriately requires understanding about when it is working properly. Psychiatrists and other mental health professionals, however, rarely consider what appropriate psychological functioning involves. For example, the standard psychiatric textbook devotes a half page out of 4,500 pages to a discussion of normal emotions.[147] "It is indeed astonishing," neurologist Antonio Damasio exclaims, "to realize that [medical] students learn about psychopathology without ever being taught normal psychology."[148]

Yet, neuroscientists often interpret findings that link symptoms to brain operations as indicating some condition must be a disorder. For example, Eric Kandel asserts: "All mental processes are brain processes, and therefore all disorders of mental functioning are biological diseases."[149] Because all human conditions, those that are healthy as well as those that are disordered, have some connection to the brain, such statements are tautologous.

Although researchers often associate neurogenetic influences with abnormality, this need not be the case. Genetic studies of homosexuality provide an example.[150] Before 1973, psychiatric manuals considered homosexuality to be a mental disorder because in this condition "sexual interests are directed primarily toward objects other than people of the opposite sex [or] toward sexual acts not usually associated with coitus."[151] Indeed, it was one of only fourteen conditions in the Feighner criteria.[152] Had a putative gene for homosexuality been discovered at that time, it would have been considered as strong evidence that being gay was a mental illness.

In the 1990s, when homosexuality was no longer viewed as a mental disorder, articles about the "gay brain" became front-page news as researchers claimed to identify specific biological bases of homosexual behavior. In 1991 neuroscientist Simon LeVay published a study in *Science* reporting that the hypothalamus in the brains of gay men who had died from AIDS tended to be smaller than in the brains of a comparison group, suggesting that gay men might have a different neurophysiology than heterosexual men.[153] Two years later, Dean Hamer, a geneticist at the National Cancer Institute, reported in *Science*: "it appears that [the chromosomal region] Xq28 contains a gene that contributes to homosexual orientation in males."[154] Hamer's study received widespread publicity, and Xq28 soon became widely known as "the gay gene." Yet, both LeVay and Hamer believed that rooting homosexuality in physiological and genetic processes showed it was *normal,* not disordered. They became public advocates for the notion that homosexuality is a biologically based "normal variant in human behavior."[155]

Since the 1990s the notion that many homosexuals are "born gay" has attained increasing credibility and acceptance. The vast majority of publicly gay men assert that they have always been gay. Widespread support for biological explanations of homosexuality also exists in the general population. In many writings about homosexual identity, the presumed genetic basis of homosexuality came to replace more voluntary notions of "chosen lifestyle," not to mention the idea that homosexuality is a mental disorder. The perception that homosexuality is rooted in biology has been at the heart of the normalization of this phenomenon.[156]

As the example of homosexuality indicates, showing a correlation between any condition and some gene does not resolve the problem of how to separate normal from abnormal brain structures and functions. The basic problem is simply that biological processes underlie nondisordered human traits as much as disordered ones. The classification system used to distinguish normal from abnormal brains—at present, the DSM system—supplies the answer to the question of whether or not some condition is a mental disorder. When homosexuality was classified as a disorder, a genetic grounding would have been viewed as evidence for why carriers were mentally ill. Once homosexuality was normalized, the same gene was seen as indicating a normal genetic variant.

The same difficulty arises in the interpretation of images of brain activity. For example, some neuroscientists claim that people can become addicted to video games, and the APA is considering classifying Internet gaming disorder as a mental disorder.[157] They assert that gaming arouses the same brain areas associated with heightened dopamine production as do addicting drugs such as methamphetamines. Yet, all pleasurable activities—having a good meal, getting sexual pleasure, enjoying a movie, and so on—can increase dopamine levels. Brain scans, in themselves, cannot distinguish normal gratifications from addictive disorders. Researchers must infer the presence of a dysfunction from the assumption that the brains they scan are ones of people who are already presumed to have some disorder.

Depression provides another example. Genetic influences partly explain why people naturally grieve when an intimate dies. The fact that there is a genetic basis for grief does not show that grief is a disorder—rather, bereavement is biologically grounded in normal, not in defective, genetic processes. Brain scans used in studies that induce states of sadness in normal participants indicate biological changes comparable to those found among persons with depressive disorders.[158] This result suggests that these regions are related to normal sad moods and pathological depression alike. If scientists consider only the correlation between biological markers and DSM criteria without taking into account the context in which they occur, they would wrongly conclude that a marker indicates a disorder rather than some phenomenon common to normal sadness and disorder.

Research that shows a biological or genetic correlate of a psychic condition says nothing in itself about whether or not that condition is a disorder. Therefore, it is premature to conclude, as Kandel does:

> [W]e do know that all psychiatric disturbances result from specific changes in how neurons and synapses function, and we also know that insofar as psychotherapy works, it works by acting on brain functions, creating

physical changes in the brain. Thus, we now know that psychiatric illnesses, like neurological disorders, arise from abnormalities in the brain.[159]

Imaging and other techniques that claim to show that psychiatric illnesses "arise from abnormalities in the brain" cannot in themselves reveal disorders. Instead, they must rely on the DSM criteria to separate who is disordered from who is not. To the extent that DSM definitions do not adequately distinguish normal from abnormal conditions, neurobiological studies will misleadingly claim to be examining some disordered condition.

The neurobiological approach has yet to produce a coherent answer to Felix Brown's question of how to define when a disordered condition is present. "Functional neuroimaging," philosopher and psychologist Derek Bolton explains, "is critical for many purposes, it has many scientific and clinical uses, but diagnosing mental disorder—in the sense of telling whether a psychological condition is a mental disorder or is normal is not one of them."[160] As the next chapter discusses, many neuropsychiatrists have come to realize that the DSM system has become a liability for their efforts to separate mental disorders from normal brain responses.

## Conclusion

Throughout history, understandings of mental illness have moved back and forth between a focus on brains or minds, on nature or nurture.[161] The most recent period has been marked by a reductionist turn toward brains and away from the external world. The pendulum of the dominant scientific thought community has swung from a denial of biological effects on human behavior to a primary concentration on the brain.[162] Brains have also attained iconic cultural status. Both scientific and popular reports associate the presumed demonstration of a brain-based influence with the primary cause of some behavior. They assume that cells, molecules, and genes are the basic level of reality to which other factors can be reduced.[163] This assumption carries the risk that current understandings will suffer the same fate as 19th-century brain anatomists who "failed so miserably because they focused on the brain at the expense of the mind."[164]

It is undoubtedly the case that knowledge about the structural and functional qualities of the brain has soared in the past decade.[165] Likewise, methods of assessing brain structure and function have grown far more advanced and precise. The sequencing of the human genome provides extraordinary opportunities for understanding the mechanisms underlying many human behaviors. Neuroscientific studies demonstrate beyond a doubt that brains are associated with all forms of human behavior. At the same time, the discovery of particular neurotransmitters and receptors has created the potential for sophisticated manipulation of psychiatric symptoms.

As yet, however, despite rhetoric to the contrary, these advances have not led to significant advances in knowledge about the causes, prognoses, or treatments for any mental disorder. Instead, the findings from neuroscientific studies are actually more congruent with the nondistinct, overlapping diagnostic model used before the DSM-III than to the specific current DSM categories. One consistent result is that many genes are associated with any single condition; a particular gene contributes only a small amount of variance to the emergence of some diagnosis. Another is that each of the central DSM categories—for example, depression, anxiety, schizophrenia, or bipolar disorder—displays both great internal heterogeneity and high overlap with other diagnoses. Many of the symptoms that characterize mental disorders are widely distributed across entities and are not localized within particular diagnoses. A third is that the central mental disorders are related to a small number of general vulnerabilities that might be called "psychoses," "internalized neuroses," and "externalized neuroses."[166] The major drug treatments, as well, target broad rather than specific conditions. Each of these major conclusions thoroughly *contrasts* with the DSM model of discrete diagnoses.

In the 21st century, suspicion grew that the DSM classification system itself might be one reason for the inability of neuroscience to produce any clinical breakthroughs.[167] In a reversal of the enthusiasm that marked the initial period of biological psychiatry's resurgence, many leading proponents of this perspective have acknowledged serious defects that might be uncorrectable without a basic change in the extant diagnostic system. Steven Hyman, a former director of the NIMH, recognizes a "gaping disconnect" between the immense progress in neuroscientific research and the total absence of translating these findings into clinical practice.[168] A prominent European psychiatrist bemoans the use of "star wars technology on bow and arrow diagnoses."[169] Thomas Insel, who directed the NIMH between 2002 and 2015, laments: "In the rest of medicine, [using clusters of symptoms] would be equivalent to creating diagnostic systems based on the nature of chest pain or the quality of fever. Indeed, symptom-based diagnosis, once common in other areas of medicine, has been largely replaced in the past half century as we have understood that symptoms alone rarely indicate the best choice of treatment."[170]

Many researchers have come to attribute the failure of the neuroscientific revolution to produce new understandings and treatments for mental illness to the basic flaws of the DSM itself. They proposed replacing the current categories with a new classificatory system that focused on understanding the neurobiological circuits that cut across traditional diagnostic boundaries. Their disappointment with the DSM set in motion a process that they hoped would fundamentally transform the symptom-based diagnostic nosology.

## Acknowledgments

Portions of this chapter are adapted from Horwitz, 2002, Horwitz and Wakefield, 2007, and Horwitz, 2016.

## Notes

1. Crick, 1994, 3.
2. Kandel, 2018, 8.
3. Lieberman, 2015, 292.
4. Sharfstein, 2005.
5. Grob, 1973; Shorter, 1997. There were, of course, some differences between the current and the original biomedical views (Harrington, 2019). The latter focused on brain anatomy and had no knowledge of biochemical processes. It also used hereditarian theories of degeneration that are foreign to today's approaches. Yet, given the drastic improvements in available technologies between the 19th and 21st centuries, the similarities in outlooks are remarkable.
6. E.g. Darwin, 1872/1998.
7. The term itself did not appear until 1900.
8. See, e.g., Degler, 1991.
9. Degler, 1991, 41.
10. Turkheimer, 2015, 228.
11. https://www.nytimes.com/2018/01/25/science/children-parents-genes-education.html.
12. Degler, 1991.
13. Durkheim, 1893/1984, 144–45.
14. Mead, 1928/2001, 280.
15. Wilson, 1998, 266.

16. E.g. Dawkins, 1976; Dennett, 1995; Pinker, 1997.
17. Segalowitz, 1999.
18. Kallmann, 1946; Pardes et al., 1989; Wender et al., 1986; Bertelson, Harvald, & Hauge, 1977.
19. Stein, Jang, & Livesley, 1999; Horwitz, 2002, 132–57.
20. E.g. Heston, 1966; Polderman et al., 2015; Plomin, 2018.
21. Rasic et al., 2014.
22. Cancro, 2000; Eisenberg, 1995.
23. E.g. Dumit, 2004. Without question, knowledge about the ways in which the brain works has soared since 1980. Neuroscientific findings have decisively refuted the localized conceptions of brain functioning and the notion of specific, mechanism-based disorders that dominated biological approaches since the 19th century. The field has demonstrated the highly interconnected nature of various brain regions: all mental processes depend on coactivation and cooperation of multiple brain circuits.
24. E.g., the cover of the July 3, 2000, issue of *Time*.
25. https://www.linkedin.com/company/societyforneuroscience/.
26. Jasanoff, 2018, 20.
27. Panofsky, 2014, 35.
28. http://braininitiative.nih.gov/about.htm. Internal Neuroethics Society, 2013, quoted in Whooley, 2019, 235.
29. Andreasen, 1984, 30, italics in original.
30. For examples see Andreasen, 1984; Guze, 1989; Sabshin 1990.
31. Vaughan, 1997, 4. Andreasen, (2001 31) too, observes: “Psychotherapy, sometimes denigrated as ‘just talk,’ is in its own way as ‘biological’ as the use of drugs.”
32. Herman, 2015, 256. In fact, brain imaging has no method to distinguish when memories are true or false. See Sacks, 2017.
33. Conrad, 1997. One sociologist estimates that from 2013 to 2014 alone there was a tenfold increase in media coverage of genetically related topics (Bliss, 2018, 193).
34. Clarke & Gawley, 2009.
35. http://www.slate.com/articles/health_and_science/science/2013/07/what_is_dopamine_love_lust_sex_addiction_gambling_motivation_reward.html. The impact of Internet use is attributed to “short-term, dopamine-driven feedback loops” (Osnos, 2018, 35). One neuroscientist connects the use of smartphones to the production of dopamine, which is “the same system that is implicated in addictions and drugs” (Rubin & Peltier, 2018, A4). Katrina brain: https://www.nytimes.com/2017/08/29/us/new-orleans-katrina-houston.html.
36. https://www.usatoday.com/story/news/nation/2018/09/27/christine-blasey-ford-brett-kavanaugh-sequelae-hippocampus-norephinephrine-epinephrine/1442969002/.
37. The information these companies receive can also be shared with pharmaceutical companies with the goal of developing drugs targeted to particular genomes. See https://www.wired.com/story/23andme-glaxosmithkline-pharma-deal/?mbid=social_fb.
38. Grob, 1998, 204.
39. Brown, 1942.

40. Some forms of dementia, Alzheimer's disease, and Huntington's disease are exceptions, although the first can only be conclusively identified after death. Some psychiatrists believe that the dexamethasone suppression test can often detect melancholic forms of depression. See Shorter, 2013a, 84.
41. van Dongen & Boomsma, 2013.
42. Schaffner, 2013, 1009.
43. Schizophrenia Working Group, 2014; Gratten, 2016.
44. Kandel, 2018, 55.
45. McGuffin & Rivera, 2015, 161–62; Major Depressive Disorder Working Group, 2013; Boraska et al., 2014; Border et al., 2019.
46. Kendler, 2005, 1250. Another factor contributing to the disillusionment with genetic findings is the failure to replicate results. A recent review concludes: "In neuropsychiatric genetics, despite a considerable body of work, studies of candidate genes largely failed to lead to broadly reproducible results" (Purcell, 2018, 5).
47. Nesse, 2019, 9.
48. Nestler, 2018, 382.
49. Levinson & Nichols, 2018, 301, 299.
50. Davidson, 2013, 206–7.
51. Hyman, 2018, 944.
52. Hyman, 2008.
53. Merikangas, Risch, & Weissman, 1994, 75.
54. Gershon & Nurnberger, 1995; Weissman et al., 2006.
55. All of these conditions remain in the fifth edition of the DSM although the latest manual removes obsessive-compulsive and post-traumatic stress disorders from the anxiety category.
56. LeDoux, 1998, 230.
57. Goldberg, 2015.
58. Plomin, 2018, 67. See also Middeldorp et al., 2005.
59. Akiskal, 1998; Cloninger, 1986.
60. Esquirol, 1838/1965.
61. Sprooten et al., 2017.
62. Anttila et al., 2018.
63. https://www.sciencedaily.com/releases/2018/06/180621141059.htm.
64. Van Dam, Iacoviello, & Murrough, 2018, 290.
65. Barlow, 1988.
66. Krueger, 1999.
67. Kendler et al., 2003.
68. Kendler et al., 1987, 457.
69. Breier, Charney, & Heninger, 1985.
70. Krueger & Markon, 2006; Krueger & South, 2009. See also Thapar & McGuffin, 1997; McKeon & Murray, 1987; Kendler, 1996; Andrews et al., 1990; Stein, Jang, & Livesley, 1999.
71. Krueger et al., 2005.
72. Kupfer, First, & Regier, 2002, 139.

73. McGlashan et al., 2000; Skodol et al., 2002; Zachar & Krueger, 2013 President Donald Trump, who has been the subject of considerable diagnostic interest, provides an example of the confusion that marks the diagnoses of personality disorders. A book, *The Dangerous Case of Donald Trump: 27 Psychiatrists and Mental Health Experts Assess a President*, tries to answer the question of whether he is mentally ill. All contributors agree that he has some mental illness: different assessors consider Trump to have, among others, a narcissistic personality disorder, antisocial personality disorder, paranoid personality disorder, and histrionic personality disorder (Lee, 2017).
74. Lichtenstein et al., 2009.
75. Duan, Sanders, & Gejman, 2010; Cross-Disorder Group, 2013. Craddock & Owen, 2010; Lichtenstein et al., 2009; Tamminga et al., 2013. Famed mid-19th-century German psychiatrist Wilhelm Griesinger anticipated these findings, asserting that there was but one underlying type of psychosis (Harrington, 2019, 15).
76. Chaste & Leboyer, 2012. In addition, neuroscientific evidence is also accumulating that indicates bipolar and unipolar depressions are not distinct entities (Duffy et al., 2014; Shorter, 2015, 68–98).
77. Charney & Sklar, 2018, 162.
78. Fanous & Kendler, 2008.
79. Gottesman & Gould, 2005.
80. Caspi & Moffitt, 2018, 835.
81. Meyer, Walsh-Mesinger, & Malaspina, 2018, 149.
82. Hyman, 2010, 171.
83. Caspi & Moffitt, 2018.
84. Hyman, 2018, 943; Krueger, 1999.
85. Nesse, 2019, 25.
86. Robins & Guze, 1970.
87. https: //www.reuters.com/article/ims-uspharmaceuticals/us-prescription-drug-sales-hit-300-bln-in-2009-idUSN3122364020100401.
88. Sarpal & Malhotra, 2018, 284.
89. Krystal & Charney, 2018, 387.
90. Nestler, 2018, 377.
91. van Dongen & Boomsma, 2013.
92. Marques & Kapur, 2018, 267.
93. Taylor, 2013, 86, 110.
94. Tyrer & Kendall, 2009, 4–5; Davidson, 2013, 209.
95. Hyman, 2012, 1.
96. Taylor, 2013, 78.
97. Kirsch et al., 2008; Moncrieff, Wessely, & Hardy, 2004; Pigott et al., 2010.
98. Taylor, 2013, 76.
99. Healy, 2008, 129.
100. Angell, 2011.
101. Ioannidis, 2008.
102. Gueorguieva, Mallinckrodt, & Krystal, 2011; Thase, Larsen, & Kennedy, 2011.

103. Kramer, 2016, 174.
104. Charland, 2013, 171.
105. Healy, 2008, 89.
106. Wong et al., 2010, 1276.
107. E.g. Shorter, 2015.
108. Kandel, 2018, 70. See also: https://www.nytimes.com/2018/11/30/opinion/sunday/suicide-ketamine-depression.html.
109. https://www.nytimes.com/2018/10/27/style/cbd-benefits.html.
110. Pollan, 2018.
111. https://www.nytimes.com/2019/09/04/science/psychedelic-drugs-hopkins-depression.html
112. In March 2019 the FDA approved the use of a ketamine-based nasal spray for people with severe depression. https://www.nytimes.com/2019/02/12/health/depression-drugs-ketamine.html.
113. Conley & Fletcher, 2017, 160.
114. Spector, 2012.
115. Kandel, 2018, 253.
116. https://www.nytimes.com/2013/08/20/health/a-dry-pipeline-for-psychiatric-drugs.html.
117. Kramer, 2016, 229.
118. Quoted in Miller, 2010, 502.
119. Andreasen, 2001, xi.
120. Krueger et al., 2008.
121. Pescosolido et al., 2008.
122. Sullivan, Dal, & O'Donovan, 2012.
123. Brown, 2002.
124. Bromet et al., 2011.
125. Merikangas et al., 1996.
126. Helzer & Canino, 1992.
127. Warner et al., 1995.
128. Miller, 1998, 23.
129. Bouchard, 1994, 1700.
130. Plomin, 2018, viii.
131. The term "missing heritability" has arisen to refer to the failure to find specific alleles that account for any mental disorder. See Conley & Fletcher, 2018, for a good discussion of the variety of genetic and environmental interactions on human behavior in general.
132. Guttmacher & Collins, 2003, 997.
133. Andreasen, 2001, 27; Andreasen, 1984, 8.
134. Parnas & Bovet, 1995.
135. The centrality of context creates serious problems in interpreting the results of brain scans, which take place in contexts that do not resemble any naturally occurring settings. Brain activity is measured while people lie flat and still in a narrow tube situated in an unfamiliar room surrounded by unfamiliar people. Stimuli that emerge

from this highly artificial setting are not comparable to those in the social world outside of the laboratory (Kagan, 2017, 14–16).

136. Burton, 1621/2001, 143.
137. Horwitz & Wakefield, 2012, 25.
138. MacDonald, 1981, 156.
139. Eaton & Weil, 1955.
140. E.g. Horwitz & Davies, 1994.
141. Grandparents and parents of a child with anorexia have higher than average rates of some mental disorder, but that disorder is very unlikely to be anorexia (Kagan, 2017, 27–28).
142. Schwartz & Corcoran, 2017.
143. O'Dea, 1957; Mail, 1989.
144. Gilger, 2000.
145. Kagan, 2017, 101–2.
146. Hecht, 2015, 246.
147. Sadock et al., 2009; See Nesse, 2019, 50.
148. Damasio, 1994. 255. Sociologist Nikolas Rose (1998, 26) generalizes this observation: "Our vocabularies and techniques of the person, by and large, have not emerged in a field of reflection on the normal individual, the normal character, the normal personality, the normal intelligence, but rather, the very notion of normality has emerged out of a concern with types of conduct, thought, expression deemed troublesome or dangerous."
149. http://www.apa.org/monitor/2012/06/roots.aspx. Kandel's statement echoes psychiatrist Seymour Kety's (1974, 961) often quoted and equally tautologous 1974 proclamation: "if schizophrenia is a myth, it is a myth with a significant genetic component!"
150. Shostak, Conrad, & Horwitz, 2008.
151. APA, 1968, 44.
152. "This diagnosis is made when there are persistent homosexual experiences beyond age 18" (Feighner et al., 1972, 61).
153. LeVay, 1991.
154. Hamer et al., 1993.
155. LeVay & Hamer, 1994.
156. Much earlier, prominent sexologist Havelock Ellis posited that the inborn nature of homosexuality meant that it was a natural rather than defective behavior (Bayer, 1987, 21).
157. https://www.nytimes.com/2017/04/01/opinion/sunday/video-games-arent-addictive.html; https://www.nytimes.com/2018/09/07/opinion/sunday/teenager-anxiety-phones-social-media.html.
158. Mayberg et al., 1999.
159. Kandel, 2018, 29.
160. Bolton, 2008, 62.
161. Makari, 2015, 510.
162. Eisenberg, 1995.

163. Kandel, 1998; Pinker, 1997.
164. Harrington, 2019, 13.
165. Schwartz & Corcoran, 2017.
166. Psychiatrist Paul McHugh (1999) groups all mental illnesses into four categories based on what factors cause them: brain diseases, vulnerable constitutions, dysfunctional lifestyles, and stressful life events.
167. Parnas, 2015, 181.
168. Hyman, 2007, 725. See also Hyman & Nestler, 1993.
169. Katschneg, 2010.
170. https://www.nimh.nih.gov/about/directors/thomas-insel/blog/2013/transforming-diagnosis.shtml.

# 9

# The Successes and Failures of the DSM Revolution

> Every generation, moreover, insisted that the specialty stood on the threshold of fundamental breakthroughs that would revolutionize the ways in which mental disorders were understood and treated. In the nineteenth century the instrument of change was the mental hospital. In the mid-twentieth century, psychodynamic and psychoanalytic psychiatry became the vehicle by which the mysteries of normal and abnormal behavior would be revealed. At present the road to salvation is presumably through biological psychiatry, neuroscience, and genetics.
>
> —Gerald Grob, "Psychiatry's Holy Grail" (1998, 217)

The DSM-III solved the psychiatric profession's deep crisis of legitimacy in the 1970s through firmly equating mental with physical illnesses. It did not, however, ground its many specific diagnoses in any causal system. The neuroscientific revolution that began in the following decade took the next step, supplanting the DSM's theory neutral approach with a brain-based focus on psychic problems. Having created the impression that the DSM diagnoses had neurobiological causes, a key unresolved problem was how to distinguish pathological from normal conditions.

Despite its brief attempt to provide a general definition of mental disorder, which many of the criteria sets for particular diagnoses disregarded, the DSM-III revolution had ignored issues of validity—for example, whether diagnostic criteria accurately distinguish symptoms that result from a dysfunction from those that are appropriate responses to given contexts, if dimensions or categories best represent psychiatric conditions, whether general or specific vulnerabilities produce disorders—in favor of developing reliable categories of measurement. Neuroscientific research, too, neglected questions about validity and simply accepted the DSM criteria. During the 21st century, previously overlooked concerns about whether the various diagnostic sets were valid measures of mental disorder came to the fore.

## A Successful Revolution

In an incredibly short period of time after 1980, diagnostic psychiatry gained almost unchallenged professional and cultural dominance. The notion that the mentally ill suffer from discrete diseases rapidly became so thoroughly embedded in the knowledge system and organization of the psychiatric profession that these conditions seemed to be the only possible subject matter of the field.[1] Researchers had to use the DSM diagnoses to obtain funding and to publish articles in respectable journals. Textbooks were organized through reference to these entities. Despite fierce initial criticism from psychoanalysts and behavioral psychologists, the DSM diagnoses also quickly came to dominate curricula and practices in psychology, social work, and all other mental health disciplines.[2] After 1980, all mental health professionals played by the psychiatric profession's diagnostic rules. Although many, or even most, clinicians were skeptical about the authenticity of the diagnostic system, the vast majority, including dynamic practitioners, promptly adopted the new nosology because of its practical benefits in garnering both reimbursement and legitimacy.[3]

Government agencies, professional organizations, pharmaceutical companies, advocacy groups, and political figures sponsored educational campaigns that widely promoted the idea that mental illnesses were genuine diseases.[4] These efforts assured sufferers that they had medical conditions that professionals could effectively treat. The general public swiftly came to take the existence of the DSM's distinct mental disorders for granted. Experiences with them became the themes of many talk shows, television programs, popular news stories, and best-selling books.[5] The Internet, as well, features voluminous information, advice, and sources of interaction about these disorders. The belief that mental illnesses are distinct medical conditions has such great cultural resonance at present that it is difficult to imagine that they are anything but natural and unchangeable entities.

The cultural success that followed the DSM-III revolution led surging numbers of patients to seek mental health treatment. In the early 1980s, only about 10% of individuals that community surveys diagnosed with some mental disorder had contact with mental health professionals.[6] By the early 1990s, service use had increased to encompass about a quarter of such persons. The amount of help-seeking continued to increase; in the early 2000s about 40% of presumed cases received some professional treatment and nearly 30% received care from a psychiatrist or other mental health specialist.[7] From 1987 to 2007, the number of Americans obtaining Supplemental Security Income or Social Security Disability Insurance for some mental disorder increased by almost 250% from 1 in 184 to 1 in 76.[8]

The DSM revolution was also followed by the rise of a new class of antidepressant drugs, the selective serotonin reuptake inhibitors (SSRIs). After entering the market in the late 1980s, the SSRIs quickly became as successful as the tranquilizers had been in the 1950s and 1960s. Their use quadrupled in just a decade from the early 1990s to the early 2000s. By 2000, they were the best-selling drug class of any sort; fully 10% of the U.S. population was taking them. By 2008, the numbers of antidepressant users had grown by almost 400% compared to the period from 1988 to 1994.[9] Remarkably, by 2012 about a quarter of all adult women in their thirties and forties were taking antidepressants at any given time; nearly a third of female patients in general medical practice received a prescription for an SSRI.[10]

In addition to the new antidepressants, an array of recently patented antipsychotic drugs arose as treatments for a much greater range of conditions than ever before. Although they were not more effective than older, off-patent medications, Zyprexa, Abilify, Seroquel, and others generated huge profits for the pharmaceutical industry, which marketed them for the DSM's newly expanded category of bipolar disorder and as add-on treatments for depression. Overall, revenue from sales of psychotropic drugs rose by an amazing 600% from 1987 to 2001.[11]

While high rates of psychotropic drug use have characterized American society since the 19th century, several differences mark the present explosion. One is that long-term use is now typical; in contrast, prior drug classes such as the barbiturates or benzodiazepines could be taken on an as-needed basis. Another is that increasing proportions of patients use more than one drug simultaneously. The rate of psychiatric outpatients who received two or more medications grew from 43% in 1996–1997 to 60% ten years later.[12] Finally, in the past children and adolescents were rarely targeted for drug treatments; now, by far the highest growth rates in prescription drug use is occurring in this population. The period from 1994 to 2001 witnessed a 250% increase in the number of adolescents who received a prescription for some psychotropic medication.[13] Recent years have seen an especially steep rise in rates of stimulants that treat attention-deficit/hyperactivity disorder (ADHD); over two million youths now receive these drugs each year.[14] The numbers of persons twenty years old and younger who take the strongest kind of medication, the antipsychotics, soared from about 200,000 in 1993–1995 to about 1,225,000 in 2002, a more than six-fold increase in less than a decade.[15]

As an economic, social, and cultural phenomenon, the DSM revolution was a tremendous success. Nevertheless, some serious cracks have emerged in the foundation of the field's current knowledge system. Despite the DSM's unquestioned institutional achievements and the rapidly rising rates of mental health service use, especially for drug treatments, around the turn of the century a new crisis arose over psychiatric classification that involved questions of how to define valid cases of mental disorder. In contrast to the basic challenges to the credibility of psychiatry itself that spawned the DSM-III, the 21st-century troubles were limited to issues that concerned researchers. This group, who had initiated the insurrection that gave rise to diagnostic psychiatry, had come to recognize fundamental flaws in the DSM nosology. Most clinicians, however, were satisfied with the extant classificatory system but indifferent to the questions about validity that instigated the field's next quandary. The conflict between researchers and clinicians had a very different outcome in the DSM-5 revision process in 2013 than the resounding victory of academic investigators over analysts that marked the creation of the DSM-III.

## Questions of Validity Arise

Gerald Grob observes that every generation of psychiatrists has proclaimed that the profession is "on the threshold" of fundamental breakthroughs in understandings of mental illness.[16] In the few cases where a clear organic grounding for some condition was discovered—syphilis, epilepsy, dementia, Huntington's disease—authority was historically transferred from psychiatry to neurology. Through using explicit symptom-based conditions that could be reliably measured, the DSM-III initially offered the possibility of finally overcoming psychiatry's recurrent series of disappointments to find distinct causes, prognoses, and outcomes of specific mental disorders. "It seems that we are about to move into a period when genetics will define disease entities in psychiatry," a prominent review published in the 1980s predicted.[17] This optimistic view assumed that a reliable system of classification would be the surest pathway to etiologically derived, valid categories. As long as this assurance was accepted, psychiatry's position within the ecology of medical professions would be secure.

Yet, a striking contradiction was at the heart of the DSM classifications: research conducted after 1980 indicated that these diagnoses bore little resemblance to the entities that they were supposed to categorize. The DSM-III was established to realize Eli Robins and Samuel Guze's goals for valid mental disorders: each condition should be precisely described, identified through biological markers, and validated through outcome and family studies. After these procedures confirmed the usefulness of some condition, it would generate etiologically homogeneous groups that would respond to specific treatments.

Research in the two decades after 1980, however, consistently found that diagnostic psychiatry failed to accomplish any of these objectives.[18] The previous chapter discussed how findings from neuroscientific and epidemiological studies demonstrated that, far from differentiating clear and distinct conditions, diagnoses were heterogeneous and overlapping. For example, it was difficult to disentangle presumably distinct psychotic conditions—bipolar, schizophrenic, and schizoaffective—from one another.[19] At the extreme, about 80% of people with a personality disorder also met criteria for some Axis I condition.[20] In addition, research indicated that common general processes such as internalization, externalization, or psychosis accounted for a broad array of conditions.[21] This explained why most people who met the criteria for one DSM condition also met the criteria for at least one additional DSM diagnosis.[22]

Family and genetic explorations had not identified biological or other markers for any mental disorder. In 2002, the group charged with developing an agenda to revise the DSM-IV concluded that "the field of psychiatry has thus far failed to identify a single neurobiological phenotypic marker or gene that is useful in making a diagnosis of a major psychiatric disorder or for predicting response to psychopharmacologic treatment." The same group noted: "Today there is only rudimentary knowledge of the genetic and nongenetic factors that cause the common psychotic, affective, anxiety, and substance use disorders that constitute the large majority of serious psychiatric disturbances."[23] Follow-up studies, as well, indicated that diagnoses were as likely to change as to remain stable over time. One such study of five hundred patients indicated that after seven years, less than half retained the same diagnosis.[24] Finally, the DSM had not led to any distinct treatments for different disorders: "Rather than classes of psychopharmacologic agents matching up with particular diagnoses, we have moved into an era of pharmacologic promiscuity in which many agents are being found to be effective for a variety of disorders," one psychiatrist summarized.[25]

Observers were also alarmed by the high rate of nonspecific diagnoses—"not otherwise specified" (NOS)—found in clinical practice. The DSM, despite its hundreds of conditions, was not helpful in classifying a high proportion of patients: estimates indicated that more than a third of patients received NOS diagnoses.[26] In some cases, such as the eating disorders, the rate of NOS diagnoses approached or exceeded the number of specific diagnoses.[27] Extraordinarily, NOS was the single most common classification of the personality disorders.[28] Despite all of these flaws, psychiatric classification had undergone only relatively minor changes since 1980.

Unlike the reliability problems that drove psychiatry's crisis of legitimacy in the 1970s, problems of validity were at the heart of 21st-century concerns.[29] The discrepancy between the findings from empirical research and psychiatry's classification scheme created the predicament that the field's symptom-based, categorical diagnoses did not seem to fit the phenomena they attempted to describe, explain, and treat. Three decades after the paradigm shift to diagnostic psychiatry and two decades after President George H. W. Bush proclaimed that the 1990s would be the "Decade of the Brain," the DSM's scientific accomplishments had not achieved the hope that reliability would lead to validity. Indeed, from the point of view of psychiatric researchers at the turn of the century, the DSM was a catastrophe.

In contrast to the events leading up to the DSM-III, the leading opponents of the DSM-5 revision process were psychiatric insiders. The most prominent critic, Allen Frances, had edited the DSM-IV. A former director of the National Institute of Mental Health (NIMH), Steven Hyman, regarded the DSM as "an absolute scientific nightmare," and "an unintended epistemic prison that was palpably impeding scientific progress."[30] In 2013 one of Hyman's successors, Thomas Insel, proclaimed "there's no reality" to the DSM diseases.[31] For Insel, the DSM's chief "weakness is its lack of validity. Unlike our definition of ischemic heart disease, lymphoma or AIDS, the DSM diagnoses are based on a consensus about clusters of clinical symptoms, not any objective laboratory measure."[32] In a 1997 article in *Science*, Nancy Andreasen spoke of the DSM as providing "objective criterion-based assessment techniques that produce reliable and precise diagnoses." Ten years later, Andreasen proclaimed that the DSM was "not useful for research because of a lack of validity."[33] Prominent biological psychiatrist Michael Alan Taylor provided the bluntest statement: "Put [the DSM] in a shit solvent and all that's left is the binding."[34] For many academic psychiatrists, the field faced a crisis of Kuhnian proportion; its dominant knowledge paradigm bore little resemblance to the nature of mental disorders.

Researchers, however, found that translating their understandings of the problems of using symptom-based measures of mental disorder into a more valid diagnostic system was extremely difficult. Three examples illustrate different aspects of the problems involved in developing a valid classification of mental disorders: epidemiological estimates of the amount of mental disorder, the use of dimensional measurements, and how to handle the distress that accompanies bereavement.

## Epidemiology

After the DSM-III was published, the NIMH turned its attention toward demonstrating the pervasiveness of mental disorders in the population. This interest in many ways echoed the agency's postwar move away from the psychoses toward concern with the extent of pathology among the general public. The agency funded large epidemiological studies that diffused the DSM-III's symptom-based illnesses from psychiatric practice and research to community-based studies. Because these categories were based on overt symptoms, researchers could apply diagnoses developed for clinical patients to surveys of large groups. One factor facilitating the triumph of the DSM revolution was the enormous, and unexpected, amounts of putative mental illness that community-based surveys uncovered.[35]

### An Epidemic of Mental Disorder

The first major community study to employ the DSM-III diagnoses, the Epidemiologic Catchment Area (ECA) study, was published in 1980. The ECA estimated that about 16% of the population had at least one current psychiatric disorder and about double that figure reported a lifetime history of some disorder.[36] When the results of a second survey conducted with the same subjects one year after the original survey were taken into account, estimates of lifetime prevalence increased to 44% of the population.[37] The second main project, the National Comorbidity Survey (NCS), was carried out in the early 1990s. It produced even higher rates of mental disorder, estimating that 29% of the population had some disorder over the past year and that nearly half—48%—had some lifetime disorder.[38] When the NCS was redone in the early 2000s, it indicated that 60% of respondents had experienced some mental disorder.[39]

The most common categories of mental illnesses were widespread. At some point in their lives, about 31% of the population experienced an anxiety disorder, 21% a mood disorder, 25% some impulse control disorder, and 35% a substance use disorder.[40] Even conditions such as bipolar disorder, which before the DSM-III were thought to be rare, affected nearly six million Americans—over 4% of the population—each year.[41] These findings are now widely cited in the scientific and popular literature, in pharmaceutical advertisements, and in advocacy documents.[42] For example, in 1999, the Surgeon General issued a highly publicized report stating that fifty million Americans suffered from some mental disorder each year.[43] Perhaps the best-known epidemiological claim became the World Health Organization's (WHO) assertion that depression was the single leading cause of disability for people in midlife and for women of all ages.[44] This claim is now found, often in the first paragraph, of thousands of articles regarding depression. By 2013, the WHO declared that depression caused more years of disability than any other disease.[45]

The epidemic of mental illness spread from adults to children. When the DSM-III was published in 1980, less than 1% of school-age children received ADHD diagnoses.[46] By 2011, 11% had this condition.[47] Autism, as well, seemed to be growing at an alarming rate. The prevalence of autism in U.S. children, which was not even measured in the initial studies, increased by 119% from 2000 (1 in 150) to 2010 (1 in 68). By 2014, the Centers for Disease Control and Prevention reported that one in every fifty-nine children had an autism spectrum disorder.[48] Intermittent explosive disorder (IED) provides another example. In 1987 the DSM-IIIR stated that IED was "apparently very rare."[49] Twenty-five years later the NCS reported that two-thirds of adolescents had "anger attacks that involved destroying property, threatening violence, or engaging in violence."[50] Nearly 8% met full criteria for an IED. Most spectacularly, rates of childhood bipolar disorder increased *forty-fold* between 1994–1995 and 2002–2003.[51] In the latter period, over

1.6 million children and adolescents received a bipolar diagnosis. Overall, by 2010, half of adolescents reported some mental disorder by the time they were age eighteen.[52]

Because memories of unpleasant experiences fade over time, people do not remember many episodes of possible mental disorder when asked at a later period.[53] The small number of studies that assessed prevalence among the same people at relatively short intervals found drastically higher amounts of mental illness than cross-sectional research. A major study of this kind in New Zealand found that over 40% of subjects had experienced at least one episode of major depressive disorder (MDD) by age thirty-two.[54] Another study found that over half of children in Oregon followed longitudinally from childhood to age thirty reported such episodes.[55] When results of research on first onset of cases after the early thirties are considered, it is likely that accurate recall of lifetime prevalence of MDD would encompass over two-thirds of the population. Mental illness was the statistical norm; the absence of pathology was atypical. The application of the DSM's symptom-based measures in community surveys seemed to fulfill Karl Menninger's vision: "Gone forever is the notion that the mentally ill person is an exception. It is now accepted that most people have some degree of mental illness at some time, and many of them have a degree of mental illness most of the time."[56]

## Creating Mental Illness in Survey Research

The finding that such vast numbers of people had mental disorders disquieted a number of prominent researchers. They began to question whether the huge amount of mental disorder that community surveys uncovered was genuine or, instead, a mislabeling of contextually explicable distress. The nature of survey research itself might account for the enormous amount of mental illness that community studies supposedly revealed.[57]

Issues concerning the validity of the DSM categories depend on the settings in which they are used. They are less consequential in clinical practice. Patients who seek help are self-selected and typically have already made decisions that their conditions go beyond ordinary distress to warrant professional attention. They generally seek treatment only after months or even years of enduring their problems.[58] In addition to lay judgments, clinical decisions provide a second level of screening over what constitutes a legitimate mental health problem. Clinicians inevitably use contextual appraisals about the meaning of symptoms their patients report. Psychiatrists John Wing, John Cooper, and Norman Sartorius put this well in regard to an anxious patient: "In clinical terms, an individual may worry because he has something to worry about, because he is a worrier, because he has phobias, because he has depressive preoccupations, [or] because he has persecutory delusions.[59] Therapists must use their judgment to separate symptoms that arise because of realistic situations, personality dispositions, or physical conditions on the one hand and mental disorders on the other. Clinicians spend years of training learning how to distinguish disordered from benign symptoms. In treated groups, the symptom-based logic of the DSM is applied within the context of both lay and professional decisions about whether particular symptoms are signs of mental disorders.

The contextual nature of decisions over the presence of a mental disorder in clinical groups thoroughly contrasts with the a-contextual nature of assessments of symptoms in the general population. Studies of mental illness in the community translate the DSM diagnostic criteria into survey questions as exactly as possible. They assume that the DSM measures have the same meaning in untreated and in clinical populations. Yet, conditions that neither respondents nor clinicians would ever consider to be pathological nonetheless are often taken to indicate mental disorder in community surveys.

Unlike the situation in clinical practice, which involves both patient self-evaluation and clinician discretion to decide what symptoms are contextually appropriate or inappropriate, survey interviewers are required to strictly adhere to the literal wording of the symptom questions without using flexible probes. Because community evaluations require standardization, different interviewers must ask these questions in exactly the same way. As one study notes, "The interviewer reads specific questions and follows positive responses with additional prescribed questions. Each step in the sequence of identifying a psychiatric symptom is fully specified and does not depend upon the judgment of the interviewers."[60] Without such procedures, even minor variations in wording or in the interviewer's instructions or probes can lead to different results. The rigid approach of structured interviews improves the consistency of symptom assessment across interviewers and research sites and thus the reliability of diagnostic decisions.

The necessity to standardize responses means that epidemiological studies exclude processes that relate reported symptoms to their context, and so they treat all positive responses as possible signs of a mental disorder. Respondents might recall depressed moods, insomnia, loss of appetite, or diminished pleasure in usual activities that lasted for two weeks or more after the breakup of a romantic relationship, the diagnosis of a serious illness in an intimate, or the unexpected loss of a job. Although these symptoms might have dissipated as soon as a new relationship developed, the intimate recovered, or another job was found, these individuals would join the twenty million people who suffer from the "disorder" of depression each year. Others who suffer from common medical conditions such as influenza or typical pregnancy-related symptoms often meet the MDD criteria, despite not having a depressive disorder.

For example, in one major survey, the most common symptoms of depression were "trouble falling asleep, staying asleep, or waking up early" (33.7%), being "tired out all the time" (22.8%), and "thought a lot about death" (22.6%).[61] College students during exam periods, people who must work overtime, or respondents who take a survey around the time when a famous person has died would all naturally report some of these symptoms. Moreover, the duration criteria require that the symptom last for only a two-week period, ensuring that many transient and self-correcting symptoms are counted as disordered. In contrast to treated groups, whose symptoms have often endured for years, surveys lump brief episodes with persistent cases. Or, a student who is questioned during the week before she finds out whether or not she has been admitted to medical school might report enough symptoms of anxiety to qualify for a diagnosis of generalized anxiety disorder. She is not, however, mentally disordered despite the presence of these symptoms if her "disorder" immediately disappears once she finds out she has gained admission. Nevertheless, she would be counted as among the twenty-three million Americans who suffer from generalized anxiety disorders.

The key flaw of survey measures of anxiety, depression, and other common conditions is that they assume positive responses to symptom-based measures indicate the presence of a *disorder*. Because many affirmative responses can indicate context-driven symptoms, surveys that rely on a-contextual questions about symptoms produce large numbers of false positive diagnoses of people who are wrongly counted as having mental disorders. When all symptoms are counted as disordered, ordinary distress is hopelessly confounded with genuine dysfunction. The "epidemic" of various types of mental illness is largely an artifact of the application of measures that cannot distinguish normal distress from pathological disorder.[62] This situation generates the false, but typical, conclusion that epidemiological studies show "mental disorders are very frequent disorders of the brain, affecting almost every other person over his/her life course."[63]

The commonly cited huge estimates of mental disorder in the population are grounded in an ideologically useful, but scientifically questionable, conflation of normal and disordered conditions. The inflated estimates of the amount of mental disorder that community studies report justify the efforts of national and international agencies to amplify the scope of the "public health epidemic" they must deal with and garner political support to eradicate the extremely widespread problems they confront. For example, the WHO's well-known claim that depression is the world's most disabling condition stems from taking the large number of people who meet the depressive criteria in community studies and then considering all of them as having severity that is comparable to paraplegia or blindness![64] While this might be justified for the relatively small number of serious and chronic cases of melancholic depression, the same can hardly be said for someone who was sad, fatigued, unable to concentrate, and had sleep and appetite problems for just two weeks after facing a loss event. What is more, community studies are considerably more likely to diagnose more cases of the latter than the former type of depression.

The equation of disability from all cases of depression with blindness and paraplegia was trumped by the rhetorical value of viewing depression as "the major scourge of mankind."[65] High estimates of the numbers of people who are mentally ill can also facilitate advocacy groups' attempts to destigmatize mental illness through citing estimates of how many people they affect. Likewise, they can assist investigators in securing funding to study widely prevalent conditions: "Researchers always give maximal prevalence for the disorders that they have a particular interest in. In other words, if you're really interested in panic disorder, you're going to say it's very common. You never hear an expert say, 'My disorder is very rare.' Never. They always tend to see it as more common," Robert Spitzer shrewdly observed.[66]

Moreover, despite the great increase in the number of people receiving mental health treatment in the decades following 1980, nearly two-thirds of people that surveys claimed had mental disorders went untreated. It seemed, therefore, that a vast unmet need for psychiatric services existed, which served as a useful marketing tool for mental health professionals. "The business appeal of tens of millions of people needing a product is transparent," psychiatrist John Sadler notes.[67] Huge rates of mental disorder also provide drug companies immense untapped markets of presumably mentally ill people who have not yet sought medical help.[68] Ignoring the evidence discussed in the previous chapter that there were few signs that common drug treatments actually worked better than placebos, as well as the potentially grim consequences of long-term psychotropic drug use, professional, governmental, and advocacy groups joined in the pharmaceutical industry's efforts to bring treatment to ever greater numbers of the distressed.

The magnitude of the numbers of people with untreated mental illnesses led some leading scholars to rethink the way that their application in the population led the DSM categories to mistake contextually appropriate symptoms as signs of mental disorders.[69] "Based on the high prevalence rates identified in both the ECA and the NCS, it is reasonable to hypothesize that some syndromes in the community represent transient homeostatic responses to internal or external stimuli that do not represent true psychopathologic disorders," prominent psychiatric researcher Darrell Regier noted.[70]

Nevertheless, most researchers stressed a different finding of community surveys: when they ignored DSM diagnostic thresholds and focused on the absolute number of symptoms or on gradations of severity from mild to moderate to severe, they found no sharp cut points. Instead, as the degree of severity and number of symptoms respondents reported rose in a continuous fashion, so did their degree of impairment and need for professional help. In addition, they emphasized how mild cases of mental disorder could evolve into more serious conditions at a later date so that initiating treatment at an early time might prevent severe states from emerging.[71] Many researchers concluded that the categorical nature of the DSM with its sharp diagnostic cut points was the core problem that needed to be overcome. They proposed a dimensional system as the major innovation in the revision of the DSM-IV that began around the turn of the century.

## Dimensions

Concerns over the validity of the DSM diagnoses led the American Psychiatric Association (APA) in partnership with the NIMH to attempt a radical transformation of the whole enterprise of diagnostic psychiatry. In 1999 the APA set in motion the first basic revision of the DSM-III nosology, and eight years later it created a task force toward that end. The Task Force observed: "In the more than 30 years since the introduction of the Feighner criteria by Robins and Guze, which eventually led to DSM-III, the goal of validating these syndromes and discovering common etiologies has remained elusive."[72] It identified the DSM system of classifying mental disorders itself as the culprit: its serious flaws have become "increasingly problematic for research and clinical use."[73] Regier, the APA's director of research at the time, stated that the categorical nature of the DSM was "a root cause of many of the problems with current psychiatric diagnostic classifications."[74] The Task Force anticipated that neuroscientific findings would guide the attempt to fundamentally change the diagnostic system for classifying mental disorders.[75]

The DSM-5 Task Force was aware of the gaping problems of internal diagnostic heterogeneity, the broad and overlapping nature of extant diagnoses, the extensive use of NOS labels, and the elevated risk of false positive diagnoses that a-contextual criteria created. Nevertheless, it identified a different problem as at the heart of the DSM's flaws: the manual's failure to use continuous dimensions. It asserted that no clear boundaries, but only gradations, separated the two poles of sanity and madness. This decision was puzzling because one of the pillars of the DSM-III revolution was its rejection of the dynamic notion that symptoms ranged on a scale with the normal on one end and the severely abnormal on the other.[76]

The Task Force proposed a new dimensional system based on the principle that the major psychiatric conditions lay on continua that ranged from mild to severe:

> The single most important precondition for moving forward to improve the clinical and scientific utility of DSM-V will be the incorporation of simple dimensional measures for assessing syndromes within broad diagnostic categories and supraordinate dimensions that cross current diagnostic boundaries. Thus, we have decided that one, if not the major, difference between DSM-IV and DSM-V will be the more prominent use of dimensional measures in DSM-V.[77]

Most mental disorders, it stated, did not conform to the categorical neo-Kraepelinian DSM model but, instead, featured continuous gradations.[78] Moreover, the Task Force rejected the core DSM (and Kraepelinian) assumption of a sharp break between pathology and normality, assuming that mental illnesses were points on a continuum that also encompassed mild distress.

People were not disordered or non-disordered but displayed degrees of disturbance. Another rationale it gave, discussed more extensively in the next chapter, was that a small number of symptoms at one point in time could predict the development of a full-blown disorder at a later period. Surveys could target people who reported minor disturbances as at risk for becoming more seriously disturbed in the future.

The Task Force cited evidence that many common mental disorders have quantities of symptoms and levels of severity that run from mild to moderate through severe.[79] For example, one review of 177 studies found that continuous rather than categorical models provide better fits for symptoms of most major classes of mental disorders.[80] "Evidence has accumulated," the Task Force asserted, "that prototypical mental disorders such as major depressive disorder, anxiety disorders, schizophrenia, and bipolar disorder seem to merge imperceptibly both into one another and into normality with no demonstrable natural boundaries or zones of rarity in between."[81] For example, the defects in social cognition that mark schizophrenic conditions might run from mild impairments to full-blown psychotic symptoms.[82] Likewise, ADHD involved continuous degrees of attention deficits, not sharp distinctions with normal attention processes. The categorical model was especially unsuitable for the personality disorders: "There does not appear to be a qualitative distinction between normal personality functioning and personality disorder," the Task Force concluded.[83]

Dimensionalization became the central organizing logic behind the revisions and the rallying cry for those trying to reform the diagnostic manual. Researchers came to see the major psychiatric disorders as continuous, ranging from generally healthy individuals on one end, to mild and transient disturbances, to moderate symptoms, to severe and prolonged distress on the other end.[84] Its most extreme proponents argue, like Szasz, that mental disorders do not exist: "There are no disorders—they are just the extremes of quantitative dimensions."[85] Ironically, the major thrust of the intended DSM-5 revisions in many ways would have returned the manual to the dynamic assumptions that the DSM-III had thoroughly repudiated. Advocates also believed that a continuous model, which emphasized the similarities as opposed to the differences of mental health and mental illness, could reduce the stigma associated with psychiatric disorders.[86]

The Task Force's proposed project of dimensionalization involved the use of two types of scales. First, it would employ a broad, cross-cutting dimensional scale to screen all prospective patients. Rather than restrict assessment to the system of categories, the cross-cutting scale would examine broad syndromes that blur with one another and with normality. Results would yield numerical information on the state of patients but delay pigeon-holing them into a specific diagnosis. Instead, this dimensional measure would serve as an important screening device that could facilitate early identification and intervention.

After this initial screening, clinicians would derive a diagnosis from an interview in which they would draw on both the initial screen and some diagnostic category. A second dimensional intervention would then provide each patient with a numerical ranking of severity for whatever diagnosis he or she received. In order to tailor each severity scale to the specific disease, the Task Force granted each work group the flexibility to determine how to structure their scales. Some groups quantified particular symptoms and added them up for a composite severity score; others constructed severity scales by simply counting the number of symptom criteria a patient met. The irony of this discretion was that it promoted divergent diagnostic processes and thus undermined the standardization that was the foundation of the DSM-III revolution.

The problem the Task Force faced when it tried to implement the new paradigm shift was that, unlike the situation the promulgators of the DSM-III encountered in the 1970s, the 21st-century crisis was limited to the research community. Clinicians and others who employed the DSM diagnoses were not dissatisfied with them. While researchers believed that the current definitional system had systematic inadequacies, the manual's classifications were so embedded in clinical practice and social institutions that they were impervious to major changes.

## Clinicians Rebel

It is unsurprising that the DSM-III, which emerged because of politics, economics, and status, has fundamental flaws as a scientific model. These defects, however, only posed problems for researchers. Clinicians, who use diagnoses for practical as opposed to scientific ends, were relatively content with the DSM model. Indeed, clinical practice was virtually untouched by research that questioned the validity of psychiatric classifications.[87] Putting some DSM diagnosis on a reimbursement form did not mean that practitioners had to believe in its validity: indeed, studies showed widespread doubt among clinicians about the extant taxonomy.[88] In contrast, the implementation of a dimensional system would have forced clinicians to learn new protocols and figure out how to incorporate scales into their practices.

Not just providers but also administrators had too much invested in the current nomenclature to risk such a major disturbance to current procedures. The penetration of the manual into all facets of the workaday world of psychiatry made any changes that were more than marginal too potentially unsettling for anyone who used the classification for applied purposes. Psychiatrist Michael First outlined the potential institutional disruptions:

> Adopting a dimensional approach would likely complicate medical record keeping, create administrative and clinical barriers between mental disorders and medical conditions, require a massive retreating effort, disrupt research efforts (e.g., meta-analyses), and complicate clinicians' efforts to integrate prior clinical research using *DSM* categories into clinical practice.[89]

A dimensional approach would have required changes in the entire paperwork edifice of psychiatry—from hospital admissions procedures to insurance reimbursement charges.

The Task Force found that the DSM categories were far too entrenched in psychiatric practice to be so radically changed. The sheer success and widespread institutionalization of the DSM through all facets of the mental health system translated into great inertia when it came to altering it. At the annual meeting of the APA in May 2012, the APA Assembly, a body mostly composed of members selected by local psychiatric societies, voted unanimously to relegate all the dimensional scales to the appendix of the manual. Given that there was almost no research showing the practical utility of these measures, it argued that the "unproven severity scales" needed more testing. The APA Board of Trustees affirmed the Assembly's rejection of the Task Force's major innovation to DSM-5. Although the DSM's categorical system was ill-suited to research needs, it was highly adaptive for the everyday concerns of the psychiatric guild.

While the Assembly motion specifically addressed severity scales, it also manifested the long-standing tension within the psychiatric profession between clinicians and researchers.[90] Clinicians have a complicated relationship with the DSM—complications that dimensionality would have exacerbated. On the one hand, psychiatrists gain credibility from the DSM because it shows that they are treating legitimate medical disorders. And insofar as dimensionality would improve scientific research (an issue that is by no means resolved), it could help shore up the field's prestige. On the other hand, the more that the needs of researchers dictate the DSM and impinge on clinical practice, the more that clinical intuition and expertise is devalued. Scales may obtain more advanced statistical analyses and knowledge, but they do so at the expense of clinical wisdom. At least as important, specifying what diagnostic category a patient fits determines third-party reimbursement. Dimensional measurement created uncertainty over the cutting point between reimbursable and nonreimbursable conditions. In rejecting the severity scales the Assembly not only squashed all talk of a paradigm shift; it also struck a blow for clinical influence. The professional stakes of the DSM were simply too high to allow for anything other than tinkering around the edges. Despite the Task Force's goal of fundamentally altering the extant diagnostic system, the DSM-5 looks a lot like its predecessors.[91]

## The Research Domain Criterion

Psychiatry's diagnostic system suffered an even more severe blow than the Assembly's rejection when Thomas Insel, the director of the NIMH, disavowed the entire DSM apparatus. On April 29, 2013, days before the release of DSM-5, the NIMH announced on its website that it was prioritizing research that eschewed the DSM altogether.[92] According to Bruce Cuthbert and Insel, the DSM system, based on presenting signs and symptoms, did "not adequately reflect relevant neurobiological and behavioral systems—impeding not only research on etiology and pathophysiology but also the development of new treatments."[93] In a cutting rebuke published in *The New York Times*, Insel observed: "As long as the research community takes the D.S.M. to be a bible, we'll never make progress. People think that everything has to match D.S.M. criteria, but you know what? Biology never read that book."[94]

For Insel, the DSM categories were far too internally heterogeneous, too intertwined with other categories, and too likely to result in NOS diagnoses to be useful. "While DSM has been described as a 'Bible' for the field, it is, at best, a dictionary, creating a set of labels and defining each. . . [Its] weakness is its lack of validity. . . . Patients with mental disorders deserve better," he complained.[95] The NIMH rejection of the DSM's system was especially stinging because this agency had been in the forefront of sponsoring the initial construction of the DSM-III, institutionalizing its hegemony in research, and collaborating with the DSM-5 Task Force in developing the dimensional approach.

The NIMH had come to realize that the DSM system had not produced a single important advance in the understanding and treatment of mental illness. The limitations of this diagnostic system meant that mental health professionals were unable to capitalize on the vast improvements in technologies that explore the brain. This dire situation led the agency, the APA's long-time partner in creating a scientific diagnostic system, to simply abandon the DSM as a research tool and take steps to implement an entirely new classification grounded in neuroscientific principles. "Our resources are more likely to be invested in a program to transform diagnosis by 2020 rather than modifying the current paradigm," Insel stated.[96]

As an alternative to the DSM the NIMH is developing an independent taxonomy called the Research Domain Criterion (RDoC) that is explicitly intended for researchers.[97] This classification intends to draw upon the latest research in genomics, pathophysiology, and neuroscience to cultivate new theoretically grounded etiological theories of mental disorder. Unlike the DSM, it would not rely on symptoms or disorder categories. Instead, it directs attention to general mechanisms such as negative and positive valence systems that cut across many particular diagnostic entities. Although it is broadly framed within an environmental and neurodevelopmental context and takes cognition and behavior into account, the RDoC focuses on brain-based processes including genes, molecules, cells, and physiology that do not correspond to particular diagnoses.

The RDoC begins with three assumptions: that mental illnesses are brain disorders, that the tools of clinical neuroscience such as functional neuroimaging can identify the dysfunctional neural circuits that cause mental illnesses, and that data from genetics and neuroscience will yield biosignatures that can be used to diagnosis mental disorders. It proposes to employ "fundamental biobehavioral dimensions that cut across current heterogeneous disorder categories" to identify common underlying biological, neurological, or genetic causes.[98] The RDoC approach is fully and explicitly dimensional across both the severity range of a diagnosed disorder and the entire span of normal to abnormal functioning. It would not

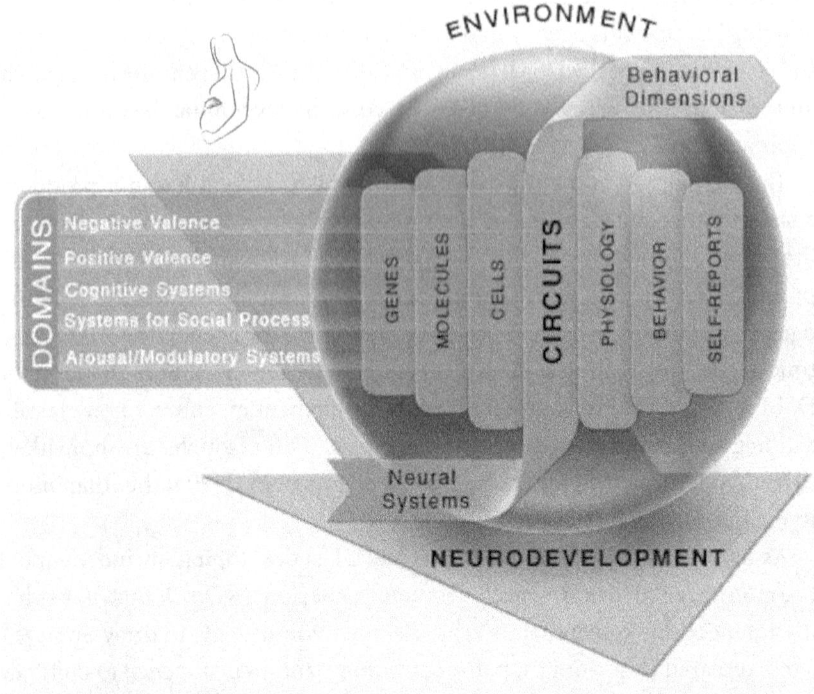


**Figure 9.1** This diagram outlines the major aspects of the National Institute of Mental Health's Research Domain Criteria, which they hope will replace the DSM in psychiatric research. From National Institute of Mental Health, Transforming the understanding and nature of mental illness. https://www.nimh.nih.gov/research/research-funded-by-nimh/rdoc/index.shtml

delineate diseased from healthy states but instead recognize their essential continuity.[99]

The RDoC is by far the most ambitious neurobiological classification scheme that has ever been developed. It is possible that this approach will overcome the serious deficiencies of the DSM and eventually provide researchers with the tools they need to accurately characterize the nature of mental disorder. Splitting classifications used in research from those applied in service provision could have a salutary impact. It is at least equally likely, however, that this initiative will become yet another failure in the long series of attempts throughout the history of psychiatry to make fundamental changes in understanding and treating mental disorder through focusing on brain-based pathologies.

## Are Dimensions the Answer?

The Task Force's plan to dimensionalize the DSM was seriously flawed. The major defect of the continuous model it proposed was that it could not make the most central distinction of any diagnostic system: distinguishing disordered from non-disordered conditions.[100] Both contextually appropriate and pathological symptoms might be dimensional, but they lie on separate dimensions, not on a single one. As Chapter 1 discussed, non-disordered conditions emerge because of specific kinds of contextual triggers, they are roughly proportionate in intensity to the provoking context, and they end about when the stressful situation that gave rise to them ends or gradually cease as natural coping mechanisms allow people to adjust to new circumstances. Even severe symptoms that arise in highly stressful contexts need not indicate disorders as long as they are proportionate to their context and do not become self-perpetuating after the stressful period ends. Conversely, symptoms that are not tied to the contexts in which they were naturally designed to arise or that emerge after a stressor but persist with disproportionate intensity or duration can constitute disorders.[101]

Depression, the most commonly cited model for a continuous approach, illustrates the problem of viewing mental disorders as gradations. A flood of studies conducted after 1980 argued that depressive disorders vary along a continuum that ranges from minor to major.[102] Using findings from epidemiological and genetic research, they maintained that no sharp boundary exists between disordered and non-disordered depression. "To accept depression as disease," psychiatrist Peter Kramer concluded, "is to see pathology or risk in minor versions."[103] At the extreme, some contended that "even one current depressive symptom may be clinically 'significant' and associated with some psychosocial dysfunction or risk."[104]

Advocates for dimensions correctly assert that the DSM cutoff of five symptoms sets an arbitrary dividing line between pathology and normality. They also make a good case that many disorders, such as depression, are naturally continuous rather than dichotomous in nature. Nevertheless, they fail to address the issue of how to distinguish contextually explicable from disordered symptoms. As the DSM's general definition of mental disorder indicates, this separation involves deciding whether some dysfunction causes a syndrome. The valid use of dimensions depends on a *prior* determination that symptoms result from a dysfunction rather than from their context. Indeed, using dimensions that run from mild through severe makes this distinction much harder to make because the probability that symptoms are normal and not pathological grows as the severity and number of symptoms declines.

In addition, the dimensional approach would exacerbate, rather than solve, the problem of heterogeneity that plagues the current categorical system. Some people with a small number of depressive symptoms will eventually develop a more serious disorder. The vast majority of mild cases, however, have causes, courses, and outcomes that are distinct from severe cases.[105] In particular, survey respondents who report a small number of transient depressive symptoms have no resemblance to serious melancholics with enduring conditions. Dimensional measurements are far more likely than current categorical criteria to mistakenly consider an "expectable or culturally approved response to a common stressor or loss" as a mental disorder. They would significantly worsen the problem of distinguishing normal from pathological conditions that the DSM's general definition of mental disorder tries to overcome.

Studies of how often people in the general population report hearing voices provide another example of the problems dimensional views confront. In conformity with the dimensional view, hallucinations and delusions are distributed continuously in the population.[106] Seventeen surveys in nine countries show that the median prevalence of hearing voices is 13.2%. To a number of observers, this finding indicates "support [for] the current movement away from pathological models of unusual experiences and towards understanding voice-hearing as occurring on a continuum in the general population."[107]

The problem, however, is that hearing voices, in isolation, is not in itself a pathological symptom. More detailed interviews indicate that most voice hearing in community populations are false positives with no clinical significance. In one study, a quarter of twenty-six-year-olds stated they experienced delusions or hallucinations, but less than 4% met criteria for a schizophrenic disorder.[108] Many reasons can account for why someone responds affirmatively to questions about hearing voices including their religious beliefs, recent bereavement, consequences of surgery, ingestion of psychedelic drugs, or participation in subcultures that validate these occurrences. Such cases are accepted by many others in the community, do not cause distress, and do not impair social behavior.[109] People who report hearing voices do not lie on one end of a spectrum where serious cases of schizophrenia are at the other end. Most do not have mild or prodromal signs of schizophrenia but, when the context of their "symptom" is taken into account, they are not disordered at all.

The basic problem with the dimensional movement is that it rejects DSM symptom thresholds but simultaneously clings to the assumption that context-free, symptom-based criteria can best diagnose mental disorders. Those who advocate viewing mental disorders as continuous disregard the problem of how to avoid misdiagnosing normal symptoms as disorders. They ignore the perennial "without cause" distinction between contextually understandable responses and possible indicators of mental disorder. Only if a more valid approach to the disorder–non-disorder distinction is first put in place will it be justifiable to encompass low-symptom conditions within the disorder category.[110]

A new dimensional DSM would vastly expand the range of mental illness. When it does not connect symptoms to the context in which they arise, it inflates the amount of presumed disorder and underestimates the extent of normal distress. Once considerations of context are abandoned, all undesirable psychic states can be seen as signs of pathology; indeed, the very possibility of normal distress is lost. The DSM-5 Task Force's dimensional thrust echoed that of the postwar community studies in the 1950s and 1960s, which viewed even the mildest symptom as one step on a ladder of presumed mental disorder.[111] This conception carries the risk of making the realm of abnormality as expansive as it was during the heyday of dynamic psychiatry. Epidemiologist Rema Lapouse's critique of postwar community studies—"If all persons who cough are counted as cases of tuberculosis, both incidence and prevalence rates will skyrocket"—holds as well for the current continuous concept of mental disorder, which makes no provision for distinguishing normal emotions from disorders.[112]

## Removing the Bereavement Exclusion

The failed attempt to introduce dimensional diagnoses was an honest, if deeply flawed, effort to overcome the deficiencies of the DSM's categorical model. Another proposal, one that succeeded, illustrates a different dynamic that discredited the Task Force. This was the elimination of the bereavement exclusion (BE) to the diagnosis of MDD. The removal of the BE might seem to be a trivial change to a single exception to one psychiatric diagnosis. In fact, however, the controversy over this issue is worth an extended discussion because it speaks to concerns that are at the heart of psychiatric diagnosis itself. This decision reduced the validity of the MDD diagnosis and made it even more heterogeneous and a-contextual than its predecessors. It is especially important because MDD is by far the most commonly diagnosed mental illness, accounting for nearly 40% of all outpatient conditions.[113] Likewise, bereavement is a common experience that virtually everyone will experience at some point in their life.

The distinction between proportionate and disproportionate responses to loss has been part of traditional psychiatric thinking for thousands of years, at least since the Hippocratics proposed the first medical definition of depression in the 5th century BCE: "if fear or sadness last for a long time it is melancholia."[114] For them, not symptoms alone but only ones of disproportionate duration ("last for a long time") to a person's circumstances indicated disorder. This distinction persisted through the DSM-II definition of neurotic depression, which had specified that only "excessive" responses to loss should be considered depressive disorders.[115] Even if people met criteria for a depressive disorder, as long as their response was of proportionate, nonexcessive severity and duration to the loss that they suffered, they did not have a disorder. Even severe losses that resulted in intense sadness responses could be non-disordered as long as the response was not "excessive" to the provoking circumstances.

In its quest to purge etiological assumptions from psychiatric classification, the DSM-III went overboard and mistakenly assumed that terms such as "excessive," which had been used for thousands of years, were also etiological and often purged these as well. This does not seem to have been an intentional decision or even one that the working group for the DSM-III had explicitly considered in its deliberations.[116] Instead, it was the inadvertent result of their effort to rid the manual of psychodynamic tenets and to improve the reliability of diagnosis. This decision, unnoticed at the time, was to have major impacts. The result in the case of depression was that *all* symptoms, whether pathological signs of dysfunctions or proportionate responses to stressful situations, were treated as indicators of mental disorder.

Nevertheless, the symptom-based DSM-III definition of MDD contained one attempt to distinguish normal sadness from depressive disorder. The text recognized that bereavement after the death of an intimate was not a psychiatric disorder, noting that: "A full depressive syndrome frequently is a normal reaction to [the death of a loved one]."[117] Patients would not receive a diagnosis of depression if their symptoms were due to what the DSM defined as "uncomplicated grief." But these criteria also recognized that—while most bereaved people who would otherwise meet the criteria for MDD do not have mental disorders—in some cases bereavement could "go wrong" and indicate a disorder. The next edition of the DSM specified that bereaved conditions that in other respects met the criteria for MDD should not be considered as disorders unless they were "complicated," involving either at least one severe symptom of marked functional impairment, morbid preoccupation with worthlessness, or psychomotor retardation or extended duration (later defined as at least two months).[118]

The problem was that there was no reason to limit this reasonable contextual exclusion to bereavement. As diagnosticians from the Hippocratics through the DSM-II recognized, the death of an intimate is not a unique loss but exemplifies symptoms that arise in response to many sorts of losses.[119] These stressors often result in depressive symptoms that, like bereavement, could be proportionate to their circumstances and so not be disordered. The DSM-IV's definition of mental disorder even used the death of a loved one as an "example" of "an expectable and culturally sanctioned response to a particular event," not as the single exclusion.[120] This characterization explicitly considered bereavement to illustrate a broader category of "expectable" reactions to events.

Moreover, a substantial body of empirical research showed that the mental health consequences of uncomplicated bereavement were similar to symptoms that stemmed from any kind of loss whether the death of a loved one, divorce, unemployment, and the like but were distinct from complicated depressive conditions. People who suffered a wide range of stress-related losses and would have qualified for the bereavement exclusion because they had no especially severe or prolonged symptoms were far more comparable to people who had *never* been depressed than to ones with serious or enduring depressive symptoms.[121] The similarity of such reactions to those of nondepressed people challenged the basic logic of symptom-based diagnoses. The critical distinction was not between bereavement and other losses but between uncomplicated, contextually appropriate conditions following losses and conditions with prolonged duration or especially severe symptoms such as suicidal thoughts, marked functional impairment, morbid preoccupation with worthlessness, or psychotic features. There was, that is, no good reason to single out bereavement as the sole exception to the diagnostic criteria.

Researchers connected to the development of the DSM-5 were attentive to these findings. Prominent psychiatric researcher and member of the DSM-5 depression work group Kenneth Kendler, used his own data set to test the contention that bereavement was a model for other stressors. His findings replicated outcomes showing depression that developed after bereavement was identical to that following other stressful life events. "The DSM-IV position is not logically defensible. Either the grief exclusion criterion needs to be eliminated or extended so that no depression that arises in the setting of adversity would be diagnosable," Kendler concluded.[122]

Then–APA president John Oldham also noted the similarity of bereavement to other losses:

> [the bereavement exclusion is] very limited; it only applies to a death of a spouse or a loved one. Why is that different from a very strong reaction after you have had your entire home and possessions wiped out by a tsunami, or earthquake, or tornado; or what if you are in financial trouble, or laid off from work out of the blue? In any of these situations, the exclusion doesn't apply. What we know is that any major stress can activate significant depression in people who are at risk for it. It doesn't make sense to differentiate the loss of a loved one as understandable grief from equally severe stress and sadness after other kinds of loss.[123]

The logical conclusion seemed to be that the BE should be *extended* to cover all loss responses that were not particularly intense or persistent. As writings since antiquity, including every edition of the DSM itself, had recognized, typical symptoms of sadness after loss were normal, not pathological.

The DSM-5 depression work group was thus faced with a stark choice. On the one hand, it could expand the BE to cover all uncomplicated responses to loss-related stressors. On the other hand, it could abolish the BE on the grounds that there was no justification for singling out bereavement for a unique exclusion. If it chose the latter, all conditions meeting the two-week, five-symptom MDD criteria, regardless of the context in which they arose, would be considered mental disorders. This was an especially consequential decision because for the past thirty years MDD was by far the most common psychiatric diagnosis among outpatients. Extending the BE, therefore, threatened both the core symptom-based logic of diagnostic psychiatry and the prevalence of its most popular condition.

It appears that the working group never even considered the option of extending the BE to other sorts of losses.[124] Instead, it chose to make the a-contextual nature of the MDD diagnosis even stronger. It maintained the extant diagnostic criteria but used the Kendler study as the reason to eliminate the BE from the text:

> The DSM-5 Mood Disorders Work-group has recommended the elimination of the bereavement exclusion criteria from major depressive episodes in light of evidence that "the similarities between bereavement related depression and depression related to other stressful life events substantially outweigh their differences."[125]

The DSM-5 added a footnote to the MDD criteria that stated:

> Responses to a significant loss (e.g., bereavement, financial ruin, losses from a natural disaster, a serious medical illness or disability) may include the feelings of intense sadness, rumination about the loss, insomnia, poor appetite, and weight loss noted in [the symptom criteria], which may resemble a depressive episode. Although such symptoms may be understandable or considered appropriate to the loss, the presence of a major depressive episode in addition to the normal response to a significant loss should also be carefully considered. This decision inevitably requires the exercise of clinical judgment based on the individual's history and the cultural norms for the expression of distress in the context of loss.[126]

This new stipulation modified the earlier exclusion in four ways. First, it removed consideration of bereavement from the diagnostic criteria and relegated it to a footnote. Second, the footnote did not contain any diagnostic criteria so that grieving people are liable to a depressive diagnosis after a two-week rather than a two-month period, which many experts believed was already far too short.[127] Third, it no longer required the presence of any especially severe symptom to override the MDD standard. Anyone who had suffered the loss of an intimate and has normal symptoms of grief such as sadness, a loss of pleasure, sleeping and eating problems, and fatigue that lasted for a two-week period following the death could meet the criteria. Finally, the textual note undermined a basic principle of the DSM system, which promotes standardized diagnostic criteria. Instead, it explicitly relied on "the exercise of clinical judgment" to decide if a bereaved case should be included or excluded from diagnosis. This returned the process of psychiatric diagnosis to the pre-1980 situation where it was dependent on the decisions of individual clinicians.

Remarkably, the abandonment of the BE even contradicted the DSM-5's *own* definition of mental disorder: "An expectable or culturally approved response to a common stressor or loss, such as the death of a loved one, is not a mental disorder."[128] This definition of mental disorder uses "the death of a loved one" to illustrate the *difference* between a painful but normal emotion and a mental disorder. The new MDD criteria refuted the manual's explicit standard for a valid mental disorder.

The DSM-5's decision to abandon contextual criteria for the MDD diagnosis also contrasted with its frequent use of such qualifiers not just in its general definition of mental disorder but also in other diagnostic sets.[129] Although the manual does not provide any systematic or explicit ways to use contextual considerations, as Wakefield and First observe, it does widely use context to distinguish normal from disordered conditions.[130] For example, criteria for specific phobia specify that the condition must be "marked" and "out of proportion to the actual danger posed by the specific object or situation." Generalized anxiety disorders involve "excessive" anxiety and worry. For ADHD, "symptoms are not solely a manifestation of oppositional behavior, defiance, hostility, or failure to understand tasks or instructions." Pedophilic disorders must persist for six months or longer "to ensure that the sexual attraction to children is not merely transient." Insomnia can only be diagnosed when it "occurs despite adequate opportunity for sleep."[131] The rigidity of the MDD's symptom-based definition, which the DSM-5 worsened, is extreme among the manual's various criteria sets.

The removal of the BE also undermined the central logic behind psychiatric diagnosis itself. The point of distinguishing one diagnosis from another is to help specify the causes, courses, outcomes, and treatments of various disorders. Yet, combining brief and uncomplicated depressive symptoms after the death of an intimate or other losses with prolonged and morbid ones does the opposite: it blends conditions that are tied to their social contexts with those that stem from internal defects, those that are transient and are unlikely to recur with ones that are more enduring, and those that are likely to disappear over time without treatment from those that might benefit from professional interventions. As diagnosticians for millennia have recognized, limited periods of uncomplicated grief represent the way that *normal* people respond to the death of a loved one. They also understood that grief is not unique but is a model for other loss responses. Research that relies on the DSM-5 criteria will hopelessly confound natural with dysfunctional reactions.

What prompted the developers of the DSM-5 to make such an unsound decision? Perhaps the best explanation stems from the nature of professional legitimacy. The BE threatened the basic rationale behind the MDD diagnosis, which has been psychiatry's most common condition since 1980, generating huge markets for psychiatry and associated interest groups. This exclusion recognized that one common loss was not pathological, but extending this logic would have also barred many others from diagnosis and treatment. Expanding the BE to other losses could have led to a major decline in the number of people who meet MDD diagnostic criteria, a substantial decrease in clients, and diminution of psychiatry's authority over distress stemming from loss events. Likewise, much of the pharmaceutical industry's enormous stake in the newer antidepressant medications involved their close association with loss-related events, which the a-contextual DSM criteria define as "disorders." In abandoning the BE, the DSM-5 work group might have overreached; time will tell if this affront to empirical evidence and intellectual coherence, not to mention common sense, will help erode the profession's credibility as the official arbiter of sanity and madness.

## Conclusion

Forty years after the DSM-III diagnostic revolution, the fundamental dilemmas that have perennially confronted psychiatry (and other mental health professions) remain unresolved. Neuroscientific and epidemiologic findings show that the current DSM system poorly characterizes the nature of mental disorder. Contrary to the intentions of the researchers who developed the DSM-III, its conditions have tremendous internal heterogeneity, artificial comorbidity, a plague of NOS diagnoses, and an inability to separate contextually appropriate from dysfunctional symptoms. These inadequacies led the DSM-5 Task Force to propose fundamental changes in the categorical system that was at the heart of these problems. Yet, the pathway they choose to remedy the situation—the introduction of dimensions—was not only ill-advised but would have made the problem of false positives much worse. The APA Assembly and Board of Trustees wisely rejected this premature upheaval in psychiatric diagnosis. The DSM-5, however, did implement other changes, in particular, the abolition of the BE, which exacerbated the confusion between normality and pathology. Far from improving previous editions of the DSM, this key change in the DSM-5 made the manual an even worse guide to the nature of mental disorder.

The DSM-5 revision process also revealed a deep split within the psychiatric profession between researchers and clinicians. The problems of validity that inevitably emerged from symptom-based diagnostic criteria only bedeviled the psychiatric research community. In contrast, the various epidemics of mental illness—for example, depression, anxiety, post-traumatic stress disorder, bipolar disorder, ADHD—that have swept the United States and a growing number of other countries since 1980 yield great benefits for many constituencies. They provide clinicians and drug companies with new potential clients for their services and products. They allow advocacy groups to advance claims that the mentally ill suffer from real diseases that are extraordinarily widespread in the population. They also let the many institutional users of the DSM maintain a highly successful administrative nosology. The current symptom-based diagnostic system is simultaneously practically invaluable and intellectually bankrupt.

## Acknowledgments

Portions of this chapter are adapted from Horwitz, 2002, Horwitz & Wakefield, 2006, Horwitz & Wakefield, 2007, Whooley & Horwitz, 2013 and Wakefield & Horwitz, 2016.

## Notes

1. Luhrmann, 2000.
2. Eysenck, Wakefield, & Friedman, 1983. A coalition of psychodynamic organizations produced an alternative diagnostic manual in 2006, but this system failed to gain any traction (Alliance of Psychodynamic Organizations, 2006).
3. Philips, 2010, 70; Whooley, 2010; Greenberg, 2013, 68; Smith, 2014.
4. E.g. U.S. Department of Health and Human Services, 1999; Hirschfeld et al., 1997; New Freedom Commission, 2003. Prominent political figures including Rosalyn Carter, Tipper Gore, and Hilary Clinton played leading roles in encouraging the public to destigmatize mental illnesses and to recognize that they were genuine medical conditions.
5. E.g. https://www.harpersbazaar.com/celebrity/latest/g15159447/celebrities-depression-anxiety-mental-health/; cover of *People*, April 26, 2011.
6. Shapiro et al., 1985.

7. Wang et al., 2005; Kessler et al., 2005.
8. Angell, 2011, 22.
9. Pratt, Brody, & Gu, 2011.
10. Raofi & Schappert, 2006; Mojtabai & Olfson, 2008a; Pratt, Brody, & Gu, 2011; Substance Abuse and Mental Health Services Administration, 2012. See also Kantor et al., 2015.
11. Harrington, 2019, 250.
12. Mojtabai & Olfson, 2010.
13. Thomas et al., 2006.
14. Zuvekas, Vitiello, & Norquist, 2006.
15. Olfson et al., 2006.
16. Grob, 1998.
17. Mullan & Murray, 1989.
18. Phillips, 2013, 144–46.
19. Clark et al., 2017, 119.
20. Kupfer et al., 147.
21. Kupfer, First, & Regier, 2002, 70; Clark et al., 2017, 105, 119. Kotov et al., 2017.
22. Clark et al., 2017, 14.
23. Kupfer, First, & Regier, 2002, 33.
24. Clayton et al., 1992.
25. Phillips, 2013, 146.
26. Taylor, 2013, 39.
27. Clark et al., 2017, 87.
28. Kupfer et al., 2002, 129.
29. This does not mean that questions of reliability have been solved. Many conditions in the DSM-5 field trials displayed appalling reliability. The kappa statistics for the core conditions of MDD and generalized anxiety disorder were only .28 and .20, well below the results for the DSM-III field trials. Incredibly, the mixed anxiety/depressive diagnosis had a reliability of 0! The report on the field trials euphemistically concluded that "We are now coming to the end of the neo-Kraepelinian era initiated in the U.S. by Robins and Guze with a renewed appreciation of both the benefits and limitations of a strict categorical approach to mental disorder diagnosis" (Regier et al., 2013, 59).
30. Belluck & Carey, 2013.
31. Belluck & Carey, 2013.
32. https://www.nimh.nih.gov/about/directors/thomas-insel/blog/2013/transforming-diagnosis.shtml.
33. Andreasen, 1997, 2007.
34. Quoted in Shorter, 2015, 1.
35. The proponents of the DSM-III revolution expected that its well-defined criteria would lead to more conservative estimates of mental illness in the population (Maxmen, 1985, 57).
36. The ECA surveyed more than 18,000 adults in the community and 2,500 persons in institutions in five sites (New Haven, Durham, Baltimore, St. Louis, and Los Angeles) to generate national estimates of prevalence (Robins et al., 1984).

37. Regier et al., 1998.
38. Kessler et al., 1994.
39. The NCS sampled about 8,100 persons meant to represent the population of the United States (Kessler et al., 2005). European studies produced comparable estimates: from one-third to 40% of residents of European Union countries had "brain disorders" each year (Gustavsson et al., 2011). See also Steel et al., 2014.
40. www.hcp.med.harvard.edu/ncs/ftpdir/NCS-R_Lifetime_Prevalence_Estimates.pdf.
41. www.hcp.med.harvard.edu/ncs/ftpdir/NCS-R_Lifetime_Prevalence_Estimates.pdf. Manuals after the DSM-III expanded the bipolar spectrum to include milder conditions: these states now afflict 5% of the population (Akiskal et al., 2000).
42. U.S. Department of Health and Human Services, 1999.
43. U.S. Department of Health and Human Services, 1999.
44. Murray & Lopez, 1996, 30.
45. Whiteford et al., 2013.
46. Olfson et al., 2003.
47. Conrad, 2018, 9.
48. https://www.cdc.gov/ncbddd/autism/data.html.
49. APA, 1987, 322.
50. McLaughlin et al., 2012, 596.
51. Olfson et al., 2006b.
52. Merikangas et al., 2010.
53. Kruijshaar et al., 2005.
54. Moffitt et al., 2007; Moffitt et al., 2010.
55. Rohde et al., 2013.
56. Menninger, 1963.
57. Narrow et al., 2002. Others wondered whether a cultural climate that emerged after 1980, which was marked by the growing availability and acceptability of therapeutic values, helped shape the rising affirmation of symptoms in survey research (Rose, 2019, 65).
58. McNally, 2011, 65.
59. Wing, Cooper, & Sartorius, 1974, 135.
60. Leaf, Myers, & McEvoy, 1991, 12.
61. Judd et al., 1994.
62. Regier et al., 1998. These rising rates most likely resulted from changing diagnostic criteria between the DSM-III and DSM-IIIR (1986) and other methodological artifacts. "Overall, one has to conclude that, despite the frequent claims to the contrary, there is little evidence from [epidemiological] studies of a systematic increase of levels of mental illness," sociologist Joan Busfield's (2011, 73) analysis determined.
63. Wittchen, 2000, 2–3.
64. Murray & Lopez, 1996.
65. Kramer, 2005, 215.
66. Quoted in Lane, 2007, 77.
67. Sadler, 2013a, 25.
68. E.g. Kohn et al., 2004.

69. E.g. Narrow et al., 2002; Regier et al., 1998; Goldberg, 2011; Maj, 2011; Wakefield & Spitzer, 2002.
70. Regier et al., 1998, 114.
71. Kessler et al., 2003.
72. Kupfer, First, & Regier, 2002, xviii.
73. Kupfer, First, & Regier, 56; Cuthbert, 2005.
74. Regier et al., 2018, 943.
75. Regier et al., 2018, 937.
76. Grob, 1991a, 6–7. Although the DSM-II was not explicitly dimensional, it did note that many of its conditions ranged from mild, to moderate, to severe (APA, 1968, 124).
77. Regier et al., 2009.
78. Not all members of the Task Force agreed with the wholesale attempt at dimensionalization. In March, 2009, Jane Costello resigned from the Child and Adolescent Disorders workgroup, circulating a resignation letter in which she expressed that she was "increasingly uncomfortable with the whole underlying principle of rewriting the entire psychiatric taxonomy at one time. I am not aware of any other branch of medicine that does anything like this." She cited the dimensional proposals as the tipping point that precipitated her resignation, because of the "possibility of doing a psychometrically careful and responsible job given the time and resources available is remote, while to do anything less is irresponsible" (Whooley & Horwitz, 2013, 39).
79. Hyman, 2010.
80. Haslam, 2013, 998–9. Autism, schizotypy, and substance use were possible exceptions.
81. Kendler & Gardner, 1998; Kupfer et al., 2002, 12. See also Watson, 2005.
82. Addington & Barbato, 2015, 199.
83. Kupfer et al., 2002, 125.
84. Clark et al., 2017, 104.
85. Plomin, 2018, 58–59.
86. Pierre, 2013, 117.
87. Taylor, 2013, 101–2.
88. Whooley, 2010; Greenberg, 2013; Smith, 2014.
89. First, 2005, 560.
90. Whooley, 2010.
91. Despite the failure of the Task Force to fundamentally alter the DSM system, it did have some successes. A new autism spectrum diagnosis yoked the previously separate autistic and Asperger's disorders, assuming a continuum where Asperger's disorder is a mild form of autistic disorder. It also moved the personality disorders and intellectual disability from their former separation on Axis II to the main section of disorders. In effect, this dismantled the former multiaxial system and thus made psychiatric classification more in line with general medical diagnoses. It merged substance abuse and substance disorder into a single category of "substance use disorder." Whether any of these changes will improve the validity of these categories

is an open question. The Task Force, however, failed in its quest to dimensionalize the personality disorders. See Wakefield, 2015, for a discussion of the full range of DSM-5 diagnostic changes.

92. https://www.nimh.nih.gov/about/directors/thomas-insel/blog/2013/transforming-diagnosis.shtml.
93. Cuthbert & Insel, 2013, 126.
94. Quoted in Belluck & Carey, 2013, A13.
95. https://www.nimh.nih.gov/about/directors/thomas-insel/blog/2013/transforming-diagnosis.shtml.
96. Greenberg, 2013, 339.
97. Insel et al., 2010.
98. Cuthbert & Insel, 2013, 126.
99. Meyer et al., 2018, 157.
100. Wakefield, 2015.
101. Height provides an analogy for why the distinction between normality and disorder does not follow simple dimensional principles. Height is a continuous property, but height disorders are not extreme variations on a single dimension; instead they are distinct conditions that result from some dysfunction. Some people are extremely short and others extremely tall, while most people cluster around the average height of a given population. All points along this continuum are normal, not disordered. In contrast, some people have disorders such as Achondroplasia or Marfan syndrome that cause them to be extraordinarily short or tall, respectively. They do not fall on different ends of a single dimension but have identifiable dysfunctions.
102. See Hyman, 2018, 944–45.
103. Kramer, 2005, 171.
104. Judd, Akiskal, & Paulus, 1997.
105. E.g. Mojtabai, 2011; Wakefield & Schmitz, 2012, 2013a, 2013b.
106. Johns & van Os, 2001; van Os et al., 2009.
107. Beavan, Read, & Cartwright, 2011.
108. Broome, Fusar-Poli, & Wuyts, 2013, 781.
109. Nestor, Choate, & Shirai, 2015, 101.
110. Livesley, 2010.
111. Srole et al., 1962/1978.
112. Lapouse, 1967.
113. Olfson et al., 2002.
114. Hippocrates, 1923–1931, 263.
115. APA, 1968, 40.
116. Mayes & Horwitz, 2005; Horwitz & Wakefield, 2007.
117. APA, 1980, 333.
118. APA, 1986, 222.
119. Robert Burton's (1621/2001, 143–44) statement that transitory melancholy arises from "every small occasion of sorrow, need, sickness, trouble, fear, grief, passion, or perturbation of the mind, any manner of care, discontent, or thought" is exemplary.
120. APA, 1994, xxi.

121. Wakefield & Schmitz, 2012a, 2013a, 2013b; Mojtabai, 2011.
122. Kendler, Myers, & Zisook, 2008. http://psychnews.psychiatryonline.org/doi/full/10.1176%2Fpn.46.20.psychnews_46_20_3_1.
123. http://www.medscape.com/viewarticle/758788.
124. Wakefield, 2013.
125. http://www.dsm5.org/ProposedRevisions/Pages/proposedrevision.aspx?rid=427#.
126. APA, 2013, 161.
127. E.g. Kleinman, 2012.
128. APA, 2013, 20.
129. Wakefield & First, 2012.
130. Wakefield & First, 2012.
131. APA, 2013, 197, 222, 59, 698, 362.

# 10
# The Past and Future of Mental Illness

Why has the study of mind and mental illness been so difficult, and why are we still uncertain about so many basic issues? By comparison with physical scientists, investigators and practitioners in the area of mental illness must appear to progress as snails and to behave as the blind men at the elephant.

—Bennett Simon, *Mind and Madness in Ancient Greece* (1987, 31)

The puzzles that mental illnesses present have perennially beguiled both professional and lay observers. Throughout history they have asked questions regarding what qualities of madness distinguish it from sanity, the extent to which mental and physical pathologies are similar or different, the kinds of factors that lead people to become mentally ill, and the sorts of treatments that might restore their sanity. Across time, knowledge about these issues does not show any steady growth or, arguably, much progress. The immense recent technological advances in brain science have not yet led to corresponding improvements in understandings of and treatments for mental illnesses. These perplexing phenomena remain almost as mysterious now as they were millennia ago.

## A Checkered Past

Over the past two centuries, general medicine and public health have made spectacular progress in combating disease. The specific mechanisms behind numerous illnesses such as tuberculosis, typhoid fever, cholera, and many more have been isolated. Penicillin and other antibiotics provide effective treatments for many of the most common ailments. Organ transplants and devices such as pacemakers extend the lives of millions of people. Vaccines have eradicated smallpox and drastically lowered rates of diseases including polio, diphtheria, tetanus, yellow fever, whooping cough, and measles. Discoveries that resulted in cleaner water and improved sanitation prevented many maladies from ever arising. The overall result has been a dramatic increase in life expectancy: for most of human history through the late 19th century, the average person died in his or her mid-thirties. At present, typical worldwide life spans exceed seventy years.[1] No sensible person would deny that present medical understandings and care vastly exceed those in prior eras. Can the same be said for psychiatric knowledge and practice?

## Defining Sanity and Madness

Issues of how to separate normal from abnormal thoughts, emotions, and behaviors have been recurrent aspects of debates about the nature of mental illness. These involve questions about the defining qualities of disorders, whether differences between sanity and madness are discrete or continuous, and where to place boundaries between pathological and healthy occurrences. Current answers to these dilemmas are no better than—and in some respects are regressions from—earlier understandings.

Ancient Greek philosophy and medicine developed contextual definitions of sanity and madness that have proven to be remarkably stable. Madness was associated with some inner defect that set the mad apart from the sane. Situations or cultural norms could not account for their actions, and so they seemed incomprehensible to others. Aristotle made this point most dramatically when he observed that killing a man and then eating his liver was a clear sign of insanity among the Greeks but could be customary and even expectable in other groups. Isolated symptoms or behaviors did not indicate disorder; instead, their explicability in given situations and particular cultures determined if they were normal or pathological. Conversely, symptoms that were not tied to the contexts in which they were naturally designed to arise or that emerged after a stressor but persisted with disproportionate intensity or duration often constituted disorders.[2]

It was relatively easy for the Greeks to define what was normal or pathological because they only considered extreme behaviors—what we now call "psychoses"—to be signs of mental illness. Greek medicine, philosophy, and literature generally equated mental illness with madness and starkly divided the mad from the sane. As Socrates noted about them: "They don't think a slight error implies madness, but just as they call a strong desire love, so they name a great delusion madness."[3] The Greek template that sharply split the mad from the sane endured in Western medicine for thousands of years. For example, writing in 1621, Robert Burton maintained the pointed distinction between contextually appropriate, and therefore not disordered, symptoms of melancholy that arose on "occasions of sorrow, need, sickness, trouble, fear, grief . . . [that are] the character of mortality" from similar presentations that emerged "without any apparent occasion." Following the ancient "without cause" dictum, Burton viewed only symptoms that could not be linked to any generating circumstance as signs of disorder.[4]

Even after the emergence of the psychiatric profession in the 19th century, proportionality to context remained the central distinguishing feature between the poles of sanity and madness. One of the founders of biological psychiatry, Wilhelm Griesinger, summarized that insanity "is distinguished from the mental pain experienced by healthy persons by its excessive degree, by its more than ordinary protraction, by its becoming more and more independent of external influences, and by the other accessory affections which accompany it."[5] The manuals from the DSM-III through the present DSM-5, too, contain general definitions of mental disorder that focus on distinguishing inner dysfunctions from contextually explicable distress, social deviance, and conflicts between the individual and society.[6]

Although contextuality can serve as a very general guide to distinctions between madness and sanity, it is not a useful way to separate different types of disorders from each other. Since the mid-19th century, the rest of medicine has used underlying causal mechanisms related to anatomy, cell pathology, and microbiology to define diseases. In contrast, psychiatric knowledge has not advanced beyond the situation that prominent 19th-century American psychiatrist Pliny Earle described in 1886 when he stated that the field's classifications were "forced to fall back upon the symptoms of the disease—the apparent mental condition, as judged from the outward manifestations."[7] While the rest of medicine distinguishes symptoms such as cough, fever, and pain from the underlying conditions that lead to them, many of the DSM's diagnostic criteria sets exemplify what evolutionary psychiatrist Randolph Nesse calls the basic error of "viewing symptoms as diseases" without regard to the situation in which they appear.[8] In contrast, Nesse urges psychiatry to develop classifications that recognize how many symptoms of common conditions such as depression and anxiety are useful responses to particular types of circumstances.

The central problem that has bedeviled psychiatric classifications since their inception—how to go beyond external symptoms to define valid diagnostic categories of mental disorders—remains unresolved. What many researchers consider the most important recent development in psychiatry, the move toward dimensionalization, threatens to exacerbate, rather than solve, problems of validity. It regards even mild distress as an incipient form of disorder and so abandons any attempt to distinguish sanity from madness.[9] The removal of the bereavement exclusion from the criteria for major depressive disorder in the DSM-5, too, thoroughly blends naturally designed responses to contexts of loss with dysfunctional conditions. It is difficult to discern how current conceptions of sanity and madness represent growing knowledge about the nature of these puzzling phenomena.

## What Is in Between Sanity and Madness?

The issue of how to define distressing conditions that fit neither pole of sanity nor madness also remains unresolved. As Richard Napier's medical practice (described in Chapter 2) suggests, before the 18th century, no explicit label captured psychic problems that fell between these poles. Distressed, but not mad, people who visited general physicians received the same mixture of herbal, folkloric, and religious responses as medically sick clients. At that time, a new type of "in-between" condition emerged that encompassed a wide variety of nervous complaints that were neither mad nor sane.

English nerve doctor George Cheyne's English Malady was the best-known example of a widespread affliction that blurred the boundaries between the normal and the pathological.[10] Cheyne distinguished nervous complaints from the psychoses and normalized them through associations with respectable social standing. Yet, they needed treatment from specialized nerve doctors. The English Malady provided a prototype for subsequent notions of nervous diseases, most prominently American neurologist George Beard's neurasthenia, which were sharply distinct from madness but which required medical attention.[11]

At the turn of the 19th century, Sigmund Freud revolutionized previous conceptions of the connections between sanity and madness.[12] Freud's major legacy was to thoroughly blur the boundaries between mental health and mental illness. Both normal and neurotic conditions stemmed from the same sources of instincts, repressed childhood memories, and social inhibitions. Health and sickness alike arose from autobiographical contingencies in responding to the conflicting demands of biology, psychology, and society. Their common roots simultaneously normalized pathology and pathologized normality. Freud's notions opened a vast array of widespread afflictions to psychiatric interpretations. Especially in the United States during the two decades that followed the end of World War II, dynamic psychiatrists embraced expansive notions of abnormality and shunned attempts to create distinct boundaries between the normal and the pathological.[13] They also expanded the range of presumed psychiatric expertise to entire populations, not just treated patients.

The fluid dynamic conceptions of normality and abnormality, however, lacked any medical authority. The psychosocial model embodied in the DSM-I and DSM-II did not value precise diagnoses, and its diagnostic categories were very general, cursory, and often etiologically based in unproven psychodynamics. In 1980, the DSM-III overthrew the dynamic view of highly overlapping spheres of normality and abnormality.[14] In its stead, it provided explicit definitions of several hundred categories of mental illness that were presumably distinct from normality. The DSM-III reformulated the amorphous psychological, behavioral, and social difficulties that dynamic psychiatry had already classified as pathological and converted them into highly specific, symptom-based, diagnoses. It did not so much expand the realm of mental disorder as carve the existing realm into far more discrete, medical-seeming entities.

Since 1980 the DSM diagnostic system has undergirded a remarkably successful remaking of psychiatry and other mental health disciplines. Its various conditions are not mere distress or deviance—and are certainly not a "myth"—but seem to be genuine disorders. This view relegitimated psychiatry as a medical discipline and, along with its allies in federal agencies, drug companies, advocacy groups, and the research community, provides it with a mission to confront a public health problem of vast proportions. Moreover, because most mental illnesses remain untreated, there seems to be an urgent need for expanded mental health services for a wide range of problems.

The inherently extensive degree of indeterminacy in definitions of mental illness provides a wide berth to interest groups who want to shape dividing lines between the normal and the pathological. In contrast to most of history, the social context of modern mental health practice insures that the scope of disorder is maximized and that of normality is minimized. Mental health professionals are only reimbursed when they treat disorders; researchers receive funding for studying disordered psychic conditions; policymakers are on safe political ground when they target accepted illnesses; and mental health advocates gain legitimacy for obtaining services for people with diseases. Moreover, patients must have disorders to get their treatments and services paid for. Perhaps most conspicuously, drug companies reap the highest profits when they market their products for the widest possible array of mental illnesses. None of these benefits accrue from calling any condition "normal."

One reason for the current expansive range of mental illness thus lies in the unprecedented number of interest groups that have stakes in considering a wide variety of behaviors as pathological. The result, in the colorful language of the DSM-5's fiercest critic, psychiatrist Allen Frances, is that

> DSM-5 will turn temper tantrums into a mental disorder. . . . Normal grief will become Major Depressive Disorder. . . . The everyday forgetting characteristic of old age will now be misdiagnosed . . . creating a huge false positive population of people. . . . Excessive eating 12 times in 3 months is no longer just a manifestation of gluttony and the easy availability of really great tasting food. DSM-5 has instead turned it into a psychiatric illness. . . . DSM-5 has created a slippery slope by introducing the concept of Behavioral Addictions that eventually can spread to make a mental disorder of everything we like to do a lot. . . . Many millions of people with normal grief, gluttony, distractibility, worries, reactions to stress, the temper tantrums of childhood, the forgetting of old age, and "behavioral addictions" will soon be mislabeled as psychiatrically sick.[15]

This vast realm of pathology does not result from better understandings of the nature of mental illness. Instead, the DSM's symptom-based definitions, coupled with the inherently porous lines between sanity and madness, provide opportunities for many groups to encompass a huge array of disturbing conditions within the boundaries of mental disorder.

## Are Mental Disorders Similar to or Different from Physical Disorders?

Another perennial question about mental illness is the extent to which psychic and physical disorders are comparable or distinct. Hippocratic physicians did not sharply divide mental from organic conditions. They used the same humoral theories to explain both types of disturbances. Likewise, they rejected notions of specificity. In the case of mental illness, they defined only a small number of broad conditions—mania, melancholia, hysteria, epilepsy, and phrenitis (madness accompanied by fever)—as disorders.[16] Such generalized conceptions persisted until the 18th century when diagnosticians began to, albeit unsuccessfully, use various symptoms to specify distinctive pathologies. The theories of mental disease that emerged at this time turned away from holistic Hippocratic notions as the idea began to emerge that, comparable to bodily diseases, each mental illness was a relatively specific entity. This new view extended a naturalistic framework to the innermost aspects of humans.[17] At the same time, Locke's empiricist theory divorced mental from organic conditions, rooting the former in disturbed thoughts and the latter in disturbed bodies.

The formation of the psychiatric profession in the 19th century was accompanied by attempts to use carefully observed symptoms to define mental disorders in analogous ways to physical ones. Initially, psychiatrists employed Gall's phrenology in their attempts to isolate separable mental illnesses. Psychiatrists developed numerous symptom-based classifications, but none succeeded before Kraepelin's observations about the distinct symptoms, courses, and outcomes of dementia praecox and manic depression.[18] Even Freud's central thrust during the 1890s involved specifying and distinguishing a variety of neurotic conditions. Before his thinking radically changed around the turn of the century, Freud remained within the biomedical tradition that viewed mental illnesses as specific in nature, organically grounded, and comparable to physical illnesses.

From 1900 onward, Freud's influential work challenged the previous equation of mental and physical illnesses, turning from a biomedical toward a psychic framework for the neuroses. In this new view, mental illnesses were distinct from biological ones in every major respect. For one thing, they were psychological as opposed to organic in composition. For another, diverse psychic symptoms did not provide direct indicators of some underlying disease. Instead, observable symptoms were symbolic and multilayered and disguised hidden meanings. American dynamic psychiatrists moved even further against specific conceptions, claiming, in Karl Menninger's words: "There is only one class of mental illness—namely mental illness."[19]

From the beginning of the 20th century through 1980, psychodynamic notions renounced the earlier efforts to define specific conditions. At that time, the DSM-III imposed a symptom-based model that assumed its many conditions were distinct, not general, in nature. This bolstered psychiatrists' medical credentials and legitimated them as responders to genuine diseases. The decades that followed coupled striking advances in observing brains and genes with troubling findings: the picture of mental illnesses that emerged from neuroscientific studies had little to do with the DSM's specific, symptom-based, and categorical entities. The manual's precise descriptions of many distinct disorders did not correspond to the phenomena that it tried to classify. Instead, neuroscientific research shows that each major condition is heterogeneous, that presumably separate disorders are highly overlapping, and that general rather than specific vulnerabilities underlie mental illnesses. In addition, it became clear that the major psychotropic drugs worked in generalized, not particular, ways. These findings led the National Institute of Mental Health (NIMH) and many biologically oriented researchers to reject the DSM's specificity model. Neuroscientific research is returning to earlier views that a small number of expansive conditions—for example, psychoses, internalized neuroses, externalized neuroses—best capture the domain of mental disorder.

In contrast to the steady progress medicine has made in defining physical diseases, one short-lived psychiatric classification has continuously displaced the previous nosology. Edward Shorter summarizes: "Every now and then, psychiatry systematically undergoes a total knowledge wipeout. All the accumulated knowledge of the past is scrapped."[20] The present DSM does not seem better able to specify the causes, predict the prognoses and outcomes, or develop particular treatments for its various entities than the many classifications that have come and gone in past centuries. Genetic and epidemiological findings question even Kraepelin's delineation of manic depression and dementia praecox. In 1963, Karl Menninger summarized the field of psychiatric nosology: "And so we are still in an era of nosological upheavals, confusion, and uncertainty."[21] Menninger's reflection aptly captures the present state of attempts to define the various mental illnesses.

## What Causes Mental Disorder?

The history of thinking about the causes of mental illness does not display any straightforward trajectory. Views that focus on internal, external, or interactive causes have waxed and waned over the course of history. Explanations fluctuate from emphasizing brains, minds, or social forces. Most eras, however, have recognized the importance of a variety of interior and environmental sources of mental disorders.

As with the other major issues surrounding mental disorders, the Ancient Greeks provided prototypical answers to the kinds of factors that lead to madness. The origins of external, internal, and interactive theories of mental illness stem from Greek literature, philosophy, and medicine, respectively. Before Hippocratic medicine arose, literary portrayals depicted divine intervention as the cause of madness among previously normal figures. Platonic philosophy, however, viewed insanity as arising from internal struggles between reason and the emotions. Although the Hippocratics assumed that the mentally ill had disordered brains, they did not sharply distinguish internal from external causes; both disrupted the holistic relationship between individuals and their surroundings and so could lead to madness.

After a long period when spiritual views of madness dominated, two diverse empirical conceptions emerged in the 17th and 18th centuries. On the one hand, medical theories replaced notions of humoral imbalance with those featuring observable interior mechanisms grounded in nerves, fibers, and tissues. On the other hand, Locke's influential psychological writings rooted mental disorder in disturbed thoughts that arose from distorted perceptions of the environment.

Despite their stark divergences, both brain-based and Lockean views agreed that mental illnesses could arise from some mixture of internal and external factors. For example, Cheyne simultaneously rooted nervous conditions in physical properties of brains and nerves and held that these afflictions stemmed from particularly English conditions related to wealth, weather, land, diet, and sedentary lifestyles. Later, American neurologist George Beard echoed Cheyne's view that, while neurasthenia was an organic condition, it arose from the stresses that accompanied modern civilization. Although Beard believed that social factors led to the ubiquity of neurasthenia, he also emphasized how heredity determined which particular people succumbed to the pressures of contemporary life. Likewise, although the dominant schools within 19th-century neurology and psychiatry focused on inherited, brain-based reasons for mental disorders, they also noted how environmental and moral considerations triggered which particular individuals became mad.[22] In contrast, Galton's hereditarian approach sharply split inherited from acquired influences: a stronger impact of one factor implied the weaker influence of the other.

Freud developed an interactive theory of mental illness that put the conflict between instinctual forces and the social sources of their repression at the heart of his explanations. Although his views changed considerably over the course of his career, Freud consistently emphasized the incompatibility between intense sexual and aggressive urges within individuals and equally powerful social needs to suppress these impulses. For Freud, internal and external forces were not distinct causes of neuroses but simultaneously worked to produce the deep struggles that inevitably bedeviled the neurotic and the normal alike.

After Freud's death in 1939, American psychiatry entered a unique period marked by a focus on psychosocial factors. World War II sparked the emergence of environmentally oriented views among military psychiatrists for why soldiers became psychologically impaired. This emphasis peaked in the postwar period when psychiatrists and other mental health professionals partnered with the NIMH to turn attention to the social conditions that evoked mental disturbances. At the same time, they downplayed the possible biological underpinnings of psychic disorders. Even the promotion of the tranquilizing drugs during the 1950s and 1960s was grounded in portrayals that the conditions they relieved were psychosocial, not biological, in nature. Organic explanations of mental disorders persisted among asylum-based psychiatrists and a few researchers but virtually disappeared from mainstream professional thinking.

The psychosocial emphasis that prevailed during the postwar period proved to be politically, economically, and culturally unsustainable. At first, the DSM-III confronted psychiatry's crisis of legitimacy by promoting a theory-neutral diagnostic system that was deliberately agnostic about the causes of its various conditions. It was thus compatible with any source of mental illness, whether internal or external. During the 1980s, however, an aggressively biological view of mental illness emerged, fueled by animosity at the previous dominance of the dynamic view, the development of powerful new methods for exploring brains, and the renaissance of biological thinking about human behavior in scientific and lay culture. As a consequence, psychiatry repudiated psychosocial perspectives and returned to its earlier biological thrust. More recently, as discussed later, many neuroscientists and psychiatrists are recognizing the inherently interactive relationship of genes and environments.

Mental disorders might be grounded in brains, yet they stubbornly resist explanations that reduce them to organic matter.[23] Forty years after the

DSM-III revolution, historical sociologist Andrew Scull summarizes, the "origins of major forms of madness remain almost as mysterious as ever."[24]

## Are People with Severe Mental Illnesses Better Off Now?

The arc of psychiatric history in responses to mental disorders sharply contrasts with that found in other branches of medicine. Since the mid-19th century, organic fields display steady progress and, often, dramatic breakthroughs in understandings of and treatments for somatic conditions. Despite expectations to the contrary, the rise of neuropsychiatric explanations of mental illness has not led to any discernible benefits for people with these conditions.

Hippocratic physicians employed an eclectic array of treatments. Some encouraged restrained modes of living as ways to redress imbalances between individuals and their environments.[25] Others involved alcohol and drugs, particularly opium, to relieve mental distress. Ancient Greek and Roman therapies also employed cognitive regimes modeled on Stoic or Epicurean philosophies that taught people to control their disturbing emotions through willfully ignoring outside stressors. Overall, responses to mental unrest were not dogmatic but used whatever techniques worked in individual cases. This practical approach endured. Before a dedicated profession of psychiatry arose in the 19th century, typical responses to the mad embodied a potpourri of potions, spells, drugs, and external restraints in the quest to restore mental health and control disruptive behavior.

Locke's theory that mental illnesses resulted from mistaken association of ideas that are correctable through environmental changes justified dramatic changes in inpatient treatment. The rise of asylums based on the principles of moral treatment followed from his emphasis on changing circumstances rather than brains. Moral treatment used individualistic approaches that strove to take into account the unique psychosocial circumstances of its generally well-off clients. It was, however, a short-lived phenomenon.[26]

From the mid-19th century to the onset of the Second World War, the emphasis on changing milieus waned as neglect and coercion typified asylums. One after another, the various organic theories that arose in this era failed to deliver on their promises to improve treatments for madness. Theories based on hereditary degeneration resulted in the sterilization of thousands of mental patients in the United States; hundreds of thousands were murdered in Nazi Germany.[27] Twentieth-century efforts that removed patients' organs, placed them in insulin-induced comas, or lobotomized their brains all came to naught. Electroconvulsive therapy, which can help people with depression and bipolar conditions who have not benefitted from less drastic interventions, is the only asylum-era procedure that persists [28]

The initial movement of psychiatrists from mental institutions to outpatient practices during the postwar period was marked by psychotherapies that relied on verbal interactions between therapists and their patients, often supplemented with a prescription for some drug. Since the 1980s, however, the reimbursement policies of managed care organizations render the provision of drugs far more cost-effective than time-intensive modes of talk therapies. The result has been a plunging number of psychiatrists who engage in any kind of psychotherapy. By 2004–2005, less than 30% of visits to psychiatrists involved these treatments, and only about 10% of psychiatrists used them with all their patients.[29]

At the same time as psychiatry turned from inpatient to outpatient practices, the combined effects of federal policies that would not finance care in mental hospitals, legal rulings that restricted involuntary commitments, and the widespread employment of antipsychotic drugs in the community resulted in the virtual abandonment of mental institutions. Despite the drastic growth in the U.S. population between the 1950s and the present, the mean number of residents in mental hospitals at any particular time dropped from about 550,000 to less than 40,000.[30] Yet, no adequate treatment system for persons with serious mental illnesses emerged to replace the now nearly empty institutions.[31]

In the absence of community treatment programs that encompass housing, jobs, recreation, counseling, and social services, incarceration in jails and prisons has become an increasingly common response to the seriously mentally ill. The grim reality is that people with severe mental illnesses are now more likely to be found in law enforcement than health-care facilities.[32] A 2017 report from the Bureau of Justice Statistics indicated that "383,000 individuals with severe psychiatric disease were behind bars in the United States in 2014 or nearly 10 times the number of patients remaining in the nation's state hospitals."[33] Indeed, the three institutions that currently contain the largest number of mentally ill people are the Riker's Island correctional facility in New York City, the Cook County prison in Chicago, and the Los Angeles county jail.[34] It is difficult to see how the rising focus on drug treatments for outpatients and the neglect or incarceration of persons with

**Figure 10.1** Far more people with serious mental illnesses are now found on the streets or in prisons than in mental institutions. David Shankbone, A man sleeping on the street in the Bowery, August 2008.

the most severe conditions represents progress in the response to mental illness.

Nor have recent developments resulted in better public perceptions of people with serious mental illnesses. Participants in the neuroscientific revolution took as an article of faith that viewing mental disorders as brain dysfunctions would improve the lives of the mentally ill, especially through reducing the stigma of mental illness. They assumed that biomedical portrayals would provide not just more accurate depictions of mental disorders but also the additional benefit of transforming public attitudes toward the mentally ill. "The recognition that mental illnesses are diseases affecting the brain," Nancy Andreasen wrote in 1984, ". . . has already done a great deal to diminish the fear, shame, and guilt attached to mental illness."[35] In 2012, Eric Kandel went even further, proclaiming without irony that "schizophrenia is a disease like pneumonia. Seeing it as a brain disorder destigmatizes it immediately."[36]

Since 1980 the number of Americans who regard mental illnesses such as schizophrenia, depression, and alcohol dependence as results of chemical imbalances or genetic factors has indeed grown. In addition, they increasingly recommend medical and psychiatric treatments, especially drugs, as the most appropriate responses to these conditions.[37] Yet, these biomedical depictions have not resulted in declining stigmatization: community surveys since 1980 show that views of psychiatric patients remain unfavorable.[38] This is because there is no simple equation of biological causes and less stigma; respondents often associate brain-based explanations with diminished ability to change the disruptive condition in question.[39] Serious mental illnesses—regardless of how they are perceived—create interpersonal challenges for responders that, say, cancer or cardiovascular diseases do not entail. Brain-based interpretations do not annul the behavioral disturbances that are intrinsic to mental disorder.

Perhaps the greatest disappointment accompanying the adoption of a biomedical view of mental disorder has been the shocking growth in disparities of life expectancy between persons with serious mental illnesses and others. In contrast to trends in the general population, death rates among this group have drastically increased in recent decades. The mortality ratio between people with schizophrenia and the general population has more than doubled, rising from 1.8 in the 1970s to 3.0 in the 1980s, to 3.2 in the 1990s to 3.7 in the 2000s.[40] Colton and Manderscheid report that between 1997 and 2000: "Public mental health clients lost decades of potential life and died at younger ages than their cohorts nationwide for the years studied."[41] Stunningly, people with serious mental illnesses now die an average of about twenty years younger than those without a mental disorder.[42] While the causes of these stark, and growing, differences are not clear, they are associated with poor living conditions marked by poverty, social isolation, homelessness, physical inactivity, tobacco use, and obesity. The long-term and massive use of antipsychotic drugs is also likely to contribute to shorter life spans.

At minimum, the turn toward neuroscience has not succeeded in improving the quality of life of persons with serious mental illnesses and in some ways might have helped to facilitate their declining longevity. In 2017, former NIMH Director Thomas Insel concluded:

> I spent 13 year at NIMH really pushing on the neuroscience and genetics of mental disorders, and when I look back on that I realize that while I think I succeeded at getting lots of really cool papers published by cool scientists

> at fairly large costs—I think $20 billion—I don't think we moved the needle in reducing suicides, reducing hospitalizations, improving recovery for the tens of millions of people who have mental illness.[43]

Rhetoric aside, the current biomedical revolution has, in historian Anne Harrington's words, "overreached, overpromised, overdiagnosed, overmedicated, and compromised its principles."[44]

Perhaps the most promising path toward bettering the lives of persons with serious mental illnesses lies in renewing the insights of Hippocratic and moral treatments. These schools recognized that changing environments is at least as important as changing brains for bettering the lives of this group. Although drug treatments are often integral aspects of successful healing programs, they must be combined with interventions that provide resources such as housing, community services, and jobs that are distant from the current concerns of biological psychiatry. It is difficult to see how current therapies represent progress in comparison to long-abandoned approaches that recognized how changes in inner processes must be accompanied by attention to external circumstances.

## The Future of Mental Illness

In certain respects, the era from 1980 to the present has been a golden age for psychiatry. The DSM and its broad range of mental illnesses have been widely embraced socially, culturally, and institutionally. The profession has the important mission of confronting and lowering rates of mental illness that seem to be not just at an all-time high but growing in each decade. Rates of treatment for psychiatric problems have soared as the willingness of the public to seek mental health care has increased during this period.[45] Aside from people with the most severe cases of psychoses, professionals rarely impose labels of mental illness on resistant patients; instead, clinicians and their clients participate in a shared voluntary culture of therapeutic values.

Although therapeutic cultures have existed in various forms since the 18th century, a key difference between earlier periods and the present is that opposing institutions and beliefs no longer challenge them. In contrast to past eras, no health professional, regardless of his or her theoretical allegiance, considers mental illness to be a blameworthy condition. The media, as well, consistently presents positive portrayals of therapeutic values. Likewise, politicians on all sides generally unite in advocating compassionate attitudes toward the mentally ill. Most organized religious groups also embrace therapeutic viewpoints. Crucially, the public has thoroughly accepted the DSM structure and its many mental illnesses. Even military, law enforcement, and athletic institutions that have traditionally made punitive responses to mental illness no longer scorn their members who claim to have mental health problems.[46]

Yet, despite the widespread affirmation of the current diagnostic model and the accompanying culture of therapy, a few cracks have arisen in psychiatry's classification system. As the controversy regarding the failed proposal to dimensionalize diagnoses in the DSM-5 indicates, accepted views of mental illness are in some peril. Many researchers have concluded that the underlying nature of most disorders do not fit the categorical system that diagnostic psychiatry imposes on them. Despite the DSM-5 Task Force's failure to capitalize on the findings from neuroscientific research, the incompatibilities between these results and the current diagnostic system might eventually make the DSM system unsustainable.

If a diagnostic system fails to adequately order its domain, it may eventually collapse.[47] The NIMH has already developed an alternative taxonomy for researchers, the Research Domain Criteria, which thoroughly rejects the DSM's symptom-based framework. An intellectual crisis that is currently confined to academics might spread as it becomes increasingly apparent that current diagnostic categories fail to provide distinct explanations, prognoses, and treatments for the conditions they ought to reflect.

Although the current categorical structure remains strongly entrenched in clinical practice and social institutions, several factors might lead to alternative ways of viewing mental disorders. It is becoming clearer that particular genes or combinations of genes do not produce specific disorders. Instead, genetic factors produce broad vulnerabilities to, for example, psychotic, internalizing, and externalizing traits. Once researchers gain greater understandings of such nonspecific genetic factors, they are not likely to correspond with the current symptom-based system. The questionable validity of the current psychiatric nosology might not be sufficient to lead to its demise, but it might help stimulate its eventual replacement.

Will the neuroscientific revolution eventually produce fundamental changes in understanding and treating mental disorders? Or will it join the long history of promised, but failed, breakthroughs? Gerald Grob observes:

> Many etiological theories (hereditarian, Freudian, neurochemical, etc.) have been proposed to great fanfare only to buckle under the weight of their initial promise. In other branches of medicine, by contrast, the demonstration of a relationship between the presence of certain symptoms and a specific bacterial organism had led to the development of a new classification system based on etiology rather than symptomatology. The inability to pursue a parallel course left psychiatry with a classification system based on external symptoms that tended to vary in the extreme.[48]

Although to date biological psychiatry, neuroscience, and genetics have not produced the promised forward leaps, the genuine advances in abilities to examine brains, neurons, and genes lend credence to the belief that significant advances are on the horizon. What are the prospects that psychiatrists and others will make genuine progress in comprehending the nature, causes, and treatments of mental illnesses? And, if they do, what problems could accompany such advances?

## Genes and Environments

One promising aspect of much current thinking about the causes of mental illness lies in its rejection of the century-old dominance of Galton's heritability paradigm, which pitted inner and outer forces against each other. The basic premise of this zero-sum view was that the greater the influence of nature, the less impact of nurture, as well as the converse. Refuting the legacy of a long fascination with twin and adoption studies that opposed genetic to environmental forces, many researchers now embrace an interactive conception that shows their mutual interdependence: favorable environmental circumstances can suppress and adverse conditions can provoke the expression of mental disorders among the genetically vulnerable. Genes and environments provide complementary, not competing, explanations for the appearance of mental disorders. The recognition that environmental and genetic forces are interactive offers hope that the Galtonian split of nature from nurture will become anachronistic. "In the current climate," sociologist Catherine Bliss observes, "gene–environment science is the number one priority for most public health agencies around the world."[49]

One spur to the integration of external and genetic effects has been research showing how genetic influences depend on the social contexts in which they are expressed. As sociologist Peter Bearman notes, environments can be viewed as petri dishes in which genetic things can happen.[50] Indeed, one of the major findings from the Human Genome Project has been the crucial role of environmental factors in either mitigating or promoting genetic effects. Just because a person possesses a certain type of gene does not mean that the gene will be expressed or that the expression of the same gene will be the same in different contexts. For example, genes for aggression among people in deprived environments might lead their carriers to jail; the identical genes can facilitate financial success among those in resource rich circumstances.[51]

Comparably, genetic factors that predispose people to develop mental disorders can become manifest only in certain kinds of environments. For example, the New Zealand study mentioned in previous chapters found that genes regulating serotonin levels had no direct effect on rates of depression. However, people who both possessed the short allele of the 5-HTT gene and experienced especially high levels of social stress were more likely to develop depression.[52] Another body of research indicates that the prevalence of schizophrenia is higher among immigrants than among the population of their home countries, possibly because of the greater discrimination they face in their new locales.[53] Genetic tendencies might only become manifest when environmental stress intensifies.

Conversely, favorable external circumstances can quash genetic liabilities. Without the right environmental trigger, genes might never be expressed at all. For example, people with genetic variants tied to alcoholism are less likely to develop it when they receive high levels of family support or adhere to religions that do not tolerate drinking.[54] Social factors even shape the prevalence of conditions that are firmly rooted in genetics. For example, de novo genetic mutations linked to autism seem related to the older ages when fathers have children, which itself is a product of profound social changes in work and family life.[55] The rising number of children with autistic spectrum disorders could in part stem from the postponement of fatherhood. "Mental phenomena can be fully understood only in the context of an organism's interacting in an environment" neuroscientist Antonio Damasio summarizes.[56]

The rise of epigenetics has been another source for the growing importance of the mutual study of genes and environments.[57] Epigenetic research examines how environmental forces turn the DNA strands that surround genes off and on. "[The] brain continually rewires itself and changes gene expression as a function of learning and life events," Thomas Insel and Bruce Cuthbert summarize.[58] Such studies counter Galtonian influences to show how brain effects are not fixed at the time of birth but display neuroplasticity in response to external stimuli. Environments, especially those acting during the first few years of life, can literally get under the skin to provoke mental illness.[59] This occurs through not only highly stressful traumatic events such as child abuse, famine, or extreme poverty but also customary experiences including cultural socialization, the degree of social and economic security, and education. These epigenetic effects can last throughout the life course. Some researchers also assert that epigenetic impacts are heritable and can be passed on to new generations, although this contention is unproven at present.[60]

Findings from epigenetic studies challenge previous assumption that genetic influences are unchangeable. Brains are intrinsically social organs that are biologically designed to be extremely sensitive to the outer world.[61] Far from being stable structures established at conception, neurons and neurochemicals, as well as genetic expressions, are attuned to respond over an individual's life course to social, cultural, and interactional factors. The central issue for many current researchers is not whether nature or nurture influences human behavior but regards the interplay between heredity and social milieu. "Gene–environment interactions," report sociologists Dalton Conley and Jason Fletcher "are the rule rather than the exception."[62] The appreciation that one-sided views focusing on either genes or environments are inherently incomplete provides hope that knowledge about mental illnesses will grow in future years.[63] Although examinations of the interactions between genes and the outside world add immense complexity to the study of mental disorders, they also provide the greatest chances of making genuine progress in understanding these perplexing phenomena.

## Early Identification of Mental Illness

More than a century ago, the mental hygiene movement strove to identify maladjusted youth before they developed full-blown disorders. Its efforts provided a model for screening efforts during World War II that aimed to prevent mentally disordered men from entering the armed services and presumably harm combat readiness.[64] The military instituted a massive attempt to detect mental illness among draftees, which resulted in the rejection of an extraordinary number of potential soldiers as mentally unfit to serve. Although the program was, at best, of limited effectiveness, it implanted the idea that screening the general population could identify embryonic signs of mental illness before they developed into serious cases. Accordingly, in the postwar era psychiatrists and the NIMH advocated for identifying and treating mental illnesses at early stages.

At the beginning of the 21st century a movement arose that echoed these earlier efforts. Taking their cue from epidemiological findings that indicated most persons who have some mental illness do not seek professional help, that some mild cases will become severe in later years, and that the amount of untreated mental illness is especially high among adolescents, mental health advocates and policymakers argued that it was imperative to identify disorders as early as possible with the goal of providing help that could prevent incipient disorders from worsening. To this end, they developed widespread programs to first screen and then take actions to prevent problems among young people from emerging or deteriorating.

Another influence intensifying calls for mass screening has been widely publicized school shootings by juveniles who presumably are mentally ill. Such killings seem preventable if their perpetrators can be recognized and treated before they commit such heinous deeds. Considerable momentum exists to institute broad programs that screen school-age populations for signs of mental illness and provide identified youths with mental health services that might prevent future homicidal or suicidal acts.[65]

Many initiatives now strive to recognize adolescents who are at risk for developing some mental disorder. In 2003, a presidential commission proclaimed: "Every child should be screened for mental illness once in their youth in order to identify mental illness and prevent suicide among youth."[66] The next year President George W. Bush signed a bill that authorized $82 million to fund programs beginning in sixth grade that would detect youth who are prone to becoming mentally ill. Most states have implemented some sort of screening in schools, and the number of students who participate in such programs is expanding rapidly. These efforts generally rely on brief, self-reported symptom measures to initially identify students who are potentially at high risk for some mental disorder. This group then undergoes a second, more intensive clinical interview to confirm the possible presence of a mental illness.

Despite their superficial appeal, the implementation of widespread screening programs and consequent mental health treatment has a number of worrisome aspects. Screening efforts to date rely on self-reports of mental health problems, and their goals are transparent. Anyone with serious mental health difficulties, not to mention potential killers, can easily avoid recognition by denying their problems. Moreover, screening identifies anywhere from 20% to more than 50% of students as potentially having mental disorders.[67] Far more of those identified turn out to be false positives than youths who will actually develop serious mental health problems in the future, not to mention the tiny number that will become homicidal. The proportion of false positive identifications of students who are deemed to be at risk for becoming psychotic could be as high as 95%.[68] Screening programs can falsely label millions of students as being at risk for becoming mentally ill with potential negative consequences for their lives.

In addition, effective mental health treatment depends on the willingness of people to participate in therapeutic efforts. Screening programs assume that those identified as being at risk for developing some mental illness will also want to get treatment, a questionable assumption. Moreover, research indicates that, aside from the most severe cases, success rates for medication therapy with even willing patients barely exceed placebo responses. There is little reason to believe that providing therapy to people who have not begun treatment on their own initiative will be of more value. Programs that do not rely on voluntary help-seeking for mental health problems are unlikely to succeed. It is especially doubtful that the miniscule number of individuals who perpetrate mass shootings would benefit from treatment, even in the unlikely case that they could be identified. Programs for this small group could only be a real deterrent if they involve a regime of involuntary treatment that would capture the innocent as well as the potentially dangerous alike. Mental health screening will likely mislabel huge numbers of people without protecting society from its deranged members.

Finally, such programs can have undesirable practical applications. So many youth screen positively that only a massive increase in mental health funding could provide even superficial treatment to those identified as at risk for some mental illness. Such wasteful employment of resources is likely to detract from the unmet needs of people with serious mentally illnesses who could genuinely use mental health services. The most shameful aspect of the current mental health system is the failure to replace the now-shuttered mental hospitals with any coherent community-based system.[69] Many individuals with psychotic conditions spend long periods of time in prisons and jails while hundreds of thousands of others are homeless. Elderly persons with serious mental disorders often reside in nursing homes that are ill-equipped to care for them. Instead of continuing the fruitless quest to expand mental health services to vast numbers of people with mild cases, policymakers should concentrate on providing help to those with severe disorders who live on the streets or languish in institutions that were not meant to hold them.

### Using Biomarkers to Identify People Who Are at Risk for Becoming Mentally Ill

While current screening programs use self-reported measures and subsequent follow-up clinical interviews, the genomic era has created the possibility of foregoing these processes and directly assessing genotypes to identify at-risk populations. It has led to a new category of people: those who neither currently have nor have ever had a mental disorder but who possess genetic traits that are thought to put them at risk for developing one in the future. Geneticists now have inexpensive, simple, and easily employed methods for assessing individuals' genomes. "Consequently," according to psychologists Stephen Shirk and Nathaniel Jungbluth, "it may turn out that genetic variables, potentially easily sampled with a mouth swab, could contribute substantially to the prediction of subsequent internalizing [and other] disorders."[70]

It is probable that diagnosticians of mental illness will increasingly make use of such markers in upcoming years, even for people who show no current symptoms of some condition. One sociologist uses attention-deficit hyperactivity disorder (ADHD) to illustrate the potential employment of genotypes:

> So the example I have been using is there is a GWAS [genome-wide association study] on ADHD; parents could have their kid genotyped to decide based on the kid's genotype and the GWAS to show a high likelihood adult in ADHD. They can perhaps request an intervention be performed with their kid in their schools before the kid was even showing symptoms.[71]

Likewise, a considerable body of research has emerged that strives to identify young people who have not yet become schizophrenic but are thought to have genes that put them at risk for doing so.[72]

Using genes related to any mental disorder can provide opportunities to identify, intervene, and therefore minimize the damage that any condition might entail long before any signs and symptoms appear. Its proponents also optimistically assume that genetic profiles can generate treatments that are tailored to personalized genomes.[73] Yet, this process also has troublesome aspects. One is the vast expansion of the population seen as having or as at risk of developing some mental illness: the number of people with at-risk genetic variants who do not ever develop a condition will greatly exceed the number that will become disordered. David Healy estimates:

> if 1 person per 100 has a disease at least 10 per 100 can be expected to carry a risk factor for this disease. Traditional medicine mandates the treatment of the 1 diseased individual, whereas the new emphasis on risk mandates the treatment of all 10 who are at risk. A further attraction is that whereas the treatment of a disease comes to a full stop when the patient is cured, the predisposing risk and the risk of relapse may go on forever.[74]

Neuroscientific research on depression provides an example. An extensive body of research on the 5-HTT gene, which controls the way that serotonin passes messages through brain cells, has emerged. The influential New Zealand study mentioned earlier found that among people who experienced four or more stressful life events, 43% with two short alleles and 33% with one short allele developed major depression compared to only 17% of those with two long alleles who did.[10] The identification of the short allele of the 5-HTT gene as a risk factor for depression among highly stressed people creates the possibility that genetic tests could identify people who are at risk of developing depression even if they do not show any actual depressive symptoms. Such findings could be used to justify incorporating genetic measures into screening efforts. Yet, two-thirds of the population seemingly has a genotype that is associated with vulnerability to depression. The use of such presumed innate susceptibilities could massively pathologize any screened population.

Another problem involves the polygenic influences on virtually all mental disorders. Any particular gene only explains a small amount of variance in who will develop some disorder. Far more people who possess a specific genetic variant will therefore *not* come to have the condition than the number who will have it. In other words, the number of false positive identifications will far exceed the number of true cases that are identified. The specification of at-risk genes creates the possibility for a vast expansion of pathology in the future to people who neither display any symptoms of disorder at a particular point in time nor will become disordered in the future. Nevertheless, the temptation exists to prescribe medication or apply other technologies to such individuals that might prevent the condition from emerging, even to those whose outcomes will be benign.[75]

The identification of at-risk individuals through genetic screening thus has several problematic aspects. Such people can be recognized extremely early in life, potentially at the time of birth or even before. Many identified individuals (and their parents) can be chronically fearful of getting some serious disorder that in fact will never appear. Such fears can last for a lifetime. People labeled as being at risk for some mental illness can also develop identities centered on their chances of becoming mentally ill. Psychologist Illina Singh and sociologist Nikolas Rose perceptively pose the basic questions:

> How will people feel about themselves given their risk profile, and will others perceive them differently? Will "risk" and "potential" eventually dominate ideas of personal identity, health status and opportunity in rigid, coercive or stigmatizing ways? Will these ideas become institutionalized within education, law, and policy? And how will such change affect the life trajectories of children identified as at risk early in life?[76]

The risk factors and consequent possibility of relapse will be omnipresent throughout the life course so that life-long medication regimens might seem necessary.

Despite these issues, biotechnology companies are making large investments in techniques for uncovering possible genes for common mental illnesses, among other conditions. Correspondingly, pharmaceutical companies might become able to develop new classes of more refined, genetically tailored medications that could be prescribed to presumably at-risk populations so disorders will never actually arise. The popular gene testing business 23andMe has partnered with the pharmaceutical company GlaxoSmithKline with the goal of developing drugs that target individualized genomes.[77] Its founder urges parents to obtain their children's genetic profiles so that they can begin preventive measures for potential problems as early as possible.[78] The market for such therapies is potentially huge.

The future is likely to be marked by a vast expansion of individuals who are identified as at risk for becoming mentally ill later in life despite not displaying any current signs of mental illness. Such persons can comprise a substantial portion, perhaps a majority, of the population. Yet, many, and probably most, people with some genetic marker will never develop a mental disorder. They are, however, attractive targets for pharmaceutical companies and mental health professionals. The potential costs of technological breakthroughs in identifying people who are at risk of developing mental disorders, however, might outweigh the hoped-for benefits.

## Can Psychiatry Survive?

The emergence of the psychiatric profession in the 19th century was associated with its mandate to provide care for institutionalized patients. This connection dissolved during the mid-20th century when mental institutions themselves began to empty. Hospital-based psychiatry is now almost dead. The profession can no longer justify itself through its caretaking for persons with the most severe forms of mental illness.[79]

During the post–World War II period, psychiatry became linked to prestigious and successful outpatient practices where it dispensed various combinations of psychotherapies and drug treatments to distressed community members. During the mid-1960s, however, a combination of antipsychiatric ideology, medical antipathy, interprofessional competition, and a shaky reimbursement structure put the field's professional viability in question. The standing of the field was so low that in 1976 the director of the NIMH asked: "What is the role of psychiatry? Does it even have a role, or is it, as some would have it, dead or dying?"[80]

In contrast, recent decades have been kind to the psychiatric profession. The DSM-III's importation of a biomedical paradigm restored the field's prestige and legitimacy, which it has maintained through the present. Its broad range of mental illnesses has gained cultural acceptance; the conditions that require its attention are widespread and growing. Therapeutic values now permeate cultural and social institutions. At the same time, the pharmaceutical industry's profits from its psychotropic products brought substantial monetary rewards to its psychiatric collaborators. Moreover, psychiatry departments are valued within medical schools because they garner considerable amounts of drug company and governmental funding. Yet, some signs indicate that the recent flourishing of the psychiatric profession might be coming to an end.[81]

Several emerging trends raise questions about psychiatry's viability as it moves into the third decade of the 21st century. Ironically, one challenge stems from the enormous use of psychotropic drugs. The public has broadly accepted the view of mental disorder as a brain disease that is best treated with medication. Antidepressants, antipsychotics, anxiolytics, mood stabilizers, and stimulants are all widely marketed, sought after, and dispensed. Yet, the growing number of drug prescriptions has primarily occurred in general medical, not psychiatric, practice.[82] Psychiatrists write just 23% of prescriptions for all types of psychotropic drugs while general physicians (GPs) account for 60% and other medical specialists the remainder.[83] GPs write about 80% of prescriptions for the most common mental illness, depression.[84] Visits to GPs are cheaper, usually less time-consuming, and more convenient than visits to psychiatrists. In the future it is likely that both insurance companies and patients will increasingly see GPs as the preferred source of medication for mental health conditions. The dispensation of drugs threatens to transfer psychiatry's grip on the outpatient market to general physicians.

Worse, the era of pharmaceutical industry sponsorship of psychiatry shows signs of ending.[85] No truly novel psychotropic drugs for general use have been discovered since the 1950s, and most of the enormously lucrative antipsychotic, antidepressant, and anxiolytic medications that came on the market in the 1980s and 1990s have lost their patents and, consequently, much of their profitability.[86] The major drug companies have largely abandoned their efforts to discover new products, a development that could have dire consequences for psychiatry and its place in the hierarchy of medical specialties. If academic psychiatrists cannot garner funding from the pharmaceutical industry, their standing within medical schools is likely to diminish.

At the same time as medication treatments increasingly take place in primary medical care and the drug industry is beginning to withdraw from the search for new products, psychotherapies are more and more becoming the province of nonmedical providers. Psychiatry has already mostly ceded the practice of psychotherapy to other clinicians including psychologists, social workers, and counselors.[87] Rates of psychiatric outpatients who received any psychotherapy plunged from roughly 70% in 1987 to about 40% in 2007.[88] Many of even this minority receive brief and cursory forms of talk therapies. Patients who desire psychotherapies are now likely to turn to nonmedical sources of help who can provide cheaper and equally efficacious care as psychiatrists.

A third factor that might threaten the future viability of the psychiatric profession, especially if the quest for biomarkers of mental disorders is successful, lies in the perennial trend for biologically established conditions to move from psychiatry to neurology. Although no such markers have yet emerged from neuroscientific research, prominent candidates include melancholic subtypes of depression, panic disorder, and catatonic forms of schizophrenia.[89] As has happened in the past with syphilis, epilepsy, dementia, and Huntington's disease, once some condition is established as a definitive brain-based defect, it becomes unlikely to be viewed as a mental disorder at all. Instead, neurologists will become the primary source of treatment for people with confirmed brain diseases.

The rise of electronic communication poses a final, and particularly profound, threat to psychiatric authority. Mental health clients, like all consumers, now increasingly use the Internet to understand and deal with their problems.[90] Cyberspace empowers individuals and the subcultures they create to bypass professional authority and to self-define what sort of condition they have and how best to treat it. The ability to instantly connect in chat rooms and websites with millions of others who are not physically co-present allows consumer groups to form and develop their own ways of thinking. Growing online social movements, echoing some of the major themes of the anti-psychiatry school, question whether any condition is a mental illness or, alternatively, a valued way of living that entails special insights. They regard many states that psychiatrists view as disorders as healthy signs of neurodiversity and demand the power to make their own definitions of their conditions and responses to their situations.[91]

The omnipresence of electronic communication can minimize the stigma of previously disvalued qualities and promote a far broader array of traits as valued ways of being human. Its impacts, however, need not be positive. For instance, many "pro-ana" websites glorify anorexia nervosa and provide information on the best ways to facilitate it.[92] For better or worse, in coming years such virtual groups are likely to grow and expand as sources of social acceptance for people who have been regarded as mentally disordered. They are also likely to profoundly challenge the authority and legitimacy of professional groups, including psychiatrists.

These trends could present a major test to the very existence of the psychiatric profession. Its control over mental institutions and the people with severe conditions who once occupied them has already dissipated. If general physicians are the primary purveyors of psychotropic medications, drug companies become less willing to fund psychiatric research, non-physicians provide psychotherapy, neurologists deal with established brain-based conditions, and the Internet becomes the chief disseminator of knowledge about mental health, what role will be left for psychiatrists in understanding and treating mental disorders in future years?

To date, the extremely sophisticated technologies that view brains and genomes have not produced fundamental breakthroughs in understandings of mental disorders. Instead, they reveal how the inherent complexities of brains might preclude the possibility of discovering simple causes of or treatments for any mental illness. Kenneth Kendler presents a good portrayal of the intricacies confronting understandings of mental disorder in the neuroscientific era:

> Psychiatric disorders are a result of multiple etiological processes impacting on many different levels and often further intertwined by mediational and moderational interactions between levels. It is not possible a priori to identify one privileged level that can unambiguously be used as the basis for developing a nosologic system.[93]

The future is far more likely to involve unravelling different particular strands of this multifaceted whole than making dramatic discoveries that alter our basic picture of mental illness. The basic puzzles that madness has posed since the earliest medical, philosophical, and literary writings—how to separate sanity from insanity, how to distinguish mental and bodily illnesses, how to specify the variety of internal and external forces and interactions between them that lead people to become mentally ill, and how to effectively treat mental disorders—are likely to remain unresolved for the foreseeable future and perhaps forever.

## Acknowledgments

Portions of the section "Early Identification of Mental Illness" are adapted from Horwitz & Wakefield, 2009.

## Notes

1. Pinker, 2018, 53–54.
2. Aristotle, 2009, 127.
3. Xenophon, 2013, 226–27.
4. Burton, 1621/2001, 143–44.
5. Griesinger, 2000, 226.

6. APA, 2013, 20.
7. Earle, 1886.
8. Nesse, 2019, 27.
9. E.g. Kessler et al., 2003b.
10. Cheyne, 1733/1991.
11. Beard, 1881.
12. Freud, 1900/1965.
13. Menninger, 1963.
14. APA, 1980.
15. https://www.psychologytoday.com/us/blog/dsm5-in-distress/201212/dsm-5-is-guide-not-bible-ignore-its-ten-worst-changes.
16. Aside from melancholia, the Greek meanings of these conditions considerably diverged from their current usages.
17. Makari, 2015, 63.
18. David Healy (2008, 127) notes that at present it would be virtually impossible to replicate Kraepelin's method of discovering the natural course of any serious mental disorder because virtually all affected persons have been on long-term regimens of drug treatments.
19. Menninger, 1963, 9.
20. Shorter, 2013a, viii.
21. Menninger, 1963, 489.
22. Porter, 2018. Kraepelin, who almost completely disregarded environmental influences on the development and course of schizophrenia and manic depression, was an exception.
23. Jasanoff, 2018, 181.
24. Scull, 2015, 401.
25. Rosen, 1968, 132.
26. The basic principles of moral treatment persist in a few private mental hospitals that cater to wealthy clients.
27. Torrey & Yolken, 2010.
28. E.g. Khalid et al., 2008; Ross, Zivin, & Maixner, 2018.
29. Mojtabai & Olfson, 2008.
30. Mechanic, McAlpine, & Rochefort, 2014, 61.
31. Oshinsky, 2018.
32. https://law.stanford.edu/publications/the-prevalence-and-severity-of-mental-illness-among-california-prisoners-on-the-rise/.
33. https://www.treatmentadvocacycenter.org/evidence-and-research/learn-more-about/369.
34. Whooley, 2019, 222.
35. Andreasen, 1984, 8.
36. Kandel quoted in Weir, 2012, 30.
37. Pescosolido et al., 2010.
38. E.g. Phelan, 2005; Schnittker, 2008; Schomerus et al., 2012.
39. Jasanoff, 2018.

40. https://www.nytimes.com/2018/05/30/upshot/mental-illness-health-disparity-longevity.html?hp&action=click&pgtype=Homepage&clickSource=story-heading&module=second-column-region&region=top-news&WT.nav=top-news.
41. Colton & Manderscheid, 2006, A42.
42. Ilyas, Chesney, & Patel, 2017.
43. https://www.wired.com/2017/05/star-neuroscientist-tom-insel-leaves-google-spawned-verily-startup/.
44. Harrington, 2019, xiv.
45. U.S. Department of Health and Human Services, 1999.
46. E.g. https://www.nytimes.com/2018/10/01/us/las-vegas-massacre-paddock.html; https://www.nytimes.com/2019/03/14/sports/marquette-markus-howard-ncaa-tournament.html.
47. Abbott, 1988.
48. Grob, 1998, 204.
49. Bliss, 2018, 52.
50. Bearman, 2013.
51. Conley & Fletcher, 2017, 3.
52. Caspi et al., 2003. Attempts to replicate this association have not been successful. See Bolton et al., 2019.
53. Gil, Wagner, & Vega, 2000; Bourque, van der Ven, & Malla, 2011; Gold & Gold, 2014.
54. Pescosolido et al., 2008; Johnson, 2004.
55. Liu, Zerubavel, & Bearman, 2010.
56. Bearman, 2013, S12; Shostak & Moinester, 2015; Bell & Figert, 2015; Damasio, 1995, xvii.
57. Ebrahim, 2012; Landecker & Panofsky, 2013.
58. Insel & Cuthbert, 2015, 499.
59. Pescosolido, 2015, 249.
60. Shostak & Freese, 2010, 28; Conley & Fletcher, 2017, 53.
61. See the essays in Schutt, Seidman, & Keshavan, 2015; Schutt, 2015.
62. Conley & Fletcher, 2017, 139.
63. Keshavan, 2015; Segerstrale, 2000, 307. See also Guttmacher & Collins, 2003.
64. Grob, 1991a.
65. https://www.nytimes.com/2012/12/23/opinion/sunday/anatomy-of-a-murder-suicide.html.
66. New Freedom Commission, 2003.
67. Horwitz & Wakefield, 2009.
68. Pierre, 2013, 115.
69. See Torrey, 2014.
70. Shirk & Jungbluth, 2008.
71. Bliss, 2017, 185.
72. Some researchers conclude that as many as a quarter of these individuals will develop a full-blown psychotic disorder within a year and a third within two years (Addington & Barbato, 2015, 190). Such findings led the DSM-5 to include a new category of

"Attenuated Psychosis Syndrome" as a condition for further study. See also Kandel, 2018, 94.

73. Plomin, 2018, 165.
74. Healy, 2008, 237–38.
75. Broome, Fusar-Poli, & Wuyts, 2013, 791.
76. Singh & Rose, 2009, 204.
77. https://www.wired.com/story/23andme-glaxosmithkline-pharma-deal/?mbid=social_fb.
78. Plomin, 2018, 178.
79. Whooley, 2019, 217.
80. Bertram Brown quoted in Whooley, 2019, 163.
81. Taylor, 2013; Katschnig, 2010.
82. In 1974, psychiatrist E. Fuller Torrey predicted that the growing number of GPs prescribing psychotropic drugs threatened the very existence of the psychiatric profession: "The psychiatrist has become expendable, he is left standing between people who have problems in living and those who have brain disease, holding an empty bag."
83. DuBosar, 2009: http://www.acpinternist.org/archives/2009/11/national-trends.htm.
84. GPs write prescriptions for about 90% of anxiolytics, 65% of stimulants, and 50% of antipsychotics (Marcus & Olfson, 2010; Mojtabai & Olfson, 2008; DuBosar, 2009).
85. Hyman, 2012.
86. In March, 2019, the Food and Drug Administration approved a new drug specifically for the treatment of postpartum depression. It is enormously expensive and requires two days of inpatient stay when women began to take it. Nevertheless, its benefits barely exceed placebo treatment. https://www.nytimes.com/2019/03/19/health/postpartum-depression-drug.html?action=click&module=Latest&pgtype=Homepage.
87. Olfson et al., 2002.
88. Olfson et al., 2002; Marcus & Olfson, 2010.
89. Shorter, 2013a. Martin (2002, 695) notes: "It is increasingly difficult to distinguish scientifically between the disciplines of neurology and psychiatry."
90. Charland, 2013.
91. Davidson, 2013; Rose, 2019.
92. https://www.aedweb.org/advocate/press-releases/position-statements/pro-anorexia-websites.
93. Kendler, 2012, 385.

# References


Abbott, A. (1988). *The system of the professions*. Chicago: University of Chicago Press.

Addington, J. & Barbato, M. (2015). "Social cognition in those at high risk of psychosis." In *Social neuroscience: Mind, brain, and society* (R. K. Schutt, L. J. Seidman, & M. S. Kechaven, eds.), 187–207. Cambridge, MA: Harvard University Press.

Aeschylus. (1984). *Orestes* (P. Vellacott, trans.). New York: Penguin Classics.

Akiskal, H. S. (1998). "Toward a definition of generalized anxiety disorder as an anxious temperament type," *Acta Psychiatrica Scandinavica*, *98*(Suppl. 393), 66–73.

Akiskal, H. S., Bourgeois, M. L., Angst, J., Post, R., Möller, H. & Hirschfeld, R. (2000). "Re-evaluating the prevalence of and diagnostic composition within the broad clinical spectrum of bipolar disorders." *Journal of Affective Disorders*, 59(Suppl. 1), S5–S30.

Alliance of Psychodynamic Organizations. (2006). *Psychodynamic diagnostic manual*. Silver Spring, MD: Alliance of Psychodynamic Organizations.

American Psychiatric Association. (1952). *Diagnostic and statistical manual of mental disorders*. Washington DC: American Psychiatric Association.

American Psychiatric Association. (1968). *Diagnostic and statistical manual of mental disorders* (2nd ed.). Washington, DC: American Psychiatric Association.

American Psychiatric Association. (1980). *Diagnostic and statistical manual of mental disorders* (3rd ed.). Washington, DC: American Psychiatric Association.

American Psychiatric Association. (1987). *Diagnostic and statistical manual of mental disorders* (3rd ed., revised). Washington, DC: American Psychiatric Association.

American Psychiatric Association. (1994). *Diagnostic and statistical manual of mental disorders* (4th ed.). Washington, DC: American Psychiatric Association.

American Psychiatric Association. (2013). *Diagnostic and statistical manual of mental disorders* (5th ed.). Washington, DC: American Psychiatric Association.

Andreasen, N. (1984). *The broken brain: The biological revolution in psychiatry*. New York: Harper & Row.

Andreasen, N. (1997). "Linking mind and brain in the study of mental illnesses." *Science*, *275*, 1586–93.

Andreasen, N. (2001). *Brave new brain: Conquering mental illness in the era of the genome*. New York: Oxford University Press.

Andreasen, N. (2007). "DSM and the death of phenomenology in America." *Schizophrenia Bulletin*, *33*, 108–12.

Andrews, G., Stewart, G., Morris-Yates, A., Holt, P., & Henderson, S. (1990). "Evidence for a general neurotic syndrome." *British Journal of Psychiatry*, *157*, 6–12.

Angell, M. (2011). "The epidemic of mental illness: Why?" *New York Review of Books*, June 23, 20–22.

Anttila, V., Bulik-Sullivan, B., Finucane, H. K., Walters, R. K., Bras. J., Duncan, L., et al. (2018). "Analysis of shared heritability in common disorders of the brain." *Science*, *360*(6395), eaap8757. doi:10.1126/science.aap8757

Arens, K. (2013). "Wilhelm Griesinger: Philosophy as the origin of a new psychiatry." In *The Oxford handbook of philosophy and psychiatry* (K. W. M. Fulford, M.

Davies, R. Gipps, G. Graham, J. Sadler, G. Stanghellini, & T. Thornton, eds.), 53–67. New York: Oxford University Press.

Aristotle. (2009). *The Nichomacheon ethics* (W. D. Ross, trans.). Oxford World Classics, VII. New York: Oxford University Press.

Auden, W. H. (1940). "In memory of Sigmund Freud." In *W. H. Auden: Collected Poems* (E. Mendelson, ed.), 273–76. New York: Vintage.

Avicenna. (2000). "Black bile and melancholia." In *The nature of melancholy* (J. Radden, ed.), 75–78. New York: Oxford University Press.

Barber, C. (2008). *Comfortably numb: How psychiatry is medicating a nation.* New York: Pantheon.

Barlow, D. H. (1988). *Anxiety and its disorders.* New York: Guilford Press.

Battie, W. (1758/1969). *A treatise on madness.* New York: Brunner/Mazel.

Bayer, R. (1981). *Homosexuality and American psychiatry.* Princeton, NJ: Princeton University Press.

Bayer, R. & Spitzer, R. L. (1985). "Neurosis, psychodynamics, and *DSM-III*: History of the controversy." *Archives of General Psychiatry, 42*,187–96.

Beard, C. (1881). *American nervousness: Its causes and consequences.* New York: Putnam.

Bearman, P. (2013). "Genes can point to environments that matter to advance public health." *American Journal of Public Health, 103*(Suppl. 1), S11–S13.

Beavan, V., Read, J. & Cartwright, C. (2011). "The prevalence of voice-hearers in the general population: A literature review." *Journal of Mental Health, 20*, 281–92.

Beck, A. T. (1967). *Depression: Causes and treatment.* Philadelphia: University of Pennsylvania Press.

Beck, A. T. (1991). "Cognitive therapy: A 30-year retrospective." *American Psychologist, 46*, 368–75.

Beebe, G. & Apple, J. (1958). "Psychological breakdown in relation to stress and other factors." In *Variation in psychological tolerance to ground combat in World War II*, pp. 88–131. Washington, DC: National Academy of Science.

Bell, S. E. & Figert, A. E. (2015). "Moving sideways and forging ahead: Reimagining '-izations' in the twenty-first century." In *Reimagining (bio)medicalization, pharmaceuticals and genetics* (S. E. Bell & A. E. Figert, eds.), 19–40. New York: Routledge.

Belluck, P. & B. Carey, B. (2013). "Psychiatry's guide is out of touch with science, experts say." *New York Times*, May 6.

Benedict, R. (1934). "Anthropology and the abnormal." *The Journal of General Psychology, 10*, 59–82.

Berrios, G. E. (2015). "History and epistemology of psychopathology." In *Philosophical issues in psychiatry III: The nature and sources of historical change* (K. Kendler & J. Parnas, eds.), 30–50. New York: Oxford University Press.

Bertelson, A., Harvald, B. & Hauge M. (1977). "A Danish twin study of manic-depressive disorder." *British Journal of Psychiatry, 130*, 330–51.

Blackwell, B. (1973). Psychotropic drugs in use today: The role of diazepam in medical practice. *JAMA, 225*, 1637–41.

Blashfield, R. K. (1982). Feighner et al., invisible colleges, and the Matthew effect. *Schizophrenia Bulletin, 8*, 1–8.

Bleuler, E. (1978). *The schizophrenia patient: Long-term patient and family studies.* New Haven, CT: Yale University Press.

Bliss, C. (2018). *Social by nature: The promise and peril of sociogenomics.* Stanford, CA: Stanford University Press.

Bolton, D. (2008). *What is mental disorder?* New York: Oxford University Press.

Boraska, V., Franklin, C. S., Floyd, J. A., Thornton, L. M., Huckins, L. M., Southam, L, et al. (2014). "A genome-wide association study of anorexia nervosa." *Molecular Psychiatry, 19*, 1085–94.

Borch-Jacobsen, M. & Shamdasani, S. (2012). *The Freud files: An inquiry into the history of psychoanalysis*. Cambridge, England: Cambridge University Press.

Border, R., Johnson, E. C., Evans, L. M., Smolen, A., Berley, N., Sullivan, P. F. & Keller, M. C. (2019). "No support for historical candidate gene or candidate gene-by-interaction hypotheses for major depression across multiple large samples." *American Journal of Psychiatry, 176*, 376–87.

Bouchard, T. J. (1994). "Genes, environment, and personality." *Science, 264*, 1700–1.

Boudon-Millot, V. (2013). "What is a mental illness, and how can it be treated?" In *Mental disorders in the classical world* (W. V. Harris, ed.), 129–45. New York: Brill.

Bourque, F., van der Ven, E. & Malla, A. (2011). "A meta-analysis of the risk for psychotic disorders among first-and second-generation immigrants." *Psychological Medicine, 41*, 897–910.

Breier, A., Charney, D. S. & Heninger, G. R. (1985). "The diagnostic validity of anxiety disorders and their relationship to depressive illness," *American Journal of Psychiatry, 142*, 787–97.

Bromet, E., Helena Andrade, L. H., Hwang, I., Sampson, N. A., Alonso, J., de Girolamo, G., et al. (2011). "Cross-national epidemiology of DSM-IV major depressive episode." *BMC Medicine, 9*, 90.

Broome, M., Fusar-Poli, P. & Wuyts, P. (2013). "Conceptual and ethical issues in the prodromal phase of psychosis." In *The Oxford handbook of philosophy and psychiatry* (K. W. M. Fulford, M. Davies, R. Gipps, G. Graham, J. Sadler, G. Stanghellini & T. Thornton, eds.), 779–802. New York: Oxford University Press.

Brown, F. (1942). "Heredity in the psychoneuroses." *Processes of Research in Social Medicine, 35*, 785–90.

Brown, G. W. (2002). Social roles, context and evolution in the origins of depression. *Journal of Health and Social Behavior, 43*, 255–76.

Brown, N. O. (1959). *Life against death: The psychoanalytic meaning of history*. Middletown, CT: Wesleyan University Press.

Buchanan, R. (2003). "Legislative warriors: American psychiatrists, psychologists, and competing claims over psychotherapy in the 1950s." *Journal of the History of Behavioral Sciences, 39*, 225–50.

Burnham, J. (2012). *After Freud left: A century of psychoanalysis in America*. Chicago: University of Chicago Press.

Burton, R. (1621/1948). *The anatomy of melancholy* (F. Dell & P. Jordon-Smith, eds.). New York: Tudor.

Burton R. (1621/2001). *The anatomy of melancholy* (H. Jackson, ed.). New York: New York Review Books Classics.

Busfield, J. (2011). *Mental illness*. New York: Polity.

Cancro, R. (2000). "Functional psychoses and the conceptualization of mental illness." In *American psychiatry after WWII: 1944–1994* (R. W. Menninger & J. C. Nemiah, eds.), 413–29.Washington, DC: American Psychiatric Association.

Caplan, E. (2001). *Mind games: American culture and the birth of psychotherapy*. Berkeley: University of California Press.

Caspi, A. & Moffitt, T. E. (2018). "All for one and one for all: Mental disorders in one dimension." *American Journal of Psychiatry, 175*, 831–44.

Caspi, A., Sugden, K., Moffitt, T. E., Taylor, A., Craig, I. W., Harrington, H., et al. (2003). "Influence of life stress on depression: Moderation by a polymorphism in the 5-HTT gene." *Science, 301*, 386–89.

Cassidy, W. L., Flanagan, N. B., Spellman, M. & Cohen, M. E. (1957). "Clinical observations in manic-depressive disease." *JAMA, 164*, 1535–46.

Cawte, J. (1974). *Medicine is the law: Studies in psychiatric anthropology of Australian tribal societies.* Honolulu: University Press of Hawaii.

Charland, L. C. (2013). "Why psychiatry should fear medicalization." In *The Oxford handbook of philosophy and psychiatry* (K. W. M. Fulford, M. Davies, R. Gipps, G. Graham, J. Sadler, G. Stanghellini & T. Thornton, eds.), 159–75. New York: Oxford University Press.

Charney, A. & Sklar, P. (2018). "Genetics of schizophrenia and bipolar disorder." In *Charney & Nestler's neurobiology of mental illness* (5th ed., D. S. Charney, J. D. Buxbaum, P. Sklar & E. J. Nestler, eds.), 161–76. New York: Oxford University Press.

Chaste, P. & Leboyer, M. (2012). "Autism risk factors: Genes, environment, and gene-environment interactions." *Dialogues in clinical neuroscience, 14*, 281–92.

Cheyne, G. (1733/1991). *The English malady* (R. Porter, ed.). New York: Tavistock/Routledge.

Churchland, P. M. (1984). *Matter and consciousness.* Cambridge, MA: MIT Press.

Clark, L. A., Cuthbert, B., Lewis-Fernández, R., Narrow, W. E. & Reed, G. M. (2017). "Three approaches to understanding and classifying mental disorder: ICD-11, DSM-5, and the National Institute of Mental Health's Research Domain Criteria (RDoC)." *Psychological Science in the Public Interest, 72*, 72–145.

Clarke, J. & Gawley, A. (2009). "The triumph of pharmaceuticals: Portrayals of depression from 1980 to 2005." *Administration and Policy in Mental Health, 36*, 91–101.

Clayton, P. J., Guze, S. B., Cloninger, C. R. & Martin, R. L. (1992). "Unipolar depression: diagnostic inconsistency and its implications." *Journal of Affective Disorders, 26*, 111–16.

Cloninger, C. R. (1986). "A unified biosocial theory of personality and its role in the development of anxiety states." *Psychiatric Development, 4*, 167–226.

Colton, C. W. & Manderscheid, R. W. (2006). "Congruencies in increased mortality rates, years of potential life lost, and causes of death among public mental health clients in eight states." *Preventing Chronic Disease, 3*, A42.

Conley, D. & Fletcher, J. (2017). *The genome factor: What the social genomics revolution reveals about ourselves, our history & the future.* Princeton, NJ: Princeton University Press.

Conrad, P. (1997). "Public eyes and private genes: Historical frames, news constructions and social problems." *Social Problems, 44*, 139–54.

Conrad, P. (2005). The shifting engines of medicalization. *Journal of Health and Social Behavior, 46*, 3–14.

Conrad, P. (2018). *Global perspectives on ADHD: Social dimensions of diagnosis and treatment in sixteen countries.* Baltimore, MD: Johns Hopkins University Press.

Conrad, P. & Schneider, J. (1992). *Deviance and medicalization: From badness to sickness.* Philadelphia: Temple University Press.

Cooper, J., Rendell, R., Burland, B., Sharpe, L., Copeland, J. & Simon, R. (1972). *Psychiatric diagnosis in New York and London.* London: Oxford University Press.

Cooperstock, R. & Lennard, H. (1979). "Some social meanings of tranquilizer use." *Sociology of Health and Illness, 1*, 331–47.
Cosmides, L. & Tooby, J. (1999). "Toward an evolutionary taxonomy of tratable conditions." *Journal of Abnormal Psychology, 108*, 453–64.
Cowley, M. (1951). *Exile's return: A literary odyssey of the 1920s*. New York: Penguin.
Craddock, N. & Owen, M. J. (2010). "The Kraepelinian dichotomy—Going, going . . . but still not gone." *British Journal of Psychiatry, 196*, 92–95.
Crews, F. (2017). *Freud: The making of an illusion*. New York: Metropolitan Books.
Crick, F. (1994). *The astonishing hypothesis: The scientific search for the soul*. New York: Charles Scribner's Sons.
Cross-Disorder Group of the Psychiatric Genome Consortium. (2013). "Genetic relationship between five psychiatric disorders estimated from genome-wide SNPs. *Nature Genetics, 45*, 984–94.
Cuthbert, B. N. (2005). "Dimensional models of psychopathology: Research agenda and clinical utility." *Journal of Abnormal Psychology, 114*, 565–69.
Cuthbert, B. N. & Insel, T. R. (2013). "Toward the future of psychiatric diagnosis: The seven pillars of RDoC." *BMC Medicine, 11*, 126.
Dain, N. (1964). *Concepts of insanity in the United States, 1789–1865*. New Brunswick, NJ: Rutgers University Press.
Damasio, A. (1994). *Descartes' error: Emotion, reason, and the human brain*. New York: HarperCollins.
Danziger, K. (1990). *Constructing the subject: Historical origins of psychological research*. New York: Cambridge University Press.
Darwin, C. R. (1872/1998). *The expression of the emotions in man and animals*. London: HarperCollins.
Davidson, L. (2013). "Cure and recovery." In *The Oxford handbook of philosophy and psychiatry* (K. W. M. Fulford, M. Davies, R. Gipps, G. Graham, J. Sadler, G. Stanghellini & T. Thornton, eds.), 197–213. New York: Oxford University Press.
Dawkins, R. (1976). *The selfish gene*. New York: Oxford University Press.
Decker, H. (2007). "How Kraepelinian was Kraepelin?" *History of Psychiatry, 18*, 337–60.
Decker, H. S. (2013). *The making of DSM-III*. New York: Oxford University Press.
Degler, C. A. (1991). *In search of human nature: The decline and revival of Darwinism in American social thought*. New York: Oxford University Press.
Dennett, D. (1995). *Darwin's dangerous idea*. New York: Simon & Schuster.
Dodds, E. R. (1951). *The Greeks and the irrational*. Berkeley: University of California Press.
Duan, J., Sanders, A. R. & Gejman, P. V. (2010). "Genome-wide approaches to schizophrenia." *Brain Reseaerch Bulletin, 83*, 93–102.
DuBosar, R. (2009). "Psychotropic drug prescriptions by medical specialty." *ACP Internist*. http://www.acpinternist.org/archives/2009/11/national-trends.htm
Duffy, A., Horrocks, J., Doucette, S., Keown-Stoneman, C., McCloskey, S. & Grof, P. (2014). "The developmental trajectory of bipolar disorder." *British Journal of Psychiatry, 204*, 122–28.
Dumit, J. (2004). *Picturing personhood: Brain scans and biomedical identity*. Princeton, NJ: Princeton University Press.
Durkheim, E. (1893/1984). *The division of labor in society*. New York: Free Press.
Dwyer, E. (1987). *Homes for the mad*. New Brunswick, NJ: Rutgers University Press.
Earle, P. (1886). Earle to Clark Bell, April 18, 1886. Earle Papers, American Antiquarian Society, Worcester, MA.

Eaton, J. W. & Weil, R. J. (1955). *Culture and mental disorders*. Glencoe, IL: Free Press.
Ebrahim, S. (2012). "Epigenetics: The next big thing." *International Journal of Epidemiology, 41*, 1–3.
Edgerton, R. B. (1966). "Conceptions of psychosis in four East African societies." *American Anthropologist, 68*, 408–25.
Edgerton, R. B. (1969). "On the recognition of mental illness." In *Changing perspectives in mental illness* (S. C. Plog & R. B. Edgerton, eds.), 49–72. New York: Holt.
Eisenberg, L. (1995). "The social construction of the human brain." *American Journal of Psychiatry, 152*, 1563–75.
Ellenberger, H. (1970). *The discovery of the unconscious*. New York: Basic Books.
Esquirol, J. E. D. (1838/1965). *Mental maladies, a treatise on insanity*. New York: Hafner.
Eysenck, H. J. (1970). "The classification of depressive illness." *British Journal of Psychiatry, 117*, 241–50.
Euripides. (1999). *Bacchae* (P. Woodruff, trans.). Indianapolis: Hackett.
Evans, K., McGrath, J. & Milns, R. (2003). "Searching for schizophrenia in Ancient Greek and Roman literature: A systematic review." *Acta Psychiatrica Scandinavica, 107*, 323–30.
Eysenck, H., Wakefield, J. A., & Friedman, A. F. (1983). "Diagnosis and clinical assessment: The DSM-III." *Annual Review of Psychology, 34*, 167–93.
Fanous, A. H. & Kendler, K. S. (2008). "Genetics of clinical features and subtypes of schizophrenia." *Current Psychiatry Reports, 10*, 164–70.
Feighner, J. P. (1989). "The advent of the 'Feighner Criteria.'" *Citation Classics, 43*, 14.
Feighner, J. P., Robins, E., Guze, S. B., Woodruff, R. A., Winokur, G., & Munoz, R. (1972). "Diagnostic criteria for use in psychiatric research." *Archives of General Psychiatry, 26*, 57–63.
Fenichel, O. (1945/1995). *The psychoanalytic theory of neurosis, 50th anniversary edition*. New York: W. W. Norton.
Finley, E. P. (2012). *Fields of combat: Understanding PTSD among veterans of Iraq and Afghanistan*. Ithaca, NY: IRL Press.
First, M. B. (2005). "Clinical utility: A prerequisite for the adoption of a dimensional approach in DSM." *Journal of Abnormal Psychology, 114*, 560–64.
Foucault, M. (1965). *Madness and civilization: A history of insanity in the Age of Reason*. New York: Pantheon.
Foulds, G. (1976). *The hierarchical nature of personal illness*. New York: Academic Press.
Fox, R. W. (1978). *So far disordered in mind: Insanity in California, 1870–1930*. Berkeley: University of California Press.
Frank, R., McGuire, T., Regier, D., Manderscheid, R. & Woodward, A. (1994). "Paying for mental health and substance abuse care." *Health Affairs, 13*, 337–42.
Freud, S. (1893/1955). "Some points in the comparative study of organic and hysterical paralyses." In *Standard edition, Vol. 1* (J. Strachey, ed.), 157–74. London: Hogarth Press.
Freud, S. (1893/1959). "On the psychical mechanism of hysterical phenomena." In *Collected papers, Vol. 1* (J. Riviere, trans.), 24–41. New York: Basic Books.
Freud, S. (1894a/1959). "The defence neuro-psychoses." In *Collected papers, Vol. 1* (J. Riviere, trans.), 59–75. New York: Basic Books.
Freud, S. (1894b/1959). "The justification for detaching from neurasthenia a particular syndrome: The anxiety-neurosis." In *Collected papers, Vol. 1* (J. Riviere, trans.), 76–106. New York: Basic Books.

Freud S. (1895/1959). "Obsessions and phobias." In *Collected papers, Vol. 1* (J. Riviere, trans.), 128–37. New York: Basic Books.

Freud, S. (1895/1976). "Project for a scientific psychology." In *Standard edition, Vol. 1* (J. Strachey, ed. & trans.), 283–387. New York: W. W. Norton.

Freud, S. (1896a/1959). "Further remarks on the defence neuro-psychoses." In *Collected papers, Vol. 1* (J. Riviere, trans.), 155–82. New York: Basic Books.

Freud, S. (1896b/1959). "Heredity and aetiology of the neuroses." In *Collected papers, Vol. 1* (J. Riviere, trans.), 138–54. New York: Basic Books.

Freud, S. (1896c/1959). "The aetiology of hysteria." In *Collected papers, Vol. 1* (J. Riviere, trans.), 183–220. New York: Basic Books.

Freud, S. (1898/1959). "Sexuality in the aetiology of the neuroses." *Collected papers, Vol. 1* (J. Riviere, trans.), 220–48. New York: Basic Books.

Freud, S. (1900/1965). *The interpretation of dreams*. New York: Avon.

Freud, S. (1901/1990). *The psychopathology of everyday life*. New York: W. W. Norton.

Freud, S. (1905/1959). "My views on the part played by sexuality in the aetiology of the neuroses." In *Collected papers, Vol. 1* (J. Riviere, trans.), 272–83. New York: Basic Books.

Freud, S. (1905/1962). *Three essays on sexuality: Standard edition* (J. Strachey, trans.). London: Hogarth Press.

Freud, S. (1905/1989). "Fragment of an analysis of a case of hysteria ('Dora')." In *The Freud reader* (P. Gay, ed.), 172–239. New York: W. W. Norton.

Freud, S. (1909/1955). "Notes upon a case of obsessional neurosis." In *Standard edition, Vol. X: Two case histories ("Little Hans" and the "Rat Man")*, 151–318. London: Hogarth Press.

Freud, S. (1910/1989). "'Wild' Psycho-analysis." In *The Freud reader* (P. Gay, ed.), 351–56. New York: W. W. Norton.

Freud, S. (1913/1955). "Totem and taboo." In *Standard edition, Vol. XIII* (J. Strachey, ed.), 1–255. London: Hogarth Press.

Freud, S. (1914/1959). "A history of the psycho-analytic movement." In *Collected papers, Vol. 1* (J. Riviere, trans.), 287–359. New York: Basic Books.

Freud, S. (1916–1917/1989). *Introductory lectures on psycho-analysis* (P. Gay, ed.). New York: Liveright.

Freud, S. (1917/1957). "Mourning and melancholia." In *Standard edition, Vol. 14* (J. Strachey, ed. & trans.), 237–58. London: Hogarth Press.

Freud, S. (1919/1955). "Introduction to psychoanalysis and the war neuroses." In *Standard edition, Vol. 17* (J. Strachey, ed. & trans.), 205–10. London: Hogarth Press.

Freud, S. (1920/1953). *A general introduction to psychoanalysis* (J. Riviere, trans.). New York: Pocket Books.

Freud, S. (1920/1989). *Beyond the pleasure principle*. New York: W.W. Norton.

Freud, S. (1925/1959). "An autobiographical study." In *Standard edition, Vol. XX* (J. Strachey, trans.), 7–75. London: Hogarth Press.

Freud, S. (1926/1989). *Inhibitions, symptoms, and anxiety*. New York: W. W. Norton.

Freud, S. (1926/1990). *The question of lay analysis*. New York: W. W. Norton.

Freud, S. (1930/1962). *Civilization and its discontents*. New York: W. W. Norton.

Freud, S. (1933/1965). *New introductory lectures on psychoanalysis* (J. Strachey, trans.). New York: W. W. Norton.

Freud, S. (1933/1989). "New introductory lectures on psychoanalysis, lecture XXXII." In *The Freud reader* (P. Gay, ed.), 773–83. New York: W. W. Norton.

Freud, S. (1936/1963). *The problem of anxiety*. New York: W. W. Norton.

Freud, S. (1937). "Analysis terminable and interminable." *International Journal of Psychoanalysis, 18*, 373–405.

Freud, S. (1940/1989). *An outline of psychoanalysis*. New York: W. W. Norton.

Freud, S. & Breuer, J. (1895/1974). *3 studies on hysteria* (J. Strachey, trans.). New York: Pelican.

Friedan, B. (1963/2001). *The feminine mystique*. New York: W. W. Norton.

Friedman, L. J. (1990). *Menninger: The family and the clinic*. New York: Knopf.

Fromm, E. (1941/1969). *Escape from freedom*. New York: Avon Books.

Gabbard, G. O. & Gabbard, K. (1999). *Psychiatry and the cinema* (2nd ed.). Washington, DC: American Psychiatric Press.

Gabe, J. (1990). "Towards a sociology of tranquillizer prescribing." *British Journal of Addiction, 85*, 41–48.

Gabe, J. & Bury, M. (1991). "Tranquillisers and health care in crisis." *Social Science and Medicine, 32*, 449–54.

Garbowsky, M. M. (1989). *The house without the door*. Madison, NJ: Fairleigh Dickinson University Press.

Gay, P. (1988). *Freud: A life for our time*. New York: W. W. Norton.

Gay, P. (2002). *Schnitzler's century: The making of middle-class culture 1815–1914*. New York: W. W Norton.

Gershon, E. & Nurnberger, J. I. (1995). "Bipolar Illness." *Review of Psychiatry, 14*, 339–423. Washington, DC: American Psychiatric Press.

Ghaemi, S. N. (2013). "Understanding mania and depression." In *The Oxford handbook of philosophy and psychiatry*. (K. W. M. Fulford, M. Davies, R. Gipps, G. Graham, J. Sadler, G. Stanghellini & T. Thornton, eds.), 803–19. New York: Oxford University Press.

Gil, A. G., Wagner, E. F. & Vega, W. A. (2000). "Acculturation, familism, and alcohol use among Latino adolescent males: Longitudinal relations." *Journal of Community Psychology, 28*, 443–58.

Gilger, J. W. (2000). "Contributions and promise of human behavioral genetics." *Human Biology, 72*, 229–55.

Gitlin, T. (1993). *Years of hope, days of rage*. New York: Bantam.

Glas, G. (1996). "Concepts of anxiety: A historical reflection on anxiety and related disorders." In *Advances in the neurobiology of anxiety disorders* (H. G. M. Westerberg, J. A. Des Boer & D. L. Murphy, eds.), 3–20. New York: John Wiley.

Glas, G. (2003). "A conceptual history of anxiety and depression." In *Handbook of depression and anxiety: A biological approach* (J. A. den Boer & J. M. Sitsen, eds.), 1–44. New York: Marcel Dekker.

Goffman, E. (1961). *Asylums: Essays on the social situation of mental patients and other inmates*. New York: Anchor Books.

Goffman, E. (1971). *Relations in public: Microstudies of the public order*. New York: Harper.

Gold, J. & Gold, I. (2014). *Suspicious minds: How culture shapes madness*. New York: Free Press.

Goldberg, D. (2011). "The heterogeneity of 'major depression.'" *World Psychiatry, 10*, 226–28.

Goldberg, D. (2015). "Psychopathology and classification in psychiatry." *Social Psychiatry and Psychiatric Epidemiology, 50*, 1–5.

Goldman, H., Neal, H., Adams, N. H. & Taube, C. A. (1983). "Deinstitutionalization: The data demythologized." *Hospital and Community Psychiatry, 34*, 129–34.

Goldstein, J. (1987). *Console and classify: The French psychiatric profession in the nineteenth century*. New York: Cambridge University Press.
Goodwin, D. W. & Guze, S. B. (1996). *Psychiatric diagnosis* (5th ed.). New York: Oxford University Press.
Gottesman, I. I. & Gould, T. D. (2005). "The endophenotype concept in psychiatry." In *Research advances in genetics and genomics* (N. Andreasen, ed.), 63–84. Washington, DC: APA Publishing.
Grant, M. (1978). *The rise of the Greeks*. New York: Charles Scribner's Sons.
Gratten, J. (2016). "Rare variants are common in schizophrenia." *Nature Neuroscience, 19*, 1426–28.
Green, R. (2002). "Is pedophilia a mental disorder?" *Archives of Sexual Behavior, 31*, 467–71.
Greenberg, D. (1980). Reimbursement wars. *New England Journal of Medicine, 303*, 538–40.
Greenberg, G. (2013). *The book of woe*. New York: Blue Rider Press.
Greenberg, G. (2019). "Psychiatry's incurable hubris." *The Atlantic*, April, 30–32.
Greene, J. & Herzberg, D. (2010). "Hidden in plain sight: The popular promotion of prescription drugs in the 20th century." *American Journal of Public Health, 100*, 793–803.
Griesinger, W. (1867/2000). "Hypochondriasis and melancholia." In *The nature of melancholy: From Aristotle to Kristeva* (J. Radden, ed.), 223–29. New York: Oxford University Press.
Grinker, R. R. & Spiegel, J. P. (1943). *War neuroses in North Africa*. New York: Air Surgeon, Army Air Forces.
Grob, G. N. (1973). *Mental institutions in America: Social policy to 1875*. New York: Free Press.
Grob, G. N. (1987). "The forging of mental health policy in America: World War II to the New Frontier." *Journal of the History of Medicine & Allied Sciences, 42*, 410–46.
Grob, G. N. (1991a). *From asylum to community: Mental health policy in modern America*. Princeton, NJ: Princeton University Press.
Grob, G. N. (1991b). "Origins of DSM-I: Appearances and reality." *American Journal of Psychiatry, 148*, 421–31.
Grob, G. N. (1994). *The mad among us: A history of the care of America's mentally ill*. New York: Free Press.
Grob, G. N. (1995). "The paradox of deinstitutionalization." *Society, 32*, 51–59.
Grob G. N. (1998). "Psychiatry's holy grail: The search for the mechanisms of mental diseases." *Bulletin of the History of Medicine, 72*, 189–219.
Grob, G. N. & Goldman, H. H. (2006). *The dilemma of federal mental health policy: Radical reform or incremental change?* New Brunswick, NJ: Rutgers University Press.
Gronfein, W. (1985). "Psychotropic drugs and the origins of deinstitutionalization." *Social Problems, 32*, 437–54.
Group for the Advancement of Psychiatry. (1950). "The social responsibility of psychiatry: A statement of orientation." In *Published reports: Special volume—Reports #1–#34 and Symposium #1*. New York: Group for the Advancement of Psychiatry.
Gueorguieva, R., Mallinckrodt, C., & Krystal, J. H. (2011). "Trajectories of depression severity in clinical trials of duloxetine: Insights into antidepressant and placebo responses." *Archives of General Psychiatry, 68*, 1227–37.
Gustavsson, A., Svensson, M., Jacobi, F., Allgulander C., Alonso, J., Beghi, E., et al. (2011). *European Neurpsychopharmacology, 21*, 718–79.

Guttmacher, A. E. & Collins, F. S. (2003). "Welcome to the genomic era." *New England Journal of Medicine, 349*(10), 996–98.
Guven, F. (2005). *Madness and death in philosophy*. Albany, NY: SUNY Press.
Guze, S. (1989). "Biological psychiatry: Is there any other kind?" *Psychological Medicine, 19*, 315–23.
Hackett, T. (1977). "The psychiatrist: In the mainstream or on the banks of medicine?" *American Journal of Psychiatry, 134*, 432–35.
Hale, N. (1995). *The rise and crisis of psychoanalysis in the United States*. New York: Oxford University Press.
Hale, N. G. (2000). "American psychoanalysis since World War II." In *American psychiatry after World War II: 1944–1994* (R. W. Menninger & J. C. Nemiah, eds.), 77–102. Washington, DC: American Psychiatric Press.
Halliwell, M. (2013). *Therapeutic revolutions: Medicine, psychiatry, and American culture, 1945–1970*. New Brunswick, NJ: Rutgers University Press.
Hamer, D. H., Hu, S., Magnuson, V. L., Hu, N. & Pattatucci, A. (1993). "A linkage between DNA markers on the X chromosome and male sexual orientation." *Science, 261*, 321–27.
Harrington, A. (2019). *Mind fixers: Psychiatry's troubled search for the biology of mental illness*. New York: W. W. Norton.
Hartmann, H. (1939). *Ego psychology and the problem of adaptation*. New York: International Universities Press.
Haslam, N. (2013). "Reliability, validity, and the mixed blessings of operationalism." In *The Oxford handbook of philosophy and psychiatry* (K .W. M. Fulford, M. Davies, R. Gipps, G. Graham, J. Sadler, G. Stanghellini & T. Thornton, eds.), 987–1002. New York: Oxford University Press.
Healy, D. (1997). *The anti-depressant era*. Cambridge, MA: Harvard University Press.
Healy, D. (2008). *Mania: A short history of bipolar disorder*. Baltimore, MD: Johns Hopkins University Press.
Hecht, D. K. (2015). "Commentary and reflections: The lure of the gene." In *Reimagining (bio)medicalization, pharmaceuticals, and genetics* (S. Bell & A. Figert, eds.), 239–47. New York: Routledge.
Helzer, J. E. & Canino, G. (1992). "Comparative analysis of alcoholism in 10 cultural regions." In *Alcoholism in North America, Europe, and Asia*, (J. E. Helzer & G. Canino, eds.), 289–308. New York: Oxford University Press.
Helzer, J. E., Robins, L. N. & Davis, D. H. (1976). "Depressive disorders in Vietnam returnees." *Journal of Nervous and Mental Disorders, 168*, 177–85.
Herman, E. (1995). *The romance of American psychology*. Berkeley: University of California Press.
Herman, J. (2015). *Trauma and recovery*. New York: Basic Books.
Herodotus. (1996). *The histories, revised* (J. M. Marincola, ed.). New York: Penguin.
Herzberg, D. (2009). *Happy pills in America: From Miltown to Prozac*. Baltimore, MD: Johns Hopkins University Press.
Herzog, D. (2017). *Cold war Freud*. New York: Cambridge University Press.
Heston, L. (1966). "Psychiatric disorders in foster-home-reared children of schizophrenic mothers." *British Journal of Psychiatry, 112*, 819–25.
Hippocrates. (1923–1931). *Works of Hippocrates, Vol. 1* (W. H. S. Jones & E. T. Withington, eds. & trans.). Cambridge, MA: Harvard University Press.

Hippocrates. (2016). "On the sacred disease." In *Neurological concepts in Ancient Greek medicine* (T. M. Walshe, ed.), 43–60. New York: Oxford University Press.

Hirschfeld, R. M., Keller, M. B., Panico, S., Arons, B. S., Barlow, D., Davidoff, F., et al. (1997). The National Depressive and Manic-Depressive Association consensus statement on the undertreatment of depression. *Journal of the American Medical Association, 277*, 333–40.

Hoekstra, R. A., Bartels, M., Verweij, C. J. & Boomsma, D. I. (2007). "Heritability of autistic traits in the general population." *Archives of Pediatric and Adolescent Medicine, 161*, 372–77.

Holmes, B. (2013). "Disturbing connections: Sympathetic affections, mental disorder, and the elusive soul in Galen." In *Mental disorders in the classical world*, (W. Harris, ed.), 147–76. New York: Brill.

Holmes, O. W. (1861). *Currents and counter-currents in medical science*. Boston: Ticknor and Fields.

Holmes, T. & Rahe, R. (1967). "The social readjustment rating scale." *Journal of Psychosomatic Research, 11*, 213–18.

Homer. (1998). *The iliad* (R. Fitzgerald, trans.). New York: Farrar, Straus, Giroux.

Homer. (1999). *The odyssey* (R. Fagles, trans.). New York: Penguin Classics.

Horney, K. (1937). *The neurotic personality of our time*. New York: W. W. Norton.

Horwitz, A. V. (1982). *The social control of mental illness*. New York: Academic Press.

Horwitz, A. V. (2002). *Creating mental illness*. Chicago: University of Chicago Press.

Horwitz, A. V. (2011). "Creating an age of depression: The social construction and consequences of the major depression diagnosis." *Society & Mental Health, 1*, 41–54.

Horwitz, A. V. (2013). *Anxiety: A short history*. Baltimore, MD: Johns Hopkins University Press.

Horwitz, A. V. (2016). *What's normal? Reconciling biology and culture*. New York: Oxford University Press.

Horwitz, A. V. (2018). *PTSD: A short history*. Baltimore, MD: Johns Hopkins University Press.

Horwitz, A. V. & Davies, L. (1994). "Are emotional distress and alcohol problems differential outcomes to stress?" *Social Science Quarterly, 75*, 607–21.

Horwitz, A. V. & Wakefield, J. C. (2006). "Is There an Epidemic of Mental Illness?" *Contexts, 5*, 19–23.

Horwitz, A. V. & Wakefield, J. C. (2007). *The loss of sadness*. New York: Oxford University Press.

Horwitz, A. V. & Wakefield, J. C. (2009). "Should screening for depression among children and adolescents be demedicalized?" *Journal of the American Academy of Child and Adolescent Psychiatry, 48*, 683–87.

Horwitz, A. V. & Wakefield, J. C. (2012). *All we have to fear: Psychiatry's transformation of natural anxiety into mental disorder*. New York: Oxford University Press.

Hyman, S. (2010). "Diagnosis of mental disorder." *Annual Review of Clinical Psychology, 6*, 155–79.

Hyman, S. (2012). "Psychiatric drug discovery: Revolution stalled." *Science Translational Medicine, 4*, 155.

Hyman, S. E. (2007). "Can neuroscience be integrated into the DSM-V?" *Nature Reviews Neuroscience, 8*, 725.

Hyman, S. E. (2008). "A glimmer of light for neuropsychiatric disorders." *Nature*, *455*, 890–93.

Hyman, S. E. (2018). "The present and future of psychiatric diagnosis." In *Charney & Nestler's neurobiology of mental illness* (5th ed., D. S. Charney, J. D. Buxbaum, P. Sklar & E. J. Nestler, eds.), 941–46. New York: Oxford University Press.

Hyman, S. E. & Nestler, E. J. (1993). *The molecular foundations of psychiatry*. Washington, DC: American Psychiatric Press.

Ilyas, A., Chesney, E. & Patel, R. (2017). "Improving life expectancy in people with serious mental illness: Should we place more emphasis on primary prevention?" *British Journal of Psychiatry*, *211*, 194–97.

Insel, T., Cuthbert, B., Garvey, M., Heinssen, R., Pine, D. S., Quinn, K., et al. (2010). "Research domain criteria (RDoC): Toward a new classification framework for research on mental disorders." *American Journal of Psychiatry*, *167*, 748–51.

Insel, T. R. & Cuthbert, B. N. (2015). "Brain disorders? Precisely." *Science*, *348*, 499–500.

Ioannidis, J. P. A. (2008). "Effectiveness of antidepressants: An evidence myth constructed from a thousand randomized trials?" *Philosophy, Ethics, and Humanities in Medicine*, *3*, 3–14.

Jackson, S. (1986). *Melancholia and depression: From Hippocratic times to modern times*. New Haven, CT: Yale University Press.

Jamison, K. R. (1993). *Touched with fire: Manic-depressive illness and the artistic temperament*. New York: Free Press.

Jasanoff, A. (2018). *The biological mind*. New York: Basic Books.

Johns, L. C. & van Os, J. (2001). "The continuity of psychotic experiences in the general population." *Clinical Psychology Review*, *21*, 1125–41.

Johnson, F. N. (1984). *The history of lithium therapy*. London: Palgrave Macmillan.

Johnson, K. (2004). "By accident: Utah is proving an ideal genetic laboratory." *New York Times*, July 31, A1, A13.

Jones, E. (1953). *The life and work of Sigmund Freud, Vol. 1*. New York: Basic Books.

Jones, E. & Wessely, S. (2005). *Shell shock to PTSD: Military psychiatry from 1900 to the Gulf War*. New York: Psychology Press.

Jouanna, J. (2013). "The typology and aetiology of madness in ancient Greek medical and philosophical writing." In *Mental disorders in the classical world* (W. V. Harris, ed.), 97–118. New York: Brill.

Judd, L. J., Akiskal, H. S. & Paulus, M. P. (1997). "The role and clinical significance of subsyndromal depressive symptoms (SSD) in unipolar major depressive disorder." *Journal of Affective Disorders*, *45*, 5–18.

Judd, L. L., Rapaport, M. H., Paulus, M. P. & Brown, J. L. (1994). "Subsyndromal symptomatic depression: A new mood disorder?" *Journal of Clinical Psychiatry*, *55*, 18–28.

Kadushin, C. (1969). *Why people go to psychiatrists*. New York: Atherton Press.

Kagan, J. (2017). *Five constraints on predicting behavior*. Cambrige, MA: MIT Press.

Kallmann, F. J. (1946). "The genetic theory of schizophrenia: An analysis of 691 schizophrenic twin index families." *American Journal of Psychiatry*, *103*, 309–22.

Kandel, E. R. (1998). "A new intellectual framework for psychiatry." *American Journal of Psychiatry*, *155*, 457–69.

Kandel, E. R. (2018). *The disordered mind: What unusual brains tell us about ourselves*. New York: Farrar, Straus and Giroux.

Kantor, E. D., Rehm, C. D., Haas, J. S., Chan, A. T. & Giovannucci, E. L. (2015). "Trends in prescription drug use among adults in the United States, from 1999–2012." *JAMA*, *314*, 1818–31.

Kardiner, A. (1977). *My analysis with Freud: Reminiscenses*. New York: W. W. Norton.
Katschnig, H. (2010). "Are psychiatrists an endangered species?" *World Psychiatry, 9*, 21–28.
Keegan, J. (1976). *The face of battle*. New York: Penguin.
Kendell, R. E. (1976). "The classification of depressions: A review of contemporary confusion." *British Journal of Psychiatry, 129*, 15–28.
Kendell, R. E. (1986). "What are mental disorders?" In *Issues in psychiatric classification: Science, practice and social policy* (A. M. Freedman, R. Brotman, I. Silverman & D. Hutson, eds.), 23–45. New York: Human Sciences Press.
Kendler, K. S. (1990). "Toward a scientific psychiatric nosology: Strengths and limitations." *Archives of General Psychiatry, 47*, 969–73.
Kendler, K. S. (1996). "Major depression and generalised anxiety disorder: Same genes, (partly) different environments—revisited," *British Journal of Psychiatry, 168*(Suppl. 30), 68–75.
Kendler, K. (2005). "A gene for . . . : The nature of gene action in psychiatry disorders." *American Journal of Psychiatry, 162*, 1243–52.
Kendler, K. S. (2008). "The loss of sadness." *Psychological Medicine, 38*, 148–50.
Kendler, K. S. (2012). "The dappled nature of causes of psychiatric illness: Replacing the organic-functional/hardware-software dichotomy with empirically based pluralism." *Molecular Biology, 17*, 377–88.
Kendler, K. S. (2015). "Introduction." In *Philosophical issues in psychiatry III: The nature and sources of historical change* (K. Kendler & J. Parnas, eds.), 145–48. New York: Oxford University Press.
Kendler, K. S. & Gardner, C. O. (1998). "Boundaries of major depression: An evaluation of DSM-IV criteria." *American Journal of Psychiatry, 155*, 172–77.
Kendler, K. S., Heath, A. C., Martin, N. G. & Eaves, L. J. (1987). "Symptoms of anxiety and symptoms of depression: Same genes, different environments?" *Archives of General Psychiatry, 44*, 457.
Kendler, K. S., Myers, K. & Zisook, S. (2008). "Does bereavement related major depression differ from major depression associated with other stressful life events?" *American Journal of Psychiatry, 165*, 1449–55.
Kendler, K. S., Prescott, C. A., Myers, J. & Neale, M. C. (2003). "The structure of genetic and environmental risk factors for common psychiatric and substance use disorders in men and women." *Archives of General Psychiatry, 60*, 929–37.
Kernberg, O. (1995). *Love relations: Normality and pathology*. New Haven, CT: Yale University Press.
Keshavan, M. S. (2015). "The evolution, structure, and functioning of the social brain." In *Social neuroscience: Mind, brain, and society* (R. K. Schutt, L. J. Seidman & M. S. Kechaven, eds.), 29–40. Cambridge, MA: Harvard University Press.
Kessler, R. C., Beglund, P., Demler, O., Jin, R., Koretz, D., Merikangas, K. R., et al. (2003a). The epidemiology of major depressive disorder: Results from the National Comorbidity Survey replication. *Journal of the American Medical Association, 289*, 3095–105.
Kessler, R. C., Demler, O., Frank, R. G., Olfson, M., Pincus, H. A., Walters, E. E., et al. (2005). Prevalence and treatment of mental disorders, 1990–2003. *New England Journal of Medicine, 352*, 2515–23.
Kessler, R. C., McGonagle, K. A., Zhao, S., Nelson, C. B., Hughes, M., Eshelman, S., et al. (1994). Lifetime and 12-month prevalence of DSM-III-R psychiatric disorders in the United States. *Archives of General Psychiatry, 51*, 8–19.

Kessler, R. C., Merikangas, K. R., Berglund, P., Eaton, W. W., Koretz, D. S. & Walters, E. E. (2003b). "Mild disorders should not be eliminated from the DSM-V." *Archives of General Psychiatry, 60*, 1117–22.
Kety, S. (1974). "From rationalization to reason." *American Journal of Psychiatry, 131*, 957–63.
Khalid, N., Atkins, M., Tredget, J., Giles, M., Champney-Smith, K. & Kirov, G. (2008). "The effectiveness of electroconvulsive therapy in treatment-resistant depression: A naturalistic study." *The Journal of ECT, 24*, 141–45.
King, H. (2013). "Fear of flute girls, fear of falling." In *Mental disorders in the classical world* (W. V. Harris, ed.), 265–82. New York: Brill.
Kirk, S. A. (1999). "Instituting madness: The evolution of a federal agency." In *Handbook of the sociology of mental health* (C. A. Aneshensel & J. C. Phelan, eds.), 539–62. New York: Plenum.
Kirk, S. A. & Kutchins, H. (1992). *The selling of DSM: The rhetoric of science in psychiatry*. New York: Aldine de Gruyter.
Kirsch, I., Deacon, B. J., Huedo-Medina, T. B., Scoboria, A., Moore, T. J. & Johnson, B. (2008). "Initial severity and antidepressant benefits: A meta-analysis of data submitted to the Food and Drug Administration." *PLoS Medicine, 5*, e45. doi:10.1371/journal.pmed.0050045
Klein, D. F. (1999). "Harmful dysfunction, disorder, disease, illness, and evolution." *Journal of Abnormal Psychology, 108*, 421–29.
Klein, D. F. & Davis, J. M. (1969). *Diagnosis and drug treatment of psychiatric disorders*. Baltimore, MD: Williams & Wilkins.
Kleinman, A. (2012). "Culture, bereavement and psychiatry." *The Lancet, 379*, 608–9.
Klerman, G. L. (1978). "The evolution of a scientific nosology." In *Schizophrenia: Science and practice* (J. C. Shershow, ed.), 99–121. Cambridge, MA: Harvard University Press.
Klerman, G. L., Vaillant, G. E., Spitzer, R. L. & Michels, R. (1984). "A debate on DSM-III." *American Journal of Psychiatry, 141*, 539–51.
Kohn, R., Saxena, S., Levav, I. & Saraceno, B. (2004). "The treatment gap in mental health care." *Bulletin of the World Health Organization, 82*, 858–66.
Kohut, H. (1971). *The analysis of the self: A systematic approach to the psychoanalytic treatment of narcissistic personality disorders*. New York: International Universities Press.
Kohut, H. (1977). *The restoration of the self*. Chicago: University of Chicago Press.
Kotov, R., Krueger, R. F., Watson, D., Achenbach, T. M., Althoff, R. R., Bagby, R. M., et al. (2017). "The hierarchical taxonomy of psychopathology (HiTOP): A dimensional alternative to traditional nosologies." *Journal of Abnormal Psychology, 126*, 454–77.
Kraepelin, E. (1904/1917). *Lectures on clinical psychiatry* (3rd English ed., T. Johnstone, ed. & trans.). New York: Wood.
Kraepelin, E. (1907/1915). *Clinical psychiatry: A text-book for students and physicians abstracted and adapted from the seventh German edition of Kraepelin's Lehrbuch der Psychiatrie* (2nd ed., A. Ross Diefendorf, ed. & trans.). New York: Macmillan.
Kraepelin, E. (1910). *Psychiatrie, Vol. 2* (8th ed.). Leipzig: Barth.
Kraepelin, E. (1919). *Dementia praecox and paraphrenia*. (R. M. Barclay, trans.). Edinburgh, England: ES Livington.
Kraepelin, E. (1921/1976). *Manic depressive insanity and paranoia* (R. M. Barclay, trans.). New York: Arno Press.
Kramer, P. D. (1993). *Listening to Prozac: A psychiatrist explores antidepressant drugs and the remaking of the self*. New York: Viking.

Kramer, P. D. (2005). *Against depression*. New York: Penguin.
Kramer, P. D. (2016). *Ordinarily well: The case for antidepressants*. New York: Farrar, Straus and Giroux.
Krueger, R. F. (1999). "The structure of common mental disorders." *Archives of General Psychiatry*, *56*, 921–26.
Krueger, R. F. & Markon, K. E. (2006). "Reinterpreting comorbidity." *Annual Review of Clinical Psychology*, *2*, 111–33.
Krueger, R. F., Markon, K. E., Patrick, C. J. & Iacono, W. G. (2005). "Externalizing psychopathology in adulthood: A dimensional-spectrum conceptualization and its implications for DSM-V." *Journal of Abnormal Psychology*, *114*, 537–50.
Krueger, R. F. & South, S. C. (2009). "Externalizing disorders." *Psychological Medicine*, *3*, 2061–70.
Krueger, R., South, S., Johnson, W. & Iacono, W. (2008). "The heritability of personality is not always 50%." *Journal of Personality*, *76*, 1485–522.
Kruijshaar, M. E., Barendregt, J., Vos, T., de Graaf, R., Spijker, J. & Andrews, G. (2005). "Lifetime prevalence estimates of major depression: An indirect estimation method and a quantification of recall bias." *European Journal of Epidemiology*, *20*, 103–11.
Krystal, J. H. & Charney, D. S. (2018). "Current treatments for depression." In *Charney & Nestler's neurobiology of mental illness* (5th ed., D. S. Charney, J. D. Buxbaum, P. Sklar & E. J. Nestler, eds.), 387–405. New York: Oxford University Press.
Kuhn, T. (1970). *The structure of scientific revolutions* (2nd ed.). Chicago: University of Chicago Press.
Kupfer, D., First, M. & Regier, D. A. (2002). *A research agenda for DSM-V*. Washington, DC: American Psychiatric Association.
Laing, R. D. (1970). *The politics of experience and the bird of paradise*. New York: Penguin.
Landecker, H. & Panofsky, A. (2013). "From social structure to gene regulation, and back: A critical introduction to environmental epigenetics for sociology." *Annual Review of Sociology*, *39*, 333–57.
Lane, C. (2007). *Shyness: How normal behavior became a sickness*. New Haven, CT: Yale University Press.
Lapouse, R. (1967). "Problems in studying the prevalence of psychiatric disorder." *American Journal of Public Health*, *57*, 947–54.
Lasch, C. (1979). *The culture of narcissism*. New York: W. W. Norton.
Layton, L. (1985). "From Oedipus to Narcissus: Literature and the psychology of self." *Mosaic: An Interdisciplinary Critical Journal*, *18*, 97–105.
Layton, T. J., Barnett, M. L., Hicks, T. R. & Jena, A. B. (2018). "Attention deficit hyperactivity disorder and month of school enrollment." *New England Journal of Medicine*, *379*, 2122–30.
Leaf, P. J., Myers, J. K. & McEvoy, L. T. (1991). "Procedures used in the Epidemiological Catchment Area Study." In *Psychiatric disorders in America* (L. Robins & D. Regier, eds.), 11–32. New York: Free Press.
LeDoux, J. (1998). *The emotional brain*. New York: Simon & Schuster.
Lee, B. X. (2017). *The dangerous case of Donald Trump: 27 psychiatrists and mental health experts assess a president*. New York: Thomas Dunne Books.
Levack, B. (2013). *The devil within: Possession and exorcism in the Christian West*. New Haven, CT: Yale University Press.
LeVay, S. (1991). "A difference in the hypothalmic structure btween heterosexual and homosexual men." *Science*, *253*, 1034–37.

LeVay, S. & Hamer, D. H. (1994). "Evidence for a biological influence in male homosexuality." *Scientific American*, 270, May, 45–50.

Levinson, D. F. & Nichols, W. E. (2018). In *Charney & Nestler's neurobiology of mental illness* (5th ed., D. S. Charney, J. D. Buxbaum, P. Sklar & E. J. Nestler, eds.), 277–86. New York: Oxford University Press.

Lichtenstein, P., Yip, B. H., Björk, C., Pawitan, Y., Cannon, T. D., Sullivan, P. F., & Hultman, C. M. (2009). "Common genetic determinants of schizophrenia and bipolar disorder in Swedish families: A population-based study." *The Lancet*, *373*, 234–39.

Lieberman, J. A. (2015). *Shrinks: The untold story of psychiatry*. New York: Little, Brown.

Liu, K., Zerubavel, N., & Bearman, P. (2010). "Social demographic change and autism." *Demography*, *47*, 327–43.

Livesley, W. J. (2010). "Confusion and incoherence in the classification of personality disorder: Commentary on the preliminary proposals for DSM-5." *Psychological Injury and Law*, *3*, 304–13.

Locke, J. (1690/1996). *An essay concerning human understanding*. Indianapolis: Hackett.

Luhrmann, T. M. (2000). *Of 2 minds: The growing disorder in American psychiatry*. New York: Alfred A. Knopf.

Lunbeck, E. (1994). *The psychiatric persuasion*. Princeton, NJ: Princeton University Press.

Lunbeck, E. (2012). "Heinz Kohut's Americanization of Freud." In *After Freud left* (J. Burnham, ed.), 209–32. Chicago: University of Chicago Press.

Lunbeck, E. (2014). *The Americanization of narcissism*. Cambridge, MA: Harvard University Press.

Lutz, T. (1991). *American nervousness: 1903*. Ithaca, NY: Cornell University Press.

MacDonald, M. (1981). *Mystical bedlam: Madness, anxiety, and healing in seventeenth-century England*. New York: Cambridge University Press.

Madow, L. (1976). "The retreat from a psychiatry of people." *American Journal of Psychoanalysis*, *4*, 131–35.

Mail, P. D. (1989). "American Indians, stress, and alcohol." *American Indian and Alaskan Native Mental Health Research*, *3*, 7–26.

Maj, M. (2011). "When does depression become a mental disorder?" *British Journal of Psychiatry*, *199*, 85–86.

Major Depressive Disorder Working Group of the Psychiatric GWAS Consortium. (2013). "A mega-analysis of genome-wide association studies for major depressive disorder." *Molecular Psychiatry*, *18*, 497–511.

Makari, G. (2008). *Revolution in mind: The creation of psychoanalysis*. New York: Harper.

Makari, G. (2012). "*Mitteleuropa* on the Hudson." In *After Freud left* (J. Burnham, ed.), 111–24. Chicago: University of Chicago Press.

Makari, G. (2015). *Soul machine: The invention of the modern mind*. New York: W. W. Norton.

Malcolm, J. (1981). *Psychoanalysis: The impossible profession*. New York: Vintage.

Malcolm, J. (1983). *In the Freud archives*. New York: Alfred A. Knopf.

Manderscheid, R. W., Witkin, M. J., Rosenstein, M. J., & Bass, R. D. (1986). "The National Reporting Program for Mental Health Statistics: History and findings." *Public Health Reports*, *101*, 532–39.

Marcus, S. C. & Olfson, M. (2010). "National trends in the treatment for depression from 1998 to 2007." *Archives of General Psychiatry*, *67*, 1265–73.

Marcuse, H. (1955). *Eros and civilization*. Boston: Beacon Press.

Marques, T. R. & Kapur, S. (2018). "Novel approaches for treating psychotic disorders." In *Charney & Nestler's neurobiology of mental illness* (5th ed., D. S. Charney, J. D. Buxbaum, P. Sklar & E. J. Nestler, eds.), 267–76. New York: Oxford University Press.

Marshall, E. (1980). "Psychotherapy faces test of worth." *Science, 207*, 35–36.

Martin, J. B. (2002). "The integration of neuology, psychiatry, and neuroscience in the 21st century." *American Journal of Psychiatry, 159*, 695–704.

Maslow, A. (1962). *Toward a psychology of being*. New York: John Wiley.

Maxmen, J. (1985). *The new psychiatrists*. New York: New American Library.

Mayberg, H. S., Liotti, M., Brannan, S. K., McGinnis, S., Mahurin, R. K., Jerabek, P. A., et al. (1999). "Reciprocal limbic-cortical function and negative mood: Converging PET findings in depression and normal sadness." *American Journal of Psychiatry, 156*, 675–82.

Mayes, R. & Horwitz, A. V. (2005). "DSM-III and the revolution in the classification of mental illness." *Journal of the History of the Behavioral Sciences, 41*, 249–67.

McGlashan, T.H., Grilo, C. M., Skodol, A. E., McGlashan, T. H., Morey, L. C., Sanislow, C. A., et al. (2000). "The collaborative longitudinal personality disorders study." *Acta Psychiatrica Scandinavia, 102*, 256–64.

McGuffin, P. & Rivera, M. (2015). "The interaction between stress and genetic factors in the etiopathogenesis of depression." *World Psychiatry, 14*, 161–62.

McHugh, P. (1999). "How psychiatry lost its way." *Commentary*, December, 32–38.

McKeon, P. & Murray, R. (1987). "Familial aspects of obsessive-compulsive neurosis." *British Journal of Psychiatry, 151*, 528–34.

McLaughlin, K. A., Green, J. G., Hwang, I., Sampson, N. A., Zaslavsky, A. M. & Kessler, R. C. (2012). "Intermittent explosive disorder in the National Comorbidity Survey Replication." *Archives of General Psychiatry, 69*, 593–602.

McLaughlin, N. G. (1998). "Why do schools of thought fail?" *Journal of the History of Behavioral Science, 34*, 113–34.

McLean, A. (1990). "Contradictions in the social production of clinical knowledge: The case of schizophrenia." *Social Science and Medicine, 30*, 969–85.

McNally, R. (2003). *Remembering trauma*. Cambridge, MA: Harvard University Press.

McNally, R. (2011). *What is mental illness?* Cambridge, MA: Harvard University Press.

Mead, M. (1928/2001). *Coming of age in Samoa*. New York: Harper.

Mechanic, D. (1998). "Emerging trends in mental health policy and practice." *Health Affairs, 17*, 82–98.

Mechanic, D., McAlpine, D. & Rochefort, D. (2014). *Mental health and social policy* (6th ed.). New York: Pearson.

Menard, L. (2012). "Freud, anxiety, and the cold war." In *After Freud left* (J. Burnham, ed.), 189–297. Chicago: University of Chicago Press.

Menninger, K. (1963). *The vital balance*. New York: Viking Press.

Menninger, W. C. (1947). "The role of psychiatry in the world today." *American Journal of Psychiatry, 104*, 155–63.

Menninger, W. C. (1949). "Presidential address." *American Journal of Psychiatry, 106*, 1–12.

Merikangas, K. R., Angst, J., Eaton, W., Canino, G., Rubio-Stipec, M., Wacker, H., et al. (1996). "Comorbidity and boundaries of affective disorders with anxiety disorders and substance misuse: Results of an international task force." *British Journal of Psychiatry, 30*, 58–67.

Merikangas, K. R., He, J. P. & Burstein, M. (2010). "Lifetime prevalence of mental disorders in U.S. adolescents." *Journal of the American Academy of Child and Adolescent Psychiatry*, *49*, 980–89.

Merikangas, K. R., Risch, N. J. & Weissman, M. M. (1994). "Comorbidity and co-transmission of alcoholism, anxiety and depression." *Psychological Medicine*, *24*, 69–80.

Meyer, A. (1957). *Psychobiology: A science of man*. Springfield, IL: Charles C. Thomas.

Meyer, E., Walsh-Mesinger, J. & Malaspina, D. (2018). "Diagnosis and epidemiology of psychotic disorders." In *Charney & Nestler's neurobiology of mental illness* (5th ed., D. S. Charney, J. D. Buxbaum, P. Sklar & E. J. Nestler, eds.), 149–60. New York: Oxford University Press.

Meynert, T. (1884/2015). *Psychiatry: A clinical treatise on diseases of the fore-brain based upon a study of its structure, functions, and nutrition*. Wolcott, NY: Scholar's Choice.

Micale, M. S. (2008). *Hysterical men: The hidden history of male nervous illness*. Cambridge, MA: Harvard University Press.

Middeldorp, C. M., Cath, D. C., Van Dyck, R., & Boomsma, D. I. (2005). "The co-morbidity of anxiety and depression in the perspective of genetic epidemiology: A review of twin and family studies." *Psychological Medicine*, *35*, 611–24.

Miller, G. (2010). "Is pharma running out of brainy ideas?" *Science*, *329*, 502–4.

Miller, J. R. (1998). *The nurture assumption*. New York: Free Press.

Millon, T. (1983). "The DSM-III: An insider's perspective." *American Psychologist*, *38*, 804–14.

Millon, T. (1986). "On the past and future of the DSM-III: Personal recollections and projections." In *Contemporary directions in psychopathology: Toward the DSM-IV* (T. Millon & G. Klerman, eds.), 29–70. New York: Guilford Press.

Moffitt, T. E., Caspi, A. Taylor, J. Kokaua, B. Milne, G. Polanczyk & R. Poulton. (2010). "How common are common mental disorders? Evidence that lifetime prevalence rates are doubled by prospective versus retrospective ascertainment." *Psychological Medicine*, *40*, 899–909.

Moffitt, T. E., Harrington, H. L., Caspi, A., Kim-Cohen, J., Goldberg, D., Gregory, A. M. & Poulton, R. (2007). "Depression and generalized anxiety disorder: Cumulative and sequential comorbidity in a birth cohort followed prospectively to age 32 years." *Archives of General Psychiatry*, *64*, 651–60.

Mojtabai, R. (2011). "Bereavement-related depressive episodes: Characteristics, 3-year course, and implications for DSM-5." *Archives of General Psychiatry*, *68*, 920–28.

Mojtabai, R. & Molfson M. (2010). "National trends in psychotropic medication polypharmacy in office-based psychiatry." *Archives of General Psychiatry*, *67*, 26–36.

Mojtabai, R. & Olfson, M. A. (2008a). "National patterns in antidepressant treatment by psychiatrists and general medical providers: Results from the National Comorbidity Survey Replication." *Journal of Clinical Psychiatry*, *69*, 1064–74.

Mojtabai, R. & Olfson, M. (2008b). "National trends in psychotherapy by outpatient-based psychiatrists." *Archives of General Psychiatry*, *65*, 962–70.

Moncrieff, J., Wessely, S. & Hardy, R. (2004). "Active placebos versus antidepressants for depression." *Cochrane Database of Systematic Reviews*, *1*, CD003012.

Mullan, M. J. & Murray, R. M. (1989). "The impact of molecular genetics on our understanding of the psychoses." *British Journal of Psychiatry*, *154*, 591–95.

Murphy, D. (2006). *Psychiatry in the scientific image*. Cambridge, MA: MIT Press.

Murphy, D. (2014). "Natural kinds in folk psychology and in psychiatry." In *Classifying psychopathology: Mental kinds and natural kinds* (H. Kincaid & J. Sullivan, eds.), 105–22. Cambridge, MA: MIT Press.
Murray, C. J. L. & Lopez, A. D. (Eds.). (1996). *The global burden of disease*. Cambridge, MA: World Health Organization.
Narrow, W. E., Rae, D. S., Robins, L. N. & Regier, D. A. (2002). "Revised prevalence estimates of mental disorders in the United States: Using a clinical significance criterion to reconcile 2 surveys' estimates." *American Journal of Psychiatry*, *59*, 115–23.
Nemiah, J. C. (1985). "Anxiety states (anxiety neuroses)." In *Comprehensive textbook of psychiatry* (H. I. Kaplan & B. J. Sadlock, eds.), 883–94. Baltimore, MD: Williams & Wilkins.
Nesse, R. M. (2019). *Good reasons for bad feelings: Insights from the frontier of evolutionary psychiatry*. New York: Dutton.
Nestler, E. J. (2018). New approaches for treating depression. In *Charney & Nestler's neurobiology of mental illness* (5th ed., D. S. Charney, J. D. Buxbaum, P. Sklar & E. J. Nestler, eds.), 377–85. New York: Oxford University Press.
Nestor, P., Choate, V. & Shirai, A. (2015). "In search of the functional neuroanatomy of social disturbance in schizophrenia." In *Social neuroscience: Mind, brain, and society* (R. K. Schutt, L. J. Seidman & M. S. Kechaven, eds.), 88–122. Cambridge, MA: Harvard University Press.
New Freedom Commission on Mental Health. (2003). *Achieving the promise: Transforming mental health care in America*. Rockville, MD: U.S. Department of Health and Human Services.
O'Dea, T. F. (1957). *The Mormons*. Chicago: University of Chicago Press.
Ofshe, R. & Watters, E. (1994). *Making monsters: False memories, psychotherapy, and sexual hysteria*. Berkeley: University of California Press.
Olfson, M., Blaco, C., Liu, L., Moreno, C. & Laje, G. (2006). "National trends in the outpatient treatment of children and adolescents with antipsychotic drugs." *Archives of General Psychiatry*, *63*, 679–85.
Olfson, M., Gameroff, M. J., Marcus, S. C. & Jensen, P. S. (2003). "National trends in the treatment of attention deficit hyperactivity disorder." *American Journal of Psychiatry*, *160*, 1071–77.
Olfson, M., Marcus, S. C., Druss, B., Elinson, L., Tanielian, T. & Pincus, H. A. (2002). "National trends in the outpatient treatment of depression." *Journal of the American Medical Association*, *287*, 203–9.
Olfson, M., Marcus, S.C., Druss, B. & Pincus. H. A. (2002). "National trends in the use of outpatient psychotherapy." *American Journal of Psychiatry*, *159*, 1914–20.
Oppenheim, J. (1991). *Shattered nerves*. New York: Oxford University Press.
Oshinsky, D. (2018). "Should we reopen the asylums?" *New York Review of Books*, October 25, 37.
Osnos, E. (2018). "Ghost in the machine." *New Yorker*, September 17.
Ostergaard, S. D., Jensen, S. O. & Bech, P. (2011). "The heterogeneity of the depressive syndrome: When numbers get serious." *Acta Psychiatrica Scandinavica*, *124*, 495–96.
Panofsky, A. (2014). *Misbehaving science: Controversy and the development of behavior genetics*. Chicago: University of Chicago Press.
Pardes, H., Kaufmann, C. A., Pincus, H. A. & West, A. (1989). "Genetics and psychiatry: Past discoveries, current dilemmas, and future directions." *American Journal of Psychiatry*, *146*, 435–43.

Parnas, J. (2015). "Decline of psychoanalysis to the advantage of what?" In *Philosophical issues in psychiatry III: The nature and sources of historical change* (K. Kendler & J. Parnas, eds.), 180–86. New York: Oxford University Press.
Parnas, J. & Bovet, P. (1995). "Research in psychopathology: Epistemologic issues." *Comprehensive Psychiatry, 36*, 167–81.
Parry, H., Balter, M. M., Mellinger, G., Cisin, I., & Manheimer, D. (1973). National patterns of psychotherapeutic drug use. *Archives of General Psychiatry, 28*, 769–83.
Pescosolido, B. A. (2015). "Linking the social brain to the social world through network connections." In *Social neuroscience: Mind, brain, and society* (R. K. Schutt, L. J. Seidman & M. S. Kechaven, eds.), 247–79. Cambridge, MA: Harvard University Press.
Pescosolido, B. A., Martin, J. K., Long J. S., Medina T. R., Phelan J. C. & Link B. G. (2010). "'A disease like any other?' A decade of change in public reactions to schizophrenia, depression, and alcohol dependence." *American Journal of Psychiatry, 167*, 1321–30.
Pescosolido, B. A., Perry, B. L., Long, J. S., Martin, J. K, Nurnberger, J. & Hesselbrock, V. (2008). "Under the influence of genetics." *American Journal of Sociology, 114*, 171–201.
Phelan, J. C. (2005). "Geneticization of deviant behavior and consequences for stigma: The case of mental illness." *Journal of Health and Social Behavior, 46*, 307–22.
Phillips, A. (2014). *Becoming Freud: The making of a psychoanalyst*. New Haven, CT: Yale University Press.
Phillips, J. (2013). "The conceptual status of DSM-5 diagnoses." In *Making the DSM-5: Concepts and controversies* (J. Paris & J. Phillips, eds.), 143–57. New York: Springer.
Pierre, J. M. (2013). "Overdiagnosis, underdiagnosis, synthesis: A dialectic for psychiatry and the DSM." In *Making the DSM-5: Concepts and controversies* (J. Paris & J. Phillips, eds.), 105–24. New York: Springer.
Pigott, H. E., Leventhal, A. M., Alter, G. S. & Boren, J. J. (2010). "Efficacy and effectiveness of antidepressants: Current status of research." *Psychotherapy and Psychosomatics, 79*, 267–79.
Pinker, S. (1997). *How the mind works*. New York: W. W. Norton.
Pinker, S. (2018). *Enlightenment now: The case for reason, science, humanism and progress*. New York: Viking.
Plato. (1892). *Timaeus 71 D-E. Dialogues of Plato* (B. Jewett, trans.). New York: Oxford University Press.
Plato. (2005). *Phaedrus* (C. Rowe, trans.). New York: Penguin Classics.
Plomin, R. (2018). *Blueprint: How DNA makes us who we are*. Cambridge, MA: MIT Press.
Polderman, T., Benyamin, B., de Leeuw, C. A., Sullivan, P. F., van Bochoven, A., Visscher, P. M. & Posthuma, D. (2015). "Meta-analysis of the heritability of human traits based on fifty years of twin studies." *Nature Genetics, 47*, 702–9.
Pollan, M. (2018). *How to change your mind: What the new science of psychedelics teaches us about consciousness, dying, addiction, depression, and transcendence*. New York: Penguin.
Porter, R. (1987). *Mind-forg'd manacles: A history of madness in England*. Cambridge, MA: Harvard University Press.
Porter, R. (1991). "Preface." In *The English Malady* (R. Porter, ed.), vii. New York: Tavistock/Routledge.
Porter, R. (1997). *The greatest benefit to mankind: A medical history of humanity*. New York: W. W. Norton.
Porter, R. (2002). *Madness: A brief history*. New York: Oxford University Press.

Porter, T. M. (2018). *Genetics in the madhouse: The unknown history of human heredity.* Princeton, NJ: Princeton University Press.

Pratt, L. A., Brody, D. J. & Gu, Q. (2011). "Anti-depressant use in persons age 12 and Over: United States 2005–2008." *NCHS Data Brief, 76*, 1–8.

Purcell, S. M. (2018). "Genetic methodologies and application." In *Charney & Nestler's neurobiology of mental illness* (5th ed., D.S. Charney, J. D. Buxbaum, P. Sklar & E. J. Nestler, eds.), 3–16. New York: Oxford University Press.

Rangell, L. (1996). "Epilogue." In O. Feinichel *The psychoanalytic theory of neurosis, 50th anniversary edition*, E1–E12. New York: W. W. Norton.

Raofi, S. & Schappert, S. M. (2006). "Medication therapy in ambulatory medical care, United States 2003–04." *Vital Health Statistics, 13*, 1–40.

Rasic, D., Hajek, T. Alda, M. & Uher, R. (2014). "Risk of mental illness in offspring of parents with schizophrenia, bipolar disorder, and major depressive disorder: A meta-analysis of family high-risk studies." *Schizophrenia Bulletin, 40*, 28–38.

Raynes, N. V. (1979). "Factors affecting the prescribing of psychotropic drugs in general practice consultations." *Psychological Medicine, 9*, 671–79.

Redlich, F. & Kellert, S. (1978). "Trends in American mental health." *American Journal of Psychiatry, 135*, 22–28.

Regier, D. A., Kaelber, C. T., Rae, D. S., Farmer, M. E., Knauper, B., Kessler, R. C., & Norquist, G. S. (1998). "Limitations of diagnostic criteria and assessment instruments for mental disorders." *Archives of General Psychiatry, 55*, 109–15.

Regier, D. A., Morris, S. E. & Schultz, S. K. (2018). "DSM-5: Overview and goals." In *Charney & Nestler's neurobiology of mental illness* (5th ed., D. S. Charney, J. D. Buxbaum, P. Sklar & E. J. Nestler, eds.), 935–40. New York: Oxford University Press.

Regier, D. A., Narrow, W. E., Kuhl, E. A. & Kupfer, D. J. (2009). "The conceptual development of DSM-V." *American Journal of Psychiatry, 166*, 983–87.

Reiss, B. (2008). *Theatres of madness: Insane asylums & nineteenth century culture.* Chicago: University of Chicago Press.

Rennie, T. A. C. (1955). "Social psychiatry—A definition." *International Journal of Social Psychiatry, 1*, 10.

Rickels, N. K., Klein, J. J. & Bassan, M. E. (1950). "Who goes to a psychiatrist?" *American Journal of Psychiatry, 106*, 845–50.

Rieff, P. (1959). *Freud: The mind of a moralist.* Chicago: University of Chicago Press.

Rieff, P. (1966). *The triumph of the therapeutic.* Chicago: University of Chicago Press.

Roazen, P. (1992). *Freud and his followers.* Boston: DaCapo Press.

Roberts, L. M., Halleck, S. M. & Loeb, M. B. (1966). *Community psychiatry.* Madison: University of Wisconsin Press.

Robins, E. & Guze, S. (1970). "Establishment of diagnostic validity in psychiatric illness: Its application to schizophrenia." *American Journal of Psychiatry, 126*, 983–87.

Robins, L. N. & Helzer, J. E. (1986). "Diagnosis and clinical assessment: The current state of psychiatric diagnosis." *Annual Review of Psychology, 37*, 409–32.

Robins, L. N., Helzer, J. E., Weissman, M. M., Orvaschel, H., Gruenberg, E., Burke, J. D., et al. (1984). Lifetime prevalence of specific psychiatric disorders in three sites. *Archives of General Psychiatry, 41*, 949–56.

Robinson, D. (1995). *An intellectual history of psychology* (3rd ed.). Madison: University of Wisconsin Press.

Robinson, P. (1993). *Freud and his critics.* Berkeley: University of California Press.

Roccatagliata, G. (1986). *A history of ancient psychiatry.* New York: Greenwood Press.

Rohde, P., Lewinsohn, P. M., Klein, D. N., Seeley, J. R. & Gau, J. M. (2013). "Key characteristics of major depressive disorder occurring in childhood, adolescence, emerging adulthood, and adulthood." *Clinical Psychological Science, 1*, 41–53.
Rose, N. (1998). *Inventing our selves*. New York: Cambridge University Press.
Rose, N. (2019). *Our psychiatric future*. Medford, MA: Polity Press.
Rosen, G. (1968). *Madness in society*. New York: Harper.
Rosenberg, C. (1997). *No other gods: On science and American social thought*. Baltimore, MD: Johns Hopkins University Press.
Rosenberg, C. (2007). *Our present complaint: American medicine, then and now*. Baltimore, MD: Johns Hopkins University Press.
Rosenhan, D. (1973). "On being sane in insane places." *Science, 179*, 250–58.
Ross, D. (2012). "Freud and the vicissitude of modernism in the United States, 1940–1980." In *After Freud left* (J. Burnham, ed.), 163–88. Chicago: University of Chicago Press.
Ross, E. L., Zivin, K. & Maixner, D. F. (2018). "Cost-effectiveness of electroconvulsive therapy vs pharmacotherapy/psychotherapy for treatment-resistant depression in the United States." *JAMA Psychiatry, 75*, 713–22.
Roudinesco, E. (2016). *Freud in his time and ours*. Cambridge, MA: Harvard University Press.
Rubin, A. J. & Peltier, E. (2018). "France forces many pupils to put away their phones." *New York Times*, September 21, A4.
Rush, B. (1812). *Medical inquiries and observations upon the disease of mind*. Philadelphia: Kimber & Richardson.
Sabshin, M. (1990). "Turning points in twentieth-century American psychiatry." *American Journal of Psychiatry, 147*, 1267–74.
Sacks, O. (2017). *The rivers of consciousness*. New York: Alfred A. Knopf.
Sadler, J. (2013a). "Considering the economy of DSM alternatives." In *Making the DSM-5: Concepts and controversies* (J. Paris & J. Phillips, eds.), 21–38. New York: Springer.
Sadler, J. E. (2013b). "Vice and mental disorders." In *The Oxford handbook of philosophy and psychiatry* (K. W. M. Fulford, M. Davies, R. Gipps, G. Graham, J. Sadler, G. Stanghellini & T. Thornton, eds.), 451–79. New York: Oxford University Press.
Sadock, B. J., Sadock, V. A., Ruiz, P. & Kaplan, H. I. (Eds.). (2009). *Kaplan and Sadock's comprehensive textbook of psychiatry* (9th ed.). Philadelphia: Lippincott, Williams & Wilkins.
Sarpal, D. K. & Malhotra, A. K. (2018). "Current treatments for psychotic disorders." In *Charney & Nestler's neurobiology of mental illness* (5th ed., D. S. Charney, J. D. Buxbaum, P. Sklar & E. J. Nestler, eds.), 277–86. New York: Oxford University Press.
Schaffner, K. (2013). "Reduction and reductionism in psychiatry." In *The Oxford handbook of philosophy and psychiatry* (K. W. M. Fulford, M. Davies, R. Gipps, G. Graham, J. Sadler, G. Stanghellini & T. Thornton, eds.), 1003–22. New York: Oxford University Press.
Scheff, T. (1966). *Being mentally ill: A sociological theory*. Chicago: Aldine.
Schizophrenia Working Group of the Psychiatric Genomics Consortium. (2014). "Biological insights from 108 schizophrenia-associated genetic loci." *Nature, 511*, 421–27.
Schildkraut, J. J. (1965). "The catecholamine hypothesis of affective disorders: A review of supporting evidence." *American Journal of Psychiatry, 122*, 502–22.
Schnittker, J. (2008). "An uncertain revolution: Why the rise of a genetic model of mental illness has not increased tolerance." *Social Science and Medicine, 67*, 1370–81.

Schomerus, G., Schwahn, C., Holzinger, A., Corrigan, P. W., Grabe, H. J., Carta, M. G. & Angermeyer, M. C. (2012). "Evolution of public attitudes about mental illness: A systematic review and meta-analysis." *Acta Psychiatrica Scandinavica, 125*, 440–52.

Schutt, R. (2015). "The Social brain in a social world." In *Social neuroscience: Mind, brain, and society* (R. K. Schutt, L. J. Seidman & M. S. Kechaven, eds.), 231–46. Cambridge, MA: Harvard University Press.

Schwartz, S. & Corcoran, C. (2017). "Biological approaches to psychiatric disorders: A sociological approach." In *A handbook for the study of mental health* (T. Scheid & E. R. Wright, eds.), 98–125. New York: Cambridge University Press.

Scott, W. J. (1990). "PTSD in *DSM-III*: A case in the politics of diagnosis and disease." *Social Problems, 37*, 294–310.

Scull, A. (2009). *Hysteria: The biography*. New York: Oxford University Press.

Scull, A. (2015). *Madness is civilization*. Princeton, NJ: Princeton University Press.

Segalowitz, S. J. (1999). "Why twin studies really don't tell us much about human heritability." *Behavioral and Brain Sciences, 22*, 904–5.

Segerstrale, U. (2000). *Defenders of the truth: The sociobiology debate*. New York: Oxford University Press.

Selye, H. (1956). *The stress of life*. New York: McGraw-Hill.

Shakespeare, W. (1988). *Four tragedies* (D. Bevington, ed.). New York: Bantam.

Shapiro, S., & Baron, S. (1961). "Prescriptions for psychotropic drugs in a noninstitutional population." *Public Health Reports, 76*, 481–88.

Shapiro, S., Skinner E. A., Kramer, M., Steinwachs, D. M. & Regier, D. A. (1985). "Measuring need for mental health services in a general population." *Medical Care, 23*, 1033–43.

Sharfstein, S. (2005). "Big pharma and American psychiatry." *Psychiatric News*, August 19, 3.

Shephard, B. (2000). *A war of nerves*. Cambridge, MA: Harvard University Press.

Shirk, S. R. & Jungbluth, N. J. (2008). "School-based mental health checkups: Ready for practical action?" *Clinical Psychology: Science and Practice, 15*, 217–23.

Shorter, E. (1992). *From paralysis to fatigue: A history of psychosomatic illness in the modern era*. New York: Free Press.

Shorter, E. (1994). *From the mind into the body: The cultural origins of psychosomatic symptoms*. New York: Free Press.

Shorter, E. (1997). *A history of psychiatry: From the era of the asylum to the age of Prozac*. New York: Wiley.

Shorter, E. (2013a). *How did everyone became depressed: The rise and fall of the nervous breakdown*. New York: Oxford University Press.

Shorter, E. (2013b). "The history of DSM." In *Making the DSM-5* (J. Paris & J. Phillips, eds.), 3–20. New York: Springer.

Shorter, E. (2015). *What psychiatry left out of the DSM-5: Historical mental disorders today*. New York: Routledge.

Shostak, S., Conrad, P. E. & Horwitz, A. V. (2008). "Sequencing and its consequences: Path dependence and the relationships between genetics and medicalization." *American Journal of Sociology, 114* (Suppl.), S287–S316.

Shostak, S. & Freese, J. (2010). "Genetics and social inquiry." *Annual Review of Sociology, 35*, 107–28.

Shostak, S. & Moinester, M. (2015). "Beyond geneticization: Regimes of perceptibility and the social determinants of health." In *Reimagining (bio)medicalization, pharmaceuticals and genetics* (S. E. Bell & A. E. Figert, eds.), 216–38. New York: Routledge.

Simon, B. (1978). *Mind and madness in Ancient Greece*. Ithaca, NY: Cornell University Press.

Singh, I. & Rose, N. (2009). "Biomarkers in psychiatry." *Nature, 460*, 204.

Skodol, A. E., Gunderson, J. G., Pfohl, B., Widiger, T. A., Livesley, W. J., & Siever, L. J. (2002). "The borderline diagnosis I: Psychopathology, comorbidity, and personality structure." *Biological Psychiatry, 51*, 936–50.

Smith, D. (2014). "The diminished resistance to medicalization in psychiatry." *Society and Mental Health, 4*, 75–91.

Smith, M. B. (1985). *Small comfort: A history of the minor tranquillizers*. New York: Praeger.

Sophocles. (2008). *Ajax* (D. Raeburn, trans.). New York: Penguin Classics.

Spector, T. (2012). *Identically different: Why you can change your genes*. London: Weidenfeld & Nicolson.

Spitzer, R. L. (1975). "On pseudoscience in science, logic in remission and psychiatric diagnosis: A critique of Rosenhan's 'On Being Sane in Insane Places.'" *Journal of Abnormal Psychology, 84*, 442–52.

Spitzer, R. L. (1978). "The data-oriented revolution in psychiatry." *Man and Medicine, 3*, 193–94.

Spitzer, R. L. (2001). "Values and assumptions in the development of DSM-III and DSM-III-R: An insider's perspective and a belated response to Sadler, Hulgus and Agitch's 'On Values in Recent American Psychiatric Classification.'" *Journal of Nervous and Mental Disease, 189*, 357–58.

Spitzer, R. L., Endicott, J. & Robins, E. (1978). "Research diagnostic criteria: Rationale and reliability." *Archives of General Psychiatry, 35*, 773–82.

Spitzer, R. L. & Fleiss, J. L. (1974). "A re-analysis of the reliability of psychiatric diagnosis." *American Journal of Psychiatry, 125*, 341–47.

Spitzer, R. L., Sheehy, M. & Endicott, J. (1977). "DSM-III: Guiding principles." In *Psychiatric diagnosis* (V. Rakoff, H. Stancer, & H. Kedward, eds.), 1–24. New York: Bruner/Mazel.

Spitzer, R. L. & Wakefield, J. C. (2002). "Why pedophilia is a disease of sexual attraction—at least sometimes." *Archives of Sexual Behavior, 31*, 499–500.

Spitzer, R. L. & Williams, J. B. W. (1988). "Having a dream: A research strategy for DSM-IV." *Archives of General Psychiatry, 45*, 871–74.

Sprooten, E., Rasgon, A., Goodman, M., Carlin, A., Leibu, E., Lee, W. H. & Frangou, S. (2017). "Addressing reverse inference in psychiatric neuroimaging: Meta-analyses of task-related brain activation in common mental disorders." *Human Brain Mapping, 38*, 1846–64.

Srole, L., Langner, T. S., Michael, S. T., Kirkpatrick, P., Opler, M. K. & Rennie, T. A. C. (1962). *Mental health in the metropolis: The Midtown Manhattan study*. New York: McGraw-Hill.

St. Augustine. (2009). *Confessions*. New York: Oxford World Classics.

Starr, P. (1982). *The social transformation of American medicine*. New York: Basic Books.

*Statistical Manual for the Use of Hospitals for Mental Diseases* (10th ed.). (1942). Utica, NY: State Hospitals Press.

Staub, M. E. (2011). *Madness is civilization*. Chicago: University of Chicago Press.

Steel, Z., Marnane, C., Iranpour, C., Chey, T., Jackson, J. W., Patel, V. & Silove, D. (2014). "The global prevalence of common mental disorders: A systematic review and meta-analysis 1980–2013." *International Journal of Epidemiology, 43*, 476–93.

Stein, M. B., K.L. Jang, K. L. & Livesley, W. J. (1999). "Heritability of anxiety sensitivity: A twin study." *American Journal of Psychiatry, 156*, 246–51.

Substance Abuse and Mental Health Services Administration. (2012). *Mental health United States, 2010*. HHS Publication No. 12-4681. Rockville, MD: SAMHSA.

Sullivan, P. F., Dal, M. J. & O'Donovan, M. (2012). "Genetic architectures of psychiatric disorders: The emerging picture and its implications." *Nature Reviews Genetics, 13*, 537–51.

Sulloway, F. J. (1979). *Freud: Biologist of the mind*. Cambridge, MA: Harvard University Press.

Sulloway, F. J. (1991). "Reassessing Freud's case histories: The social construction of psychoanalysis." *Isis, 82*, 245–75.

Swank, R. & Marchand, W. (1946). "Combat neuroses: Development of combat exhaustion." *Archives of Neurology and Psychiatry, 55*, 236–47.

Szasz, T. (1961). *The myth of mental illness: Foundations of a theory of personal conduct*. New York: Harper & Row.

Szasz, T. S. (1960). "Civil liberties and mental illness." *Journal of Nervous and Mental Disease, 131*, 58–63.

Szasz, T. S. (1991). *Ideology and insanity*. Syracuse, NY: Syracuse University Press.

Szasz, T. S. (1997). *Insanity*. Syracuse, NY: Syracuse University Press.

Szasz, T. S. (2011). "The myth of mental illness fifty years later." *The Psychiatrist, 35*, 179–82.

Tamminga, C. A., Ivleva, E. I., Keshavan, M. S., Pearlson, G. D., Clementz, B. A., Witte, B., et al. (2013). "Clinical phenotypes of psychosis in the Bipolar-Schizophrenia Network on Intermediate Phenotypes (B-SNIP)." *American Journal of Psychiatry, 170*, 1263–74.

Taylor, M. A. (2013). *Hippocrates cried: The decline of American psychiatry*. New York: Oxford University Press.

Temerlin, M. K. (1968). "Suggestion effects in psychiatric diagnosis." *Journal of Nervous and Mental Disorders, 147*, 349–58.

Thapar, A. & McGuffin, P. (1997). "Änxiety and depressive symptoms in childhood—A genetic study of comorbidity." *Journal of Child Psychological Psychiatry, 38*, 651–56.

Thase, M. E., Larsen, K. G. & Kennedy, S. H. (2011). "Assessing the 'true' effect of active antidepressant therapy v. placebo in major depressive disorder: Use of a mixture model." *British Journal of Psychiatry, 199*, 501–7.

Thomas, C., Conrad, P., Casler, R. & Goodman, E. (2006). "Trends in the use of psychotropic medications among adolescents, 1994–2001." *Psychiatric Services, 57*, 63–69.

Tomes, N. (1984). *A generous confidence: Thomas Story Kirkbride and the art of asylum-keeping, 1840–1883*. New York: Cambridge University Press.

Tomes, N. (1994). "Feminist histories of psychiatry." In *Discovering the history of psychiatry* (M. S. Micale & R. Porter, eds.), 348–83. New York: Oxford University Press.

Tone, A. (2009). *The age of anxiety: A history of America's turbulent affair with tranquilizers*. New York: Basic Books.

Torrey, E. F. (1974). *The death of psychiatry*. Radnor, PA: Chilton.

Torrey, E. F. (2014). *American psychosis: How the federal government destroyed the mental illness treatment system*. New York: Oxford University Press.

Torrey, E. F. & Yolken, R. H. (2010). "Psychiatric genocide: Nazi attempts to eradicate schizophrenia." *Schizophrenia Bulletin, 36*, 26–32.

Trotter, T. (1807). *A view of the nervous temperament*. London: Longman.

Turkheimer, E. (2015). "The nature of nature." In *Philosophical issues in psychiatry III: The nature and sources of historical change* (K. Kendler & J. Parnas, eds.), 227–44. New York: Oxford University Press.

Tyrer, P. & Kendall, T. (2009). "The spurious advance of antipsychotic drug therapy." *The Lancet*, *373*, 4–5.

U.S. Department of Health and Human Services. (1999). *Mental health: A report of the Surgeon General*. Rockville, MD: U.S. Department of Health and Human Services.

U.S. Department of Health and Human Services, Agency for Health Care Policy and Research. (1980). *National Medical Care, Utilization, and Expenditure Survey*. Washington, DC: U.S. Department of Health and Human Services.

Van Dam, N. T., Iacoviello, B. M. & Murrough, J. W. (2018). "Diagnosis and epidemiology of depression." In *Charney & Nestler's neurobiology of mental illness* (5th ed., D. S. Charney, J. D. Buxbaum, P. Sklar & E. J. Nestler, eds.), 289–300. New York: Oxford University Press.

van Dongen, J. & Boomsma, D. I. (2013). "The evolutionary paradox and the missing heritability of schizophrenia." *American Journal of Medical Genetics*, *162B*, 122–36.

van Os, J., Linscott, R. J., Myin-Germeys, I., Delespaul, P. & Krabbendam, L. (2009). "A systematic review and meta-analysis of the psychosis continuum: Evidence for a psychosis proneness-persistence-impairment model of psychotic disorder." *Psychological Medicine*, *39*, 179–95.

Vandermeersch, P. (1994). "'Les mythes d'origine' in the history of psychiatry." In *Discovering the history of psychiatry* (M. S. Micale & R. Porter, eds.), 219–31. New York: Oxford University Press.

Vaughan, S. C. (1997). *The talking cure: The science behind psychotherapy*. New York: Grosset/Putnam.

Vogt, K. M. (2013). "Plato on madness and the good life." In *Mental disorders in the classical world* (W. V. Harris, ed.), 177–92. New York: Brill.

Wakefield, J. C. (1992). "The concept of mental disorder: On the boundary between biological facts and social values." *American Psychologist*, *47*, 373–88.

Wakefield, J. C. (1999). "Evolutionary versus prototype analyses of the concept of disorder." *Journal of Abnormal Psychology*, *108*, 374–99.

Wakefield, J. C. (2001). "The myth of DSM's invention of new categories of disorder: Houts's diagnostic discontinuity thesis disconfirmed." *Behavior Research and Therapy*, *39*, 575–624.

Wakefield, J. C. (2006). "Personality disorder as harmful dysfunction: DSM's cultural deviance criterion reconsidered." *Journal of Personality Disorders*, *20*, 157–69.

Wakefield, J. C. (2013). "DSM-5: An overview of major changes and controversies." *Clinical Social Work Journal*, *41*, 139–54.

Wakefield, J. C. (2015). "DSM-5, psychiatric epidemiology and the false positives problem." *Epidemiology and Psychiatric Science*, *24*, 188–96.

Wakefield, J. C. & First, M. B. (2012). "Placing symptoms in context." *Comprehensive Psychiatry*, *53*, 130–39.

Wakefield, J. C. & First, M. F. (2013). "Clarifying the boundary between normality and disorder: A fundamental conceptual challenge for psychiatry." *Canadian Journal of Psychiatry*, *58*, 603–5.

Wakefield, J. C. & Horwitz, A. V. (2016). "Psychiatry's continuing expansion of depressive disorder." In *Sadness or depression?* (S. Demazeux, ed.), 173–203. Dordrecht, The Netherlands: Springer.

Wakefield, J. C., Kirk, S. A. & Pottick, K. J. (2006). "The lay concept of conduct disorder: Do nonprofessionals use syndromal symptoms or internal dysfunction to distinguish disorder from delinquency?" *Canadian Journal of Psychiatry*, *51*, 210–17.

Wakefield, J. C. & Schmitz, M. F. (2013a). "Can the DSM's major depression bereavement exclusion be validly extended to other stressors? Evidence from the NCS." *Acta Psychiatrica Scandinavica, 128*, 294–305.

Wakefield, J. C. & Schmitz, M. F. (2012b). "Recurrence of depression after bereavement-related depression: Evidence for the validity of the DSM-IV bereavement exclusion from the Epidemiologic Catchment Area Study." *Journal of Nervous and Mental Disease, 200*, 480–85.

Wakefield, J. C. & Schmitz, M. F. (2013b). "Normal vs. disordered bereavement-related depression: Are the differences real or tautological?" *Acta Psychiatrica Scandinavica, 127*, 159–68.

Wakefield, J. C. & Spitzer, R. L. (2002). "Lowered estimates—but of what?" *Archives of General Psychiatry, 59*, 129–30.

Wallace, E. R. (1994). "Psychiatry and its nosology: A historico-philosophical overview." In *Philosophical perspectives on psychiatric diagnostic classification* (J. Z. Sadler, O. P. Wiggins & M. A. Schwartz, eds.), 6–86. Baltimore, MD: Johns Hopkins University Press.

Wang, P. S., Lane, M., Olfson, M., Pincus, H. A., Wells, K. B., & Kessler, R. C. (2005). "Twelve-month use of mental health services in the United States." *Archives of General Psychiatry, 62*, 629–40.

Warner, L. A., Kessler R. C., Hughes M., Anthony J. C. & Nelson C. B. (1995). "Prevalence and correlates of drug use and dependence in the United States. Results from the National Comorbidity Survey." *Archives of General Psychiatry, 52*, 219–29.

Watson, D. (2005). "Rethinking the mood and anxiety disorders: A quantitative hierarchical model for DSM-V." *Journal of Abnormal Psychology, 114*, 522–36.

Weir, K. (2012). "The roots of mental illness." *Monitor on Psychology, 125*, 30.

Weissman, M. M., Wickramaratne, P., Nomura, Y.,Warner, V., Pilowsky, D. &Verdeli, H. (2006). "Offspring of depressed parents: 20 years later." *American Journal of Psychiatry, 163*, 1001–8.

Wender, P. H., Kety, S. S., Rosenthal, D., Schulsinger, F., Ortmann, J., & Lunde, I. (1986). "Psychiatric disorders in the biological and adoptive families of adopted individuals with affective disorders." *Archives of General Psychiatry, 43*, 923–29.

Whitebook, J. (2017). *Freud: An intellectual biography*. New York: Cambridge University Press.

Whiteford, H. A., Degenhardt, L., Rehm, J., Baxter, A. J., Ferrari, A. J., Erskine, H. E., et al. (2013). "Global burden of disease attributable to mental and substance use disorders: findings from the Global Burden of Disease Study 2010." *The Lancet, 382*, 1575–86.

Whooley, O. (2010). "Diagnostic ambivalence: Psychiatric workarounds and the *Diagnostic and Statistical Manual of Mental Disorders*." *Sociology of Health & Illness, 32*, 452–69.

Whooley, O. (2019). *On the heels of ignorance: Psychiatry and the politics of not knowing*. Chicago: University of Chicago Press.

Whooley, O. & Horwitz, A. V. (2013). "The paradox of professional success: Grand ambition, furious resistance, and the derailment of the DSM-5 revision." In *Making the DSM-5: Concepts and controversies* (J. Paris & J. Phillips, eds.), 75–94. New York: Springer.

Wilson, E. O. (1998). *Consilience: The unity of knowledge*. New York: Alfred A. Knopf.

Wilson, M. (1993). "DSM-III and the transformation of American psychiatry: A history." *American Journal of Psychiatry, 150*, 399–410.

Wing, J. K., Cooper, J. E., & Sartorius, N. (1974). *Measurement and classification of psychiatric symptoms*. New York: Cambridge University Press.

Wittchen, H. U. (2000). "Epidemiological research in mental disorders: Lessons for the next decade." *Acta Psychiatrica Scandivavica, 101*, 2–10.

Wong, E. H., Yocca, F., Smith, M. A., & Lee C. M. (2010). "Challenges and opportunities for drug discovery in psychiatric disorders." *International Journal of Neuropsychopharmacology, 13*, 1269–84.

World Health Organization. (2001). *Mental health: New understanding, new hope*. Geneva, Switzerland: World Health Organization.

Xenophon. (2013). *Memorabilia. Oeconomicus. Symposium. Apology*. (E. C. Marchant & O. J. Todd, trans.). Cambridge, MA: Harvard University Press.

Zachar, P. & Krueger, R. F. (2013). "Personality disorder and validity." In *The Oxford handbook of philosophy and psychiatry* (K. W. M. Fulford, M. Davies, R. Gipps, G. Graham, J. Sadler, G. Stanghellini, & T. Thornton, eds.), 889–910. New York: Oxford University Press.

Zaretsky, E. (2004). *Secrets of the soul: A social and cultural history of psychoanalysis*. New York: Alfred A. Knopf.

Zuvekas, S. H., Vitiello, B., & Norquist, G. S. (2006). "Recent trends in stimulant use among U.S. children." *American Journal of Psychiatry, 163*, 579–85.

# Index

Figures are indicated by *f* following the page number.

*For the benefit of digital users, indexed terms that span two pages (e.g., 52–53) may, on occasion, appear on only one of those pages.*

Printed in the USA
CPSIA information can be obtained
at www.ICGtesting.com
CBHW070830310724
12241CB00038B/224

9 780190 907860